CONVECTIVE FLOW AND HEAT TRANSFER FROM WAVY SURFACES

Viscous Fluids, Porous Media, and Nanofluids

CONVECTIVE FLOW AND HEAT TRANSFER FROM WAVY SURFACES

Viscous Fluids, Porous Media, and Nanofluids

Aroon Shenoy
Mikhail Sheremet
Ioan Pop

CRC Press
Taylor & Francis Group
Boca Raton London New York

CRC Press is an imprint of the
Taylor & Francis Group, an **informa** business

CRC Press
Taylor & Francis Group
6000 Broken Sound Parkway NW, Suite 300
Boca Raton, FL 33487-2742

First issued in paperback 2019

ISBN-13: 978-0-4987-6090-4 (hbk)
ISBN-13: 978-0-367-87713-2 (pbk)

Library of Congress Cataloging-in-Publication Data

Names: Shenoy, Aroon V., 1951- author. | Sheremet, Mikhail, author. | Pop, Ioan, 1937- author.
Title: Convective flow and heat transfer from wavy surfaces : viscous fluids, porous media, and nanofluids / Aroon Shenoy, Mikhail Sheremet, and Ioan Pop.
Description: Boca Raton : Taylor & Francis, a CRC title, part of the Taylor & Francis imprint, a member of the Taylor & Francis Group, the academic division of T&F Informa, plc, [2017] | Includes bibliographical references and index.
Identifiers: LCCN 2016012377 | ISBN 9781498760904 (alk. paper)
Subjects: LCSH: Fluid mechanics. | Fluids--Thermal properties. | Heat--Convection. | Surfaces (Technology)
Classification: LCC TA357 .S4526 2017 | DDC 621.402/2--dc23
LC record available at https://lccn.loc.gov/2016012377

Visit the Taylor & Francis Web site at
http://www.taylorandfrancis.com

and the CRC Press Web site at
http://www.crcpress.com

Contents

Foreword I
by
Prof. Ali J. Chamkha
Prince Mohammad Bin Fahd University
Kingdom of Saudi Arabia

It is an extreme pleasure to write a foreword to this useful and added-value book. The authors of the book are internationally well known and have made significant pioneering contributions in the field of heat and mass transfer. Recently, the subject of wavy geometries has attracted significant research attention, and the prominence of this issue is due to its application in many engineering systems as a means of enhancing transport performance. However, there has been no attempt made, in the form of review papers or books, to highlight the systematic formulation and methods of solution for flow and heat and mass transfer for various types of engineering geometries with wavy boundaries under different physical effects and conditions. Therefore, I believe that the present book is extremely useful and inspiring for researchers who work or intend to work in this field.

The topic of the flow and heat transfer within wavy surfaces is quite new, and this is the first book that considers natural and mixed convection flows of viscous fluids and through fluid-saturated porous media with wavy surfaces. Many engineering applications are within this scientific area, e.g., cooling systems for micro-electronic devices, heat exchangers, solar collectors, underground cable systems, and so forth. Understanding the structure of wavy surfaces is really an important issue to the scientific and engineering community. The main strength and novelty of this book comes from the fact that the authors developed a generalization of the basic formulation for flow and heat transfer from wavy surfaces, channels, tubes, and cavities having wavy walls in the presence or absence of porous media and nanoparticles. The different chapters are characterized by ease of understanding and clear description of the topics and the quality of the figures.

The book starts with showing in detail the systematic formulation of the governing equations for flow from wavy surfaces. Successfully, the authors have covered almost all necessary topics in wavy geometries for both natural and mixed convection, including viscous fluid flow over a wavy vertical wall, flow in fluid-saturated porous media over a wavy vertical wall, viscous fluid flow in a wavy vertical channel, fluid flow in a wavy horizontal channel and a wavy tube, flow saturated with nanoparticles in wavy-walled cavities, and flow saturated with nanoparticles in wavy-walled porous cavities. As I go through the book, I find that the authors have adequately covered the area of convection in wavy geometries and have presented a very good literature survey on the various topics covered in the book.

Finally, I have to say that the book is strong in both theoretical fundamentals and engineering applications. I would like to thank the authors for this interesting book, which gives a very good opportunity to undergraduate and graduate students in science and engineering, and interdisciplinary researchers and academicians such as those in applied mathematics, mechanical engineering, chemical engineering, and others. I wish this book success and its readers pleasant reading, technical benefit, and further inspiration.

Foreword II
by
Prof. Oronzio Manca
Seconda Università degli Studi di Napoli
Italy

It is a great honor and pleasure to have the opportunity to write a foreword to this valuable and interesting book, which will be useful in research and education. It is an important collection of contributions on convective heat transfer.

The book presents the chapter on governing equations in a very clear way, giving a complete description of different convective heat transfer modes. In Chapter 1, the topics pass from single wavy walls to channels with different wavy geometries in natural, mixed, and forced convection, taking into account Newtonian and non-Newtonian power-law fluids. The different geometries of wavy surfaces and shapes are also highlighted.

In all chapters, the model construction, its solution, and the related results of the analysis are rigorous and very well discussed. They give, by means of different examples described in the various chapters, the right way to set up and examine a physical–mathematical problem in convective heat transfer. Although the book is about a specialized topic, it allows a deep understanding of important analytical techniques and indicates the different numerical methods to solve the assigned model in convective heat transfer. It provides a very good guide to applying the different techniques to real problems, giving very useful indications and suggestions to build physical–mathematical models. The different solutions can be useful in several engineering applications, where the convective heat transfer should be enhanced, such as electronic cooling, heat exchangers, solar components, nuclear systems, and, more in general, in energy systems. Today, heat transfer augmentation represents one of the hot topics in engineering, and several techniques are employed to realize the increase in thermal energy transfer. In this framework, the wavy wall is a passive system that allows a possible combination of extended surface and changes in fluid flow, with the result of a beneficial heat transfer enhancement.

The book extends the topic to nanofluids and porous media to present the possible techniques to enhance the heat transfer from wavy surfaces in cavities with free convection. Moreover, an excellent list of reviewed papers completes the book. This book can be taken as an important reference for researchers, industrial professionals, graduate and PhD students, not only to have whole and exhaustive information on convective heat transfer on wavy walls, but also to improve or learn analytical techniques applied to convective heat transfer.

Preface

Wavy geometries are used in many engineering systems as a means of enhancing the transport performance. Therefore, knowledge about flow and heat transfer through wavy surfaces becomes important in this context. Solar collectors, condensers in refrigerators, cavity wall insulating systems, grain storage containers, and industrial heat radiators, for example, are a few of the many applications where wavy surfaces are encountered to transfer small- or large-scale heat. The focus on the area of flow and heat transfer from wavy surfaces in complex enclosures like square, trapezoidal, and rectangular spaces has been intensifying over the years due to the increasing interest of researchers from applied mathematics, and mechanical and chemical engineering, as well as from biomechanics and engineering mechanics. A great number of technical papers have been published on this subject and these have been scattered in a number of different journals. Topics range from a variety of flow situations to the use of different mathematical techniques for the analyzes of complex flow situations involving wavy surfaces. However, our literature survey showed that no attempt has been made until now to unify this information and make it available in the form of a review paper or as a book.

The present monograph has been written with the idea of fulfilling a definite need to have a comprehensive treatise on the subject of flow and heat transfer from a wavy surface and in cavities having wavy walls. In view of efficient utilization of energy resources, it is important that the process of natural or mixed convection should be efficient while transferring heat along with minimum degradation of energy. Natural and mixed convection phenomena in cavities with plane walls has been greatly studied in the literature, in particular, due to geometrical simplicity, minimum cost, low noise, smaller size, and reliability, etc. This is also due to their relevance to many scientific issues such as nuclear reactor systems, foundry devices, heat exchangers, geophysical and astrophysical processes, electronic device cooling, chemical processing equipment, lubrication systems, food processing, solar energy collectors, and so on. The literature also contains many investigations into the heat transfer performance of natural and mixed convection in cavities bounded by vertical wavy walls. However, in some practical applications (e.g., the cooling of buildings or flush-mounted electronic heaters), the wavy surface may in fact only be a section on a side wall.

The traditional working fluids used in such systems (e.g., water, oil, or ethylene glycol), however, have a low thermal conductivity, and thus their heat transfer performance is inevitably limited. Consequently, a requirement exists for new working fluids with a higher thermal conductivity. The recent discovery of nanofluid, which is an important kind of fluid suspension consisting of uniformly dispersed and suspended nanometer-sized (10–50 nm) particles and fibers in base fluid, proposes the next approach in cooling technology. Therefore, two chapters of the book are devoted to the important topic of free convection in pure and porous wavy cavities filled by a nanofluid.

The book is designed as a comprehensive review of the field that combines the basic theory with major up-to-date findings, touches upon ongoing research, and offers some suggestions for future developments. It can also serve as a useful starting-point information base, to those who have the desire to move into this exciting field. Though the subject matter requires rather complex mathematics, a concerted effort has been made to provide simple explanations so that new entrants to this field can absorb the fundamental concepts rather easily. At the same time, a deliberate effort has been made to section the topics in such a way that the missing areas of research are automatically drawn out. It is sincerely hoped that the book will serve as a useful reference guide to all those researchers who intend to enter this pragmatic area of flow and heat transfer from wavy surfaces, and also to those who want to be creative and desire to contribute more to it. In our opinion, this book has a definite utility value to the experts in fluid mechanics, heat transfer theory, applied mathematicians, as well as for mechanical and chemical engineers interested in the investigation of flow and heat transfer processes from wavy surfaces, and their applications. The book can also be used by graduate and PhD students, and by industry professionals.

Dr. Aroon Shenoy wishes to express his very sincere thanks to Prof. Akira Nakayama (Shizuoka University, Japan) — a collaboration with whom set the first seeds for investigations into the study of natural and forced convection heat transfer from arbitrary geometrical configurations and in fluid-saturated porous media.

Prof. Mikhail A. Sheremet expresses his thanks to Prof. Sergey A. Isaev (Saint Petersburg State University of Civil Aviation, Saint Petersburg, Russia), Prof. Geniy V. Kuznetsov (Tomsk Polytechnic

University, Tomsk, Russia) and Prof. Victor I. Terekhov (Kutateladze Institute of Thermophysics, Novosibirsk, Russia) for their help and support in analysis of convective heat and mass transfer problems. Prof. Mikhail A. Sheremet also wishes to thank the Tomsk State University for the ongoing support.

Prof. Ioan Pop is honoured to pay tribute to Prof. Adrian Bejan (J.A. Jones Professor of Mechanical Engineering), Duke University, USA, whose truly pioneering, rigorous, and trend-setting spirit in tackling basic convective heat transfer problems had a growing and beneficial influence on Prof. Pop's scientific career. Prof. Ioan Pop also expresses his gratitude for the very fruitful research collaborations with Prof. Derek B. Ingham and Prof. John H. Merkin, both from the University of Leeds (UK), Prof. Akira Nakayama, Shizuoka University, Japan, Prof. Ali J. Chamkha, Prince Mohammad Bin Fahd University, Kingdom of Saudi Aradia, Prof. K. Vajravelu, University of Central Florida, USA, Prof. Tanmay Basak and Prof. S. Roy, both from Indian Institute of Technology, Madras (India), Prof. Anuar Ishak, Prof. Roslinda Nazar, Dr. Norihan Arifin, Dr. Yian Yian Lok, Dr. Syakila Ahmad, Dr. Norfifah Bachok, Dr. Fadzila Ali, and Dr. Zuki Salleh, all from different universities in Malaysia, and Dr. M. Ghalambaz (Islamic Azad University, Dezful, Iran), Dr. D.A.S. Rees (University of Bath, UK), Dr. M.M. Rahman (Sultan Qaboos University, Sultanate of Oman), Dr. T. Groşan, Dr. Radu Trîmbiţaş, Prof. Mirela Kohr, Prof. Diana Filip, Dr. Natalia C. Roşca, Dr. Alin V. Roşca, Dr. Cornelia Revnic, and Dr. Dalia Cîmpean, who have provided valuable suggestions and assistance during preparation of some of the subsections of the book. Prof. Ioan Pop is also very grateful to Prof. Adrian Petruşel, the Dean of the Faculty of Mathematics and Computer Science of the Babeş-Bolyai University, Cluj-Napoca, Romania, for his continuous support. It is also worth to be mentioned at this end that Prof. Ioan Pop has been nominated a Thomson Reuters 2015 Highly Cited Researchers.

The authors wish to appreciate the support and motivation of Jonathan Plant, Executive Editor of CRC Press, Mechanical, Aerospace, and Nuclear Engineering, USA, for making this book a reality.

Dr. Aroon Shenoy
Dr. Mikhail Sheremet
Dr. Ioan Pop

2016

Nomenclature

a	amplitude of the wavy surface (m)
a_1	dimensionless amplitude of the wavy surface defined in Fig. 2.4
a_2	dimensionless amplitude of the wavy surface in Eq. (2.23)
a_L	amplitude of lower wavy surface of horizontal channel (m)
a_U	amplitude of upper wavy surface of horizontal channel (m)
$A_1 \dots A_4$	coefficients appearing in Eq. (6.8)
A_{cyc}	per-cycle heat transfer area (m^2)
b	wave number of the wavy surface (m) in Fig. 1.3
c	adjustable parameter of the transformed coordinate system (x_2, r_2) in Eq. (6.69)
C_k	expansion coefficients in Eq. (6.70)
C_p	specific heat capacity of the fluid $(J/kg \cdot^\circ K)$
d	distance between wavy wall and parallel flat wall of half width of wavy channel (m)
d_m	mean half width between two wavy walls (m)
D_e	equivalent diameter (m) of a corrugated tube whose pressure difference and volumetric flow rate are equal to those for a straight cylindrical tube of diameter D_e
D_1	diameter (m) of small diameter segment of tube of square wave profile
D_2	diameter (m) of large diameter segment of tube of square wave profile
$D_{\max}$	maximum diameter (m) of a periodically converging–diverging tube
$D_{\min}$	minimum diameter (m) of a periodically converging–diverging tube
D^*	equivalent diameter (m) of a periodic tube defined by Eq. (6.29)

f	dimensionless velocity function
f_{PCT}	friction factor for a periodically constricted tube defined as $[(r_m/\rho\langle u\rangle^2)(-\Delta P_{PCT}/\lambda)]$
g	acceleration due to gravity (m/s^2)
g_x	component of the acceleration due to vector in the x-direction (m/s^2)
g_y	component of the acceleration due to vector in the y-direction (m/s^2)
Gr	Grashof number in the isothermal wavy wall case defined by Eqs. (2.6) and (2.24) for Newtonian fluids
h_1	dimensionless mesh length in the r-direction
h_2	dimensionless mesh length in the x-direction
h_3	dimensionless mesh length defined as $h_1 \cos\gamma$
h_s	local heat transfer coefficient based on profile length for a wavy surface
h_x	local heat transfer coefficient based on projected length for a wavy surface
k	thermal conductivity $(W/m \cdot^\circ K)$ of the fluid
K	permeability of the porous medium
L_L	length (m) of large diameter segment of tube of square wave profile
L_S	length (m) of small diameter segment of tube of square wave profile
m	distance (m) between a node on the wall and a neighbouring node
M, N	number of grid points
n	power-law index for a non-Newtonian fluid
$\overrightarrow{n}$	unit vector
n_r	number of radial expansion functions or collocation points
n_x	number of axial expansion functions or collocation points
$\overline{N}$	total number of interior collocation points
Nu	Nusselt number
$\overline{Nu}$	average Nusselt number
Nu_{PCT}	Nusselt number for the PCT
Nu_{ST}	Nusselt number for a straight tube of equivalent length and diameter D^* as the PCT

p	pressure without body force (Pa)
p'	pressure term appearing in Eq. (6.24) for the periodic tube
p_1	dimensionless pressure term
Δp_1	dimensionless pressure difference defined differently in various sections
P	total pressure including hydrostatic pressure (Pa)
P_1	dimensionless total pressure
ΔP_{1VD}	dimensionless pressure difference calculated from viscous dissipation as defined by Eq. (6.74)
Pr	Prandtl number
q	rate of heat transfer from the wall to the fluid per cycle
Q	volumetric flow rate (m^3/s)
Q_1	dimensionless volumetric flow rate defined as $Q/\pi r_m^2 U_e$
r	local distance (m) in the r-direction
r_1, r_2, r_3	dimensionless distances in the r-direction
r_L	radius (m) for large diameter segment of tube
r_{1L}	dimensionless distance in the r-direction for large diameter segment of tube
r_m	mean radius of a wavy conduit (m)
r_{1m}, r_{1mm}, r_{2m}	dimensionless radial distances
r_s	radius (m) for small diameter segment of tube
r_{1s}	dimensionless distance in the r-direction for small diameter segment of tube
r_w	radial distance (m) from the wall to the axis of symmetry
r_{1w}, r_{2w}	dimensionless radial distances from the wall to the axis of symmetry
R	radial distance (m) marking the profile of the stenosis whose boundary form is given by Eq. (6.79)
R_0	constant tube radius (m) of stenosis as shown in Fig. 6.14
Ra	Rayleigh number for Newtonian fluid
Re	Reynolds number for a Newtonian fluid defined differently in various sections

Re_0	Reynolds number defined by Eq. (6.95)
Re_1	Reynolds number for a Newtonian fluid defined as $(U_0\lambda/2\pi\nu)$
Re_{PCT}	Reynolds number for a periodically constricted tube defined as $2r_m\langle u\rangle/\nu$
s	integral defined by Eq. (2.18)
S	area of cross-section (m^2) of the separated region to the triangular area of cross-section of the corrugation
S_1	dimensionless area of cross-section of the separated region to the triangular area of cross-section of the corrugation
St_H	Strouhal number defined as $D_{\max}/U_a t_p$ or $2St_k^2/\pi Re$
St_K	Stokes number defined as $0.5D_{\max}(2\pi/\nu t_p)^{1/2}$
t	time (s)
t_1	dimensionless time
t_p	time period of oscillation (s)
T	temperature $(^\circ K)$
T_b	temperature $(^\circ K)$ of the bulk of the fluid
T_e	constant initial fluid temperature $(^\circ K)$ at distances far from the wall
T_w	wall temperature $(^\circ K)$
ΔT_{wb}	arithmetic mean temperature difference $(^\circ K)$
u	velocity (m/s) in the x-direction
u_1	dimensionless velocity in the x-direction (first transformation)
u_2	dimensionless velocity in the x-direction (second transformation)
u_a	mean velocity amplitude (m/s), which is a function of the radial distance at the section of diameter $D_{\max}$
u_{ew}	dimensionless inviscid velocity in the x-direction evaluated at the surface
u_m	mean axial velocity (m/s), which is a function of the radial distance at the section of diameter $D_{\max}$
u_C	characteristic velocity for natural convection flow

$\langle u \rangle$	mean axial velocity (m/s) in a periodically constricted tube
U	free stream velocity in the x-direction (m/s)
U_a	cross-sectional mean velocity amplitude (m/s) at the section of diameter $D_{\max}$
U_c	centerline velocity (m/s)
U_e	average or mean velocity (m/s)
U_{e0}	mean velocity (m/s) at initial conditions in the portion of constant radius R_0
U_m	cross-sectional mean axial velocity (m/s) at the section of diameter $D_{\max}$
U_C	characteristic velocity (m/s) in the definition of Reynolds number in Eq. (6.46), which takes the value of U_m in steady flow and U_a in oscillating flow
U_R	velocity ratio defined as U_m / U_a
U_0	velocity of impulsive movement of the horizontal wavy wall in the x-direction (m/s)
v	velocity (m/s) in the y-direction coordinate
v_1	dimensionless velocity in the y-direction (first transformation)
v_2	dimensionless velocity in the y-direction (second transformation)
w	spacing between protrusions (m) in Eqs. (1.10) and (1.11)
x	local distance (m) in the x-direction coordinates
x_1	dimensionless local distance in the x-direction coordinates (first transformation)
x_2	dimensionless local distance in the x-direction coordinates, which is equal to x_1
x_{10}	dimensionless axial distance at the entrance of the conduit
y	local distance (m) in the y-direction coordinates
y_1	dimensionless local distance in the y-direction coordinates (first transformation)
y_2	dimensionless local distance in the y-direction coordinates (second transformation)

Greek Symbols

α	thermal diffusivity of the fluid (m^2/s)
α^*	dimensionless heat source or sink parameter defined in Eq. (4.7)
β	coefficient of thermal expansion $(1/^\circ K)$
β'	coefficient appearing in Eq. (6.24)
β_1	constant defined in Eq. (6.28) for the case of periodic tube
χ	function defined by Eq. (6.69)
δ	boundary layer thickness (m)
ε	porosity of packed bed or porous medium
ϕ	phase angle (radians) in Eq. (1.12)
ϕ_p	represents the property being transported
γ	half taper angle $(^\circ)$
η	similarity variable
κ	dimensionless temperature difference defined by Eq. (6.41) for the periodic tube
λ	wavelength of the wavy surface (m)
μ	Newtonian fluid viscosity $(Pa \cdot s)$
μ^*	consistency index $(Pa \cdot s^n)$ of a power-law fluid
ν	kinematic viscosity of a Newtonian fluid (m^2/s) defined as (μ/ρ)
θ	dimensionless temperature distribution
θ_b	dimensionless bulk fluid temperature difference defined by Eq. (6.132)
$\theta_w(x_2)$	dimensionless surface temperature distribution $\theta(x_2, 0)$
ρ	fluid density (kg/m^3)
$\overline{\sigma}(x)$	function defining the profile of the wavy surface
$\sigma_1(x_1)$	function defining the profile of the wavy surface in Eq. (2.23)
$\sigma_r(x_1)$	function defining the profile of the wavy surface in relation to the mean radius
$\sigma_L(x_1), \sigma_U(x_1)$	function defining the profile of the lower and upper wavy surface of the horizontal channel in Eqs. (5.45)
τ	shear stress (Pa)
τ_w	wall shear stress (Pa)
$\overline{\omega}$	vorticity

ω_1	dimensionless vorticity
$\omega_1^{(0)}$	dimensionless vorticity solution at base state in the limit of $r_{2m} \to 0$ as defined by Eq. (6.147)
$\overline{\omega}_{B+1}$	value of vorticity at the point where the normal to the boundary wall at B crosses the next gridline
ξ	dimensionless temperature gradient defined by Eq. (6.35) for the periodic tube
ψ	stream function (m^2/s)
ψ_1	dimensionless stream function (first transformation)
ψ_2	dimensionless stream function (second transformation)
$\psi_1^{(0)}$	dimensionless stream function solution at base state in the limit of $r_{2m} \to 0$ as defined by Eq. (6.146)
$\psi_{1,0}, \psi_{1,1}, \psi_{1,2}$	components of the dimensionless stream function
$\overline{\psi}$	dimensionless stream function
$\overline{\psi}_0$	value of dimensionless stream function at the axis of symmetry in Eq. (6.16)
$\overline{\psi}_B$	value of dimensionless stream function at the boundary wall
$\overline{\psi}_{B+1}$	value of dimensionless stream function at the point where the normal to the boundary wall at B crosses the next gridline
$\overline{\psi}_m$	value of dimensionless stream function at the node m
$\overline{\psi}_{mm}$	value of dimensionless stream function at the node mm
$\overline{\psi}_N$	dimensionless trial stream function defined by Eq. (6.70)
$\overline{\psi}_w$	value of dimensionless stream function at the wall
Δ	deformation tensor
Φ_v	term equal to $(1/2)(\Delta : \Delta)$ and as given in Table 1.1 in rectangular coordinates (x, y)
$\overline{\Phi}_v$	term equal to $(1/2)(\Delta : \Delta)$ and defined in cylindrical coordinates (r_1, x_1) as given by Eq. (6.75)

∇ vector operator *del*

∇_1^2 operator defined by Eqs. (3.9), (3.28), (5.8)

∇^2 operator in r_1, x_1 coordinates and defined by Eq. (6.57)

$'$ prime denotes the differentiation with respect to η

Chapter 1

Governing Equations

Governing equations for flow from wavy surfaces are not truly different from those that are used when dealing with flat and smooth surfaces. The only difference lies in the use of an additional equation, which is needed for describing the profile of the wavy surface. Thus, the starting point is from the well-known Navier–Stokes equations. For example, for the simplest case of two-dimensional flow from a wavy wall, the relevant equations of continuity, momentum, and energy are written as follows:

$$\frac{\partial u}{\partial x} + \frac{\partial v}{\partial y} = 0 \tag{1.1}$$

$$u\frac{\partial u}{\partial x} + v\frac{\partial u}{\partial y} = g_x - \frac{1}{\rho}\frac{\partial p}{\partial x} - \frac{1}{\rho}\left(\frac{\partial \tau_{xx}}{\partial x} + \frac{\partial \tau_{xy}}{\partial y}\right) \tag{1.2}$$

$$u\frac{\partial v}{\partial x} + v\frac{\partial v}{\partial y} = g_y - \frac{1}{\rho}\frac{\partial p}{\partial y} - \frac{1}{\rho}\left(\frac{\partial \tau_{yx}}{\partial x} + \frac{\partial \tau_{yy}}{\partial y}\right) \tag{1.3}$$

$$u\frac{\partial T}{\partial x} + v\frac{\partial T}{\partial y} = \frac{l}{\rho C_p}\left(\frac{\partial^2 T}{\partial x^2} + \frac{\partial^2 T}{\partial y^2}\right) \tag{1.4}$$

where τ_{ij}, stress tensor, is related to the deformation tensor

$$\Delta_{ij} = [(\partial u_i/\partial x_j) + (\partial u_j/\partial x_i)]$$

through the rheological equation of state, thereby specifying the type under consideration. The stress components for two-dimensional flow of Newtonian fluids as well as non-Newtonian power-law fluids are given in Table 1.1.

Table 1.1. Shear Stress Components for Two-dimensional Flow
in x, y Coordinate System

Newtonian Fluids	Non-Newtonian Power-law Fluids
$\tau_{xx} = -\mu_0 \left(2\dfrac{\partial u}{\partial x} \right)$	$\tau_{xx} = -\mu^* \left(\left\lvert \sqrt{\phi_v} \right\rvert \right)^{n-1} \left(2\dfrac{\partial u}{\partial x} \right)$
$\tau_{yy} = -\mu_0 \left(2\dfrac{\partial v}{\partial y} \right)$	$\tau_{yy} = -\mu^* \left(\left\lvert \sqrt{\phi_v} \right\rvert \right)^{n-1} \left(2\dfrac{\partial v}{\partial y} \right)$
$\tau_{xy} = \tau_{yx} = -\mu_0 \left(\dfrac{\partial u}{\partial y} + \dfrac{\partial v}{\partial x} \right)$	$\tau_{xy} = \tau_{yx} = \mu^* \left(\left\lvert \sqrt{\phi_v} \right\rvert \right)^{n-1} \left(\dfrac{\partial u}{\partial y} + \dfrac{\partial v}{\partial x} \right)$ where $\phi_v = \dfrac{1}{2}(\Delta : \Delta)$ $= 2\left[\left(\dfrac{\partial u}{\partial x}\right)^2 + \left(\dfrac{\partial v}{\partial y}\right)^2 \right] + \left(\dfrac{\partial v}{\partial x} + \dfrac{\partial u}{\partial y} \right)^2$

The above equations are amenable to simplifications using the boundary layer theory (Bejan, 2014; Schlichting and Gersten, 2000; Pop and Ingham, 2001) wherever applicable and take similar forms to those that are commonly used for analyzing mass, momentum, and heat transport from flat and smooth surfaces. It is the following additional equation that differentiates a wavy surface from a flat, smooth one. The x-coordinate is specified along the surface while the y-coordinate is normal to it. Then, the geometric equation for a wavy surface as shown in Fig. 1.1 is defined as follows:

$$y = \overline{\sigma}(x) \tag{1.5}$$

where the function $\overline{\sigma}(x)$ takes various forms depending upon the profile of the wavy surface.

The choice of the form has been essentially dependent upon the preference of the investigator. A few expressions are cited below.

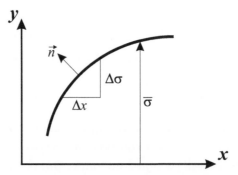

Fig. 1.1. Geometrical model for the wavy surface

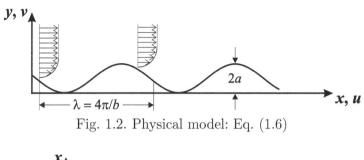

Fig. 1.2. Physical model: Eq. (1.6)

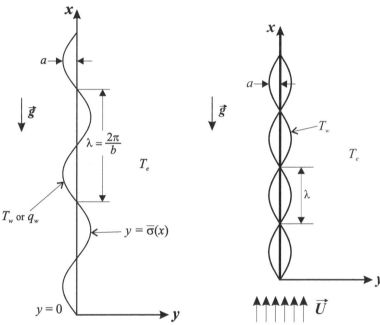

Fig. 1.3. Physical model: Fig. 1.4. Physical model:
Eq. (1.7)   Eq. (1.8a,b)

For example, during external flow past a wavy wall, the choice of the wave function could be in terms of a sine curve (Tsangaris and Potamitis, 1986; Vyas et al., 2004), Fig. 1.2,

$$y = a[1 + \sin(\lambda x)], \qquad (1.6)$$

(where λ is the wavelength of the wavy surface as shown in Figs. 1.2 and 1.4) or a sine curve (Shankar and Sinha, 1976; Yao, 1983; Moulic and Yao, 1989a,b; Kim and Chen, 1991), Fig. 1.3,

$$y = a \sin(bx) \qquad (1.7)$$

or a cosine curve (Lekoudis et al., 1970; Moulic and Yao, 1989a,b), Fig. 1.4,

$$y = a\cos(bx) \tag{1.8a}$$
$$y = (a/2)[1 - \cos(bx)] \tag{1.8b}$$

or an exponential curve (Lessen and Gangwani, 1976)

$$y = ae^{ibx}. \tag{1.9}$$

Thus, there are two additional length scales introduced for wavy surfaces. One is a, which is amplitude of the wavy surface, and b, which is the wavy number of the surface. When dealing with confined flow cases, such as flow through a channel, a number of different situations becomes obvious.

Case 1: flow between a wavy wall and a parallel wall (Fig. 1.5) (Vajravelu and Sastri, 1978; Das and Ahmed, 1992; Malashetty et al., 2001).

It can be easily seen that for this flow situation, Eq. (1.6) can be used for the wavy wall while $y = d$ would be applicable for the flat wall, where d is the distance between the wavy wall and the parallel wall.

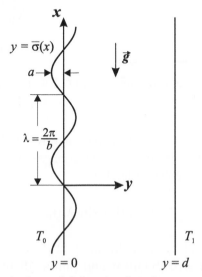

Fig. 1.5. Physical model for the flow between a wavy wall and a parallel wall

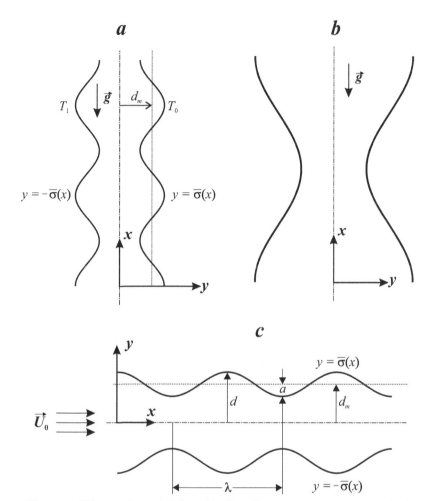

Fig. 1.6. Physical model for the flow between two symmetrical
wavy walls: a), b) vertical and c) horizontal channels

Case 2: flow between two symmetrically configured wavy walls
(Fig. 1.6) (Watson and Poots, 1971).
In this flow situation, the sinusoidal wave function is defined as follows:

$$\text{wall 1}: \quad y = d_m + a\sin(bx) \tag{1.10}$$

$$\text{wall 2}: \quad y = -d_m - a\sin(bx) \tag{1.11}$$

where $b = w/d_m$ and w, the spacing between protrusions, is defined as
$(2\pi d_m/\lambda)$ so that Eq. (1.7) holds good in this case, too. Note that d_m
is defined as the mean half width between the two walls. However, two

wavy walls may not always be configured as shown in Fig. 1.6. In fact, there could be other flow situations if asymmetry is introduced.

Case 3: flow between two asymmetrically configured wavy walls (Figs. 1.7 and 1.8) (Vajravelu, 1980 and 1989; Vajravelu and Sastri, 1980).

In such a combination of flow situations, Eq. (1.10) applicable to wall 1 is maintained; however, the equation for wall 2 is modified as follows:

$$\text{wall 2}: \quad y = -d_m - a\sin(bx + \phi) \tag{1.12}$$

where the values of ϕ equal to 0, $\pi/2$, π and $3\pi/2$ denote the changes in the orientation of the channel walls.

When dealing with flow through a conduit, a number of different situations exist depending on the chosen geometry, namely, square wave (Dullien and Azzam, 1973; Azzam and Dullien, 1977) (Fig. 1.9), conical (Sparrow and Prata, 1983) (Fig. 1.10), corrugated (Savvides and Gerrard, 1984) (Fig. 1.11), parabolic (Neira and Payatakes, 1978) (Fig. 1.12), sinusoidal (Chow and Soda, 1973) (Fig. 1.13), and the annulus (Prata and Sparrow, 1984) (Fig. 1.14).

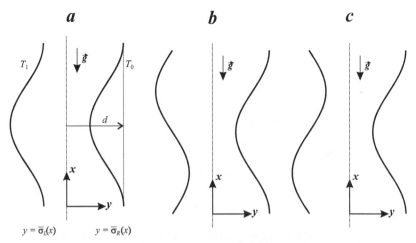

Fig. 1.7. Physical model for the flow between two vertical asymmetrical wavy walls

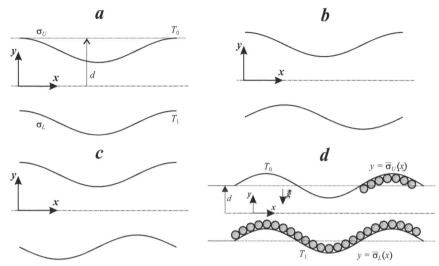

Fig. 1.8. Physical model for the flow between two horizontal
asymmetrical wavy walls

When the wall profile is sinusoidal (Chow and Soda, 1973), for
example, the following equation is used

$$r_w = r_m + a\sin(bx) \tag{1.13}$$

where r_m is mean radius of the wavy conduit. It can be seen that
Eq. (1.13) is not different from Eq. (1.10), which is applicable to a
symmetrically configured wavy channel.

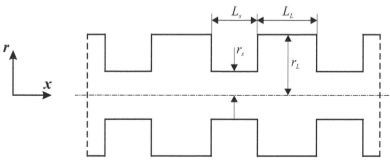

Fig. 1.9. Physical model for the square conduit

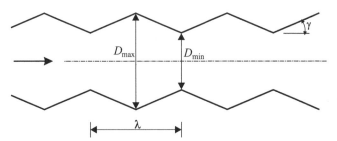

Fig. 1.10. Physical model for the conical conduit

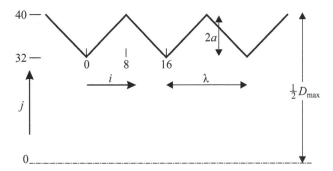

Fig. 1.11. Physical model for the corrugated conduit

During analyzes of flow and heat transfer past wavy surfaces, simple coordinate transformations are used in order to change the complex wavy geometry into a simple smooth one for which the governing equations can be solved by well-known methods. Although this procedure brings about considerable simplifications, the effort required to solve the transformed equations numerically are often just about the same as those of the original equations. However, the advantage of the transformation lies in the fact that the profile of the wavy surface can be changed during the numerical calculations without having to resort to realterations in the governing equations.

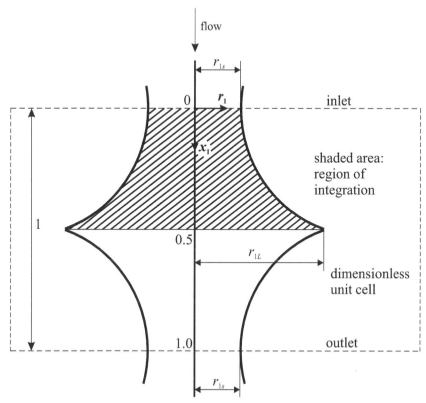

Fig. 1.12. Physical model for the parabolic conduit

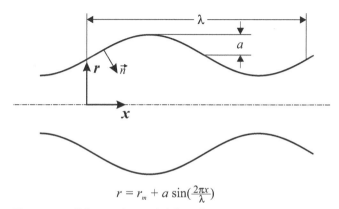

$$r = r_m + a \sin(\tfrac{2\pi x}{\lambda})$$

Fig. 1.13. Physical model for the sinusoidal conduit

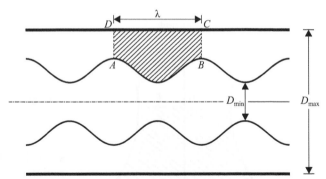

Fig. 1.14. Physical model for the annulus conduit

Chapter 2

Steady Natural and Mixed Convection Flow in Viscous Fluids over Wavy Vertical Wall

A. Natural Convection

The problem of natural convection from a wavy vertical heated surface to viscous fluids (without the presence of a porous medium) has received relatively less attention (Yao, 1983; Bhavnani and Bergles, 1991; Moulic and Yao, 1989a,b; Kishinami et al., 1990; Kim and Chen, 1991; Patel et al., 1991; Hossain and Rees, 1999; Jang et al., 2003; Jang and Yan, 2004) in comparison with the classical problem of the smooth flat vertical plate, as can be seen from the various review articles and books (Miles, 1957; Bird et al., 1960; Ostrach, 1952, 1972, 1982; Jaluria, 1980; Kakaç et al., 1987; Shenoy, 1986; Gebhart et al., 1988; Oosthuizen and Naylov, 1996; Pop and Ingham, 2001; Martynenko and Khramtsov, 2005; White, 2006; Leal, 2007; Andreev et al., 2011; Radko, 2013).

Yao (1983) was the first to analyze the Newtonian fluid natural convection flow past a wavy vertical wall. The geometry of the problem under consideration is shown in Fig. 1.3. The isothermal case was first tackled by Yao (1983) and the same procedure was used to analyze the uniform heat flux wavy vertical wall later on (Moulic and Yao, 1989a,b). Kim and Chen (1991) followed the transformation procedure of Yao (1983) and attempted a study of steady-state laminar heat transfer to non-Newtonian power-law fluids along the isothermal wavy vertical wall. Experimental work in this area has been done by Kishinami et al. (1990) and Bhavnani and Bergles (1991). Analysis and results of the above-mentioned theoretical as well as experimental works are discussed below.

Constant Temperature Case

1. Theoretical Developments

It is assumed that the wavy vertical wall is maintained at a constant temperature T_w and placed in an infinite expanse of Newtonian fluid, which is at a constant temperature T_e, such that $T_w > T_e$. It is also assumed that all physical properties of the fluid (except the density in the buoyancy term) are constant and that the natural convection flow that results from this situation is at steady-state, without heat generation and viscous dissipation.

The governing Eqs. (1.1)–(1.4) can then be written (Yao, 1983) in the non-dimensional form as follows:

$$\frac{\partial u_1}{\partial x_1} + \frac{\partial v_1}{\partial y_1} = 0 \tag{2.1}$$

$$u_1\frac{\partial u_1}{\partial x_1} + v_1\frac{\partial u_1}{\partial y_1} = -\frac{\partial p_1}{\partial x_1} + \sigma_{x_1}Gr^{1/4}\frac{\partial p_1}{\partial y_1} + (1+\sigma_{x_1}^2)\frac{\partial^2 u_1}{\partial y_1^2}$$

$$+\theta + Gr^{-1/2}\frac{\partial^2 u_1}{\partial x_1^2} - Gr^{-1/4}\left(\sigma_{x_1x_1}\frac{\partial u_1}{\partial y_1} + 2\sigma_{x_1}\frac{\partial^2 u_1}{\partial x_1\partial y_1}\right) \tag{2.2}$$

$$\sigma_{x_1x_1}u_1^2 + \sigma_{x_1}\theta = \sigma_{x_1}\frac{\partial p_1}{\partial x_1} - Gr^{1/4}(1+\sigma_{x_1}^2)\frac{\partial p_1}{\partial y_1}$$

$$-Gr^{-1/4}\left\{2\sigma_{x_1}\sigma_{x_1x_1} + u_1\frac{\partial v_1}{\partial x_1} + v_1\frac{\partial v_1}{\partial y_1} - (1+\sigma_{x_1}^2)\frac{\partial^2 v_1}{\partial y_1^2}\right\}$$

$$+Gr^{-1/2}\left\{\sigma_{x_1x_1x_1}u_1 + 2\sigma_{x_1x_1}\frac{\partial u_1}{\partial x_1} - \sigma_{x_1x_1}\partial v_1\partial y_1 - 2\sigma_{x_1}\frac{\partial^2 v_1}{\partial x_1\partial y_1}\right\} \tag{2.3}$$

$$u_1\frac{\partial\theta}{\partial x_1} + v_1\frac{\partial\theta}{\partial y_1} = \frac{1}{Pr}(1+\sigma_{x_1}^2)\frac{\partial^2\theta}{\partial y_1^2}$$

$$-\frac{1}{Pr}Gr^{-1/4}\left\{\sigma_{x_1x_1}\frac{\partial\theta}{\partial y_1} + 2\sigma_{x_1}\frac{\partial^2\theta}{\partial x_1\partial y_1}\right\} + \frac{1}{Pr}Gr^{-1/2}\frac{\partial^2\theta}{\partial x_1^2} \tag{2.4}$$

where subscript x_1 indicates differentiation with respect to x_1. The non-dimensional variables are defined as follows:

$$x_1 = \frac{x}{\lambda}; \quad y_1 = \frac{y - \overline{\sigma}}{\lambda} Gr^{1/4}; \quad p_1 = \frac{p}{\rho U_c^2};$$

$$\theta = \frac{T - T_e}{T_w - T_e}; \quad \sigma(x_1) = \frac{\overline{\sigma}(x)}{\lambda};$$

$$u_1 = \frac{u}{u_c}; \quad v_1 = \left(\frac{v - \sigma_x u}{u_c}\right) Gr^{1/4};$$

$$u_c = [\lambda g \beta (T_w - T_e)]^{1/2}. \tag{2.5}$$

Here, the Grashof number Gr is defined as follows:

$$Gr = g\beta(T_w - T_e)\lambda^3/\nu^2. \tag{2.6}$$

In what follows, it is assumed as done by Yao (1983) that the Grashof number is large enough so that heat transfer contributions are confined to a thin boundary layer along the wavy surface. However, if there is an intention to refine the analysis so as to make the results applicable to smaller values of Grashof number, then the solution ought to be obtained from the above equations based on the method of matched asymptotic expansions.

Thus, $Gr \to \infty$, boundary layer approximations can be utilized and Eqs. (2.2)–(2.4) can be rewritten in the simplified form as follows:

$$u_1 \frac{\partial u_1}{\partial x_1} + v_1 \frac{\partial u_1}{\partial y_1} = -\frac{\partial p_1}{\partial x_1} + \sigma_{x_1} Gr^{1/4} \frac{\partial p_1}{\partial y_1} + (1 + \sigma_{x_1}^2) \frac{\partial^2 u_1}{\partial y_1^2} + \theta \tag{2.7}$$

$$\sigma_{x_1 x_1} u_1^2 + \sigma_{x_1} \theta = \sigma_{x_1} \frac{\partial p_1}{\partial x_1} - Gr^{1/4}(1 + \sigma_{x_1}^2)\frac{\partial p_1}{\partial y_1} \tag{2.8}$$

$$u_1 \frac{\partial \theta}{\partial x_1} + v_1 \frac{\partial \theta}{\partial y_1} = \frac{1}{Pr}(1 + \sigma_{x_1}^2)\frac{\partial^2 \theta}{\partial y_1^2}. \tag{2.9}$$

The above equations are solved under the following boundary conditions:

$$\begin{aligned} y_1 = 0: \quad & u_1 = 0, \ v_1 = 0, \ \theta = 1 \\ y_1 \to \infty: \quad & u_1 = 0, \qquad\qquad \theta = 0. \end{aligned} \tag{2.10}$$

Since parameter $\sigma(x_1)$ characterizes the profile of the wall, when $\sigma \sim O(Gr^{-1/4})$, the effect of waviness becomes negligible and the boundary layer in Eqs. (2.1), (2.7)–(2.9) reduce exactly to those for the smooth flat plate. For finite values of $\sigma(x_1)$, the effect of waviness cannot be ignored. From Eq. (2.7), it is seen that the pressure gradient along the y_1-direction is $O(Gr^{-1/4})$, thus indicating that $Gr^{1/4}\partial p_1/\partial y_1 \sim O(1)$ from Eq. (2.8). The lowest-order pressure gradient along the x_1-direction is thus determinable from the inviscid solution and can be seen to give $\partial p_1/\partial x_1 = 0$. Consequently, elimination of $\partial p_1/\partial y_1$ between Eqs. (2.7) and (2.8) results in three equations that can be solved for u_1, v_1 and θ.

In order to obtain a numerical solution, Yao (1983) suggested the use of the following new parabolic variables:

$$x_2 = x_1; \quad y_2 = \frac{y_1}{(4x_1)^{1/4}};$$

$$u_2 = \frac{u_1}{(4x_1)^{1/2}}; \quad v_2 = (4x_1)^{1/2}v_1. \tag{2.11}$$

In terms of these variables, the appropriate boundary layer equations are written as:

$$(4x_2)\frac{\partial u_2}{\partial x_2} + 2u_2 - y_2\frac{\partial u_2}{\partial y_2} + \frac{\partial v_2}{\partial y_2} = 0 \tag{2.12}$$

$$(4x_2)u_2\frac{\partial u_2}{\partial x_2} + (v_2 - y_2u_2)\frac{\partial u_2}{\partial y_2} + \left(2 + \frac{4x_2\sigma_{x_2}\sigma_{x_2x_2}}{1+\sigma_{x_2}^2}\right)u_2^2$$

$$= \frac{\theta}{1+\sigma_{x_2}^2} + (1+\sigma_{x_2}^2)\frac{\partial^2 u_2}{\partial y_2^2} \tag{2.13}$$

$$(4x_2)u_2\frac{\partial \theta}{\partial x_2} + (v_2 - y_2u_2)\frac{\partial \theta}{\partial y_2} = \frac{1}{Pr}(1+\sigma_{x_2}^2)\frac{\partial^2 \theta}{\partial y_2^2} \tag{2.14}$$

subject to the boundary conditions (2.10). A numerical solution of Eqs. (2.12)–(2.14) has been obtained (Yao, 1983) using a finite difference scheme and the results have been presented for $\sigma = a_1\sin(2\pi x_2)$ with $a_1 = 0$ (smooth flat plate), $a_1 = 0.1$, and 0.3 for a value of $Pr = 1$. The velocity components u_2, v_2 and the temperature distribution θ at $x_2 = 1.5$ and 2.0 (nodes), and at $x_2 = 1.75$ (trough) and $x_2 = 2.25$ (crest) are shown in Figs. 2.1–2.3. It can be seen that the difference between velocity profiles at the nodes, trough, and crest are only slight and cannot be truly distinguished at the scale in the figures. However, it is obvious that the boundary layer is thicker near the nodes than near the trough and the crest.

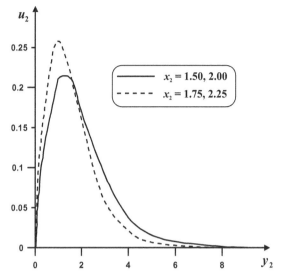

Fig. 2.1. Profiles of the velocity component u_2

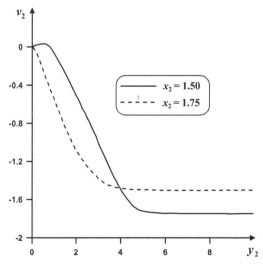

Fig. 2.2. Profiles of the velocity component v_2

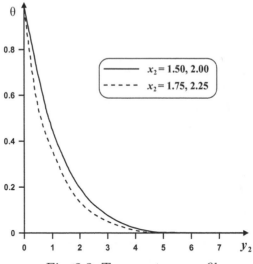

Fig. 2.3. Temperature profiles

The local Nusselt number is defined as

$$Nu = -\frac{\lambda}{k(T_w - T_e)}\overrightarrow{n} \cdot \nabla T \qquad (2.15)$$

where $\overrightarrow{n}(-\sigma_{x_2}/[1+\sigma_{x_2}^2]^{1/2}, 1/[1+\sigma_{x_2}^2]^{1/2})$ is the unit vector normal to the wavy surface as shown in Fig. 1.1. Using Eqs. (2.5) and (2.11) gives

$$Nu(4x_2/Gr)^{1/4} = -[1+\sigma_{x_2}^2]^{1/2}\left(\frac{\partial\theta}{\partial y_2}\right)\bigg|_{y_2=0}. \qquad (2.16)$$

A plot of local Nusselt number is shown in Fig. 2.4 for $a_1 = 0.1$ and 0.3, respectively.

It can be seen that for $a_1 = 0.1$, the curve actually approaches a constant that is slightly below the values of 0.5671 reported in Ostrach (1972) for the smooth vertical flat plate. Further, it is seen that near the leading edge, the magnitude of the local heat transfer rate depends on the slope of the wavy sinusoidal surface; hence, it is mainly controlled by stream motion induced by the buoyancy force parallel to the surface. On the other hand, in the downstream away from the leading edge, the heat transfer rate varies according to the orientation of the waviness. For that portion of the wavy surface that is parallel to the gravitational force, the velocity is larger and so is the heat transfer rate. The magnitude of the variation of the heat transfer rate decreases downstream, as expected, since the natural convection boundary layer thickens.

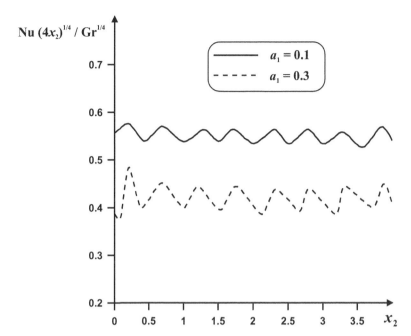

Fig. 2.4. Variation of the local Nusselt number with x_2

The average Nusselt number can be obtained by integrating Eq. (2.16) and is written as follows:

$$\overline{Nu}/Gr^{1/4} = \frac{1}{S} \int_0^{x_2} \left[\frac{1 + \sigma_{x_2}^2}{(4x_2)^{1/4}} \left(\frac{\partial \theta}{\partial y_2} \right) \right]_{y_2=0} dx_2 \qquad (2.17)$$

where

$$S = \int_0^{x_2} (1 + \sigma_{x_2}^2)^{1/2} dx_2. \qquad (2.18)$$

Figure 2.5 illustrates the results of $\overline{Nu}/Gr^{1/4}$ for $a_1 = 0$ (flat plate), 0.1, and 0.3.

The profiles clearly show that the average Nusselt number for the sinusoidal wavy vertical wall is smaller than that for the corresponding smooth vertical flat plate. For $a_1 = 0.3$, wavy variation of average Nusselt number can be observed only near the leading edge, and gradually disappears downstream.

It should be mentioned that there is only very little work published yet on flow and heat transfer of non-Newtonian fluids over wavy surfaces. In fact, the governing equations for such fluids can be obtained following the same procedure of Yao (1983) for Newtonian fluids.

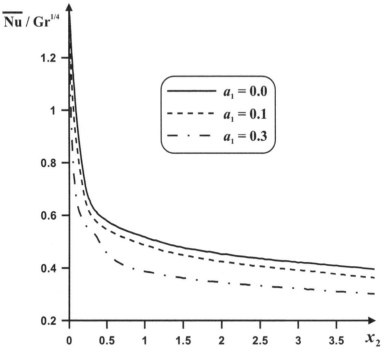

Fig. 2.5. Variation of the average Nusselt number with x_2

Kim and Chen (1991) attempted a study of steady-state laminar free convection of non-Newtonian power-law fluids along an isothermal wavy vertical wall. The viscous term for power-law fluids resulted from $(\partial \tau_{xx}/\partial x + \partial \tau_{xy}/\partial y)$ and $(\partial \tau_{yy}/\partial y + \partial \tau_{xy}/\partial x)$ by an appropriate substitution of the terms from Table 1.1. However, since the main focus of the present book is on Newtonian fluids, we omit to present here results on non-Newtonian fluids. On the other hand, since there is only scanty information in the literature on non-Newtonian fluids, we intend to present here only those results that have already been published in greater detail.

2. Experimental Efforts

Natural convection heat transfer experiments have been conducted by Kishinami et al. (1990) and Bhavnani and Bergles (1991) for Newtonian fluid flow past wavy vertical surfaces. In the study by Kishinami et al. (1990), the wavy vertical surface is formed by connecting semi-circular concave elements to semi-circular convex elements. The boundary condition imposed is that alternating adjacent sections are heated and unheated. Thus, two geometries are studied, namely one in which the concave surfaces are heated, and second, in which the convex surfaces are heated. The studies are conducted with calorimetric as well as interferometric methods. They report separation of the boundary layer and fluctuations in the flow field. Their interferograms are, however, a little difficult to interpret since a greater number of fringes are shown for the interferogram that represents lower Grashof numbers. The Nusselt numbers are correlated using a parameter that includes a ratio of the thermal conductivity of the unheated wall to the fluid conductivity and thickness of the unheated wall. The average Nusselt number is found to be lower in the case of the concave heated surfaces than that for the convex heated surfaces.

Bhavnani and Bergles (1991) present an experimental study of natural convection heat transfer characteristics of isothermal sinusoidal wavy vertical surfaces to Newtonian fluids. A Mach–Zehnder interferometer (MZI) with 100 mm optics is used in their study. The MZI consists of a 2 mW helium-neon laser light source with a wavelength of 6328 angstroms, two mirrors, and two beam splitters. All the components of the MZI are positioned in such a way that the light beam is incident on each of them at an angle of 30°. The sides of the interferometer are shrouded with clear polyethylene sheeting so as to eliminate external disturbances during observation and experiment. The interferometer is housed in a windowless room located in an interior part of the building in order to maintain a controlled atmosphere, which is essential for natural convection studied in air.

The wavy test section is made from aluminum plate with a mirror finish surface. Two types of orientations are studied in order to get two different leading-edge conditions — one is similar to that shown in Fig. 1.3, while the other is a mirror image of it. The test section length is 152.4 mm and the wavelength is 50.8 mm. Three test sections with amplitudes of 2.54 mm, 5.08 mm, and 15.24 mm are used, thereby resulting in amplitude-to-wavelength ratios of 0.05, 0.1, and 0.3, respectively.

Ten 1.58 mm-diameter holes are drilled into the rear surface of the plates at positions such that the copper-constantan thermocouples placed in them are located at the nodes of the sinusoidal profile. The isothermal boundary condition is established with four individually controlled heater assemblies and monitored by the thermocouples.

The test section assembly is suspended in the measurement path of the interferometer. Proper alignment procedures are followed in order to ensure that the optical table is leveled, the test section is vertical, and the light beam is focussed. After the alignment is completed, the heaters are set on and left for 3 to 4 hours to ensure steady-state conditions. Experiments are conducted at a temperature resulting in a light fringe being the one closest to the wall. This increases the accuracy with which the wall profile can be identified on the interferogram, to a great extent.

A composite interferogram of the test section with amplitude-to-wavelength ratio of 0.1 is shown in Fig. 2.6. The figure shows the two orientations – Fig. 2.6a is for the leading edge facing upwards and Fig. 2.6b is for the leading edge facing downwards (reversed orientation). The Grashof number based on profile length of this surface for the tests represented in Fig. 2.6 is 1.6×10^7.

This interferogram is analyzed at several different x locations using a toolmaker's microscope, and the fringe-shift information that is obtained is used to evaluate a predicted wall temperature T_w and the slope at the wall dT/dy using a curve-fitting program.

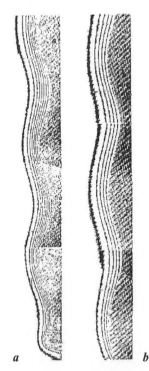

Fig. 2.6. Composite interferogram of the test section [reprinted from Bhavnani and Bergles (1991) with permission from Springer]

Local heat transfer coefficients and Nusselt numbers are calculated using conventional definitions, namely, $h_x = -k(dT/dy)/(T_w - T_e)$ and $Nu_x = h_x x/k$, respectively. In order to validate the experimental method, data is first obtained for a smooth vertical flat plate and compared with the correlation of Ostrach (1972), namely, $Nu = 0.355Gr^{0.25}$ with the Prandtl number $Pr = 0.709$. All experiments are conducted at temperatures that yielded this Prandtl number. It is found that there is a difference of 3.4% in the mean integrated heat transfer coefficient between data obtained from interferograms and that obtained using Ostrach correlation. Except for the data near the leading edge, the overall comparison with the analytical solution of Ostrach (1972) is found to be good, thereby instilling confidence in the experimental method.

The effects of two parameters are studied, namely, the amplitude-to-wavelength ratio and the orientation of the plate at the leading edge. Figure 2.7 shows the data obtained for the wavy vertical surface at the leading edge facing upwards for $a_1 = 0.05$, 0.1, and 0.3 respectively. The results from Yao (1983) converted to $Pr = 0.710$ are also shown in Figs. 2.7b and 2.7c for comparison. Although there is qualitative agreement in trends, fluctuations in the local heat transfer coefficient do not agree in terms of magnitude or frequency. Yao (1983) reports that natural convection experiments in air are often affected by low velocity forced convection flow over the test surface due to the circulation currents originating from the wall of the room. However, Bhavnani and Bergles (1991) explain that the discrepancy between the results of Yao (1983) and their own is probably due to the relatively large thickness of the boundary layer in comparison to the radius of curvature of the surface. In all cases in Fig. 2.7, the variation in heat transfer coefficient is periodic in nature and the periodicity is equal to the wavelength of the sinusoidal surface. The boundary layer is thicker at the troughs than at the crests. Hence, the local heat transfer coefficient decreases starting from the crest until the trough and then increases monotonically from the trough to reach a maximum at the crest. At the crest, the local heat transfer coefficient is greater than the corresponding plate value.

Experimental data analysis shows that the increase in heat transfer relative to the flat plate of equal projected area occurs due to an increase in surface area for the wavy surface and not due to any improvement in heat transfer mechanism. In other words, a flat plate with a surface area equal to that of the wavy surface would transfer more heat.

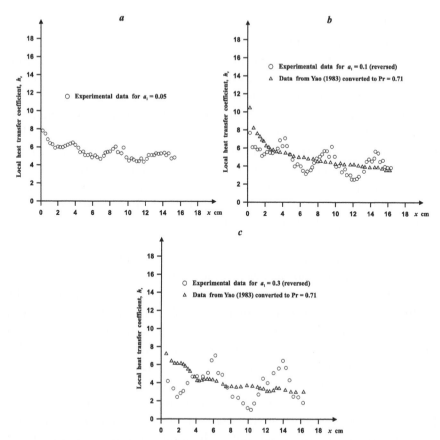

Fig. 2.7. Variation of the local heat transfer coefficient
with x for different values of a_1

It is found that the average local heat transfer coefficient decreases
as a_1 increases. This decrease is quite severe in the case of the surface
with $a_1 = 0.3$.

For this amplitude-to-wavelength ratio, it is observed that the thermal boundary layer becomes unstable and starts undergoing transition
at a location in the second cycle where the Grashof number based on
profile length is 2×10^7. This value is one to two orders of magnitude
lower than the transition Grashof number for flat vertical alignment of
the buoyancy force with respect to the solid surface, as indicated by
the fourth term on the right-hand side of Eq. (2.13).

B. Mixed Convection

The problem of mixed convection flow past a symmetric wavy vertical wall as shown in Fig. 1.4 and maintained under constant temperature conditions has been studied by Moulic and Yao (1989a,b). In the following, results of the theoretical analysis and discussion of Moulic and Yao (1989a,b) are presented. It is assumed that the axis of symmetry of the wavy vertical surface is aligned with the oncoming stream of uniform velocity U and temperature T_e. It is also assumed that the wavy vertical surface is maintained at a constant temperature T_w such that $T_w > T_e$. The symmetric wavy surface as shown in Fig. 1.4 is described by $y = \pm\sigma(x)$ and the mathematical formulation is valid for any function $\sigma(x)$ for which $\sigma(0) = \sigma_x(0) = 0$. The numerical solution is provided for the specific case of a surface described by Eq. (1.8b).

The governing Eqs. (1.1)–(1.4) can be expressed in non-dimensional form as follows:

$$\frac{\partial u_1}{\partial x_1} + \frac{\partial v_1}{\partial y_1} = 0 \tag{2.19}$$

$$u_1\frac{\partial u_1}{\partial x_1} + v_1\frac{\partial u_1}{\partial y_1} = -\frac{\partial p_1}{\partial x_1} + \sigma_{x_1}Re^{1/2}\frac{\partial p_1}{\partial y_1} + (1+\sigma_{x_1}^2)\frac{\partial^2 u_1}{\partial y_1^2} + \frac{Gr}{Re^2}\theta$$

$$+Re^{-1}\frac{\partial^2 u_1}{\partial x_1^2} - Re^{-1/2}\left(\sigma_{x_1x_1}\frac{\partial u_1}{\partial y_1} + 2\sigma_{x_1}\frac{\partial^2 u_1}{\partial x_1\partial y_1}\right) \tag{2.20}$$

$$Re^{-1/2}\left(u_1\frac{\partial v_1}{\partial x_1} + v_1\frac{\partial v_1}{\partial y_1}\right) + \sigma_{x_1x_1}u_1^2 = \sigma_{x_1}\frac{\partial p_1}{\partial x_1} - Re^{1/2}(1+\sigma_{x_1}^2)\frac{\partial p_1}{\partial y_1}$$

$$-\frac{Gr}{Re^2}\sigma_{x_1}\theta - Re^{-1/2}\left\{2\sigma_{x_1}\sigma_{x_1x_1}\frac{\partial u_1}{\partial y_1} - (1+\sigma_{x_1}^2)\frac{\partial^2 v_1}{\partial y_1^2}\right\}$$

$$+Re^{-1}\left\{\sigma_{x_1x_1x_1}u_1 + 2\sigma_{x_1x_1}\frac{\partial u_1}{\partial x_1} - \sigma_{x_1x_1}\frac{\partial v_1}{\partial y_1} - 2\sigma_{x_1}\frac{\partial^2 v_1}{\partial x_1\partial y_1}\right\}$$

$$+Re^{-3/2}\frac{\partial^2 v_1}{\partial x_1^2} \tag{2.21}$$

$$u_1\frac{\partial\theta}{\partial x_1} + v_1\frac{\partial\theta}{\partial y_1} = \frac{1}{Pr}(1+\sigma_{x_1}^2)\frac{\partial^2\theta}{\partial y_1^2}$$

$$-\frac{1}{Pr}Re^{-1/2}\left\{\sigma_{x_1x_1}\frac{\partial\theta}{\partial y_1} + 2\sigma_{x_1}\frac{\partial^2\theta}{\partial x_1\partial y_1}\right\} + \frac{1}{Pr}Re^{-1}\frac{\partial^2\theta}{\partial x_1^2} \tag{2.22}$$

where subscript x_1 indicates differentiation with respect to x_1. The non-dimensional variables are defined as follows:

$$x_1 = bx; \quad y_1 = b(y - \overline{\sigma})Re^{1/2}; \quad p_1 = \frac{p}{\rho U^2}; \quad \theta = \frac{T - T_e}{T_w - T_e};$$

$$u_1 = \frac{u}{U}; \quad v_1 = \left(\frac{v - \sigma_x u}{U}\right) Re^{1/2};$$

$$\sigma(x_1) = b\overline{\sigma}(x) = a_2 \sigma_1(x_1); \quad a_2 = 2\pi a_1 = 2\pi a/\lambda. \tag{2.23}$$

Here, the Grashof number Gr and the Reynolds number Re are defined as follows:

$$Gr = g\beta(T_w - T_e)\lambda^3/(8\pi^3\nu^2); \quad Re = U\lambda/(2\pi\nu). \tag{2.24}$$

In what follows, it is assumed as done by Moulic and Yao (1989a,b) that the term Gr/Re^2 is finite, but the Reynolds number is large enough so that the motion is confined to a thin boundary layer along the wavy surface. As $Re \to \infty$, boundary layer approximations can be utilized and Eqs. (2.20)–(2.22) can be rewritten in the simplified form as follows:

$$u_1\frac{\partial u_1}{\partial x_1} + v_1\frac{\partial u_1}{\partial y_1} = -\frac{\partial p_1}{\partial x_1} + \sigma_{x_1}Re^{1/2}\frac{\partial p_1}{\partial y_1} + (1+\sigma_{x_1}^2)\frac{\partial^2 u_1}{\partial y_1^2} + \frac{Gr}{Re^2}\theta \tag{2.25}$$

$$\sigma_{x_1 x_1}u_1^2 = \sigma_{x_1}\frac{\partial p_1}{\partial x_1} - Re^{1/2}(1+\sigma_{x_1}^2)\frac{\partial p_1}{\partial y_1} - \frac{Gr}{Re^2}\sigma_{x_1}\theta \tag{2.26}$$

$$u_1\frac{\partial \theta}{\partial x_1} + v_1\frac{\partial \theta}{\partial y_1} = \frac{1}{Pr}(1+\sigma_{x_1}^2)\frac{\partial^2\theta}{\partial y_1^2}. \tag{2.27}$$

The above equations are solved under the following boundary conditions:

$$y_1 = 0: \quad u_1 = 0, \ v_1 = 0, \quad \theta = 1$$
$$y_1 \to \infty: \quad u_1 = u_{ew}(x_1), \quad \theta = 0 \tag{2.28}$$

where $u_{ew}(x_1)$ is the x_1 component of the inviscid velocity at the wavy surface $\sigma(x_1)$ and can be approximated for small values of a_2 by the following expansion (Moulic and Yao, 1989a,b):

$$u_{ew}(x_1) = 1 + a_2\left[\frac{1}{\pi}\int_0^\infty \frac{\sigma_{x_1 x_1}(z)}{x_1 - z}dz\right] + O(a_2^2). \tag{2.29}$$

It is to be noted that when $\sigma \sim O(Re^{-1/2})$, the effect of waviness becomes negligible and the boundary layer Eqs. (2.19), (2.25)–(2.27) reduce exactly to those for the smooth flat plate.

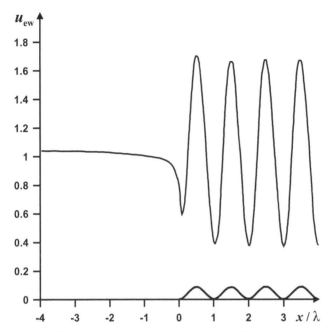

Fig. 2.8. Variation of the ideal (inviscid) velocity with x/λ

For finite values of $\sigma(x_1)$, the effect of waviness cannot be ignored. From Eq. (2.25) it is seen that the pressure gradient along the y_1-direction is $\sim O(Re^{-1/2})$, thereby indicating that $Re^{1/2}\partial p_1/\partial y_1 \sim O(1)$ from Eq. (2.26). The variation of u_{ew} with x is shown in Fig. 2.8.

The pressure gradient along the x_1-direction is thus determinable from the inviscid solution and gives the following:

$$\frac{dp_1}{dx_1} = -\left[(1+\sigma_{x_1}^2)u_{ew}\frac{du_{ew}}{dx_1} + \sigma_{x_1}\sigma_{x_1 x_1}u_{ew}^2\right]. \tag{2.30}$$

Equations (2.29) and (2.30) indicate that the inviscid solution has a singularity at the leading edge of the wavy surface. At $x_1 = 0$, du_{ew}/dx_1 is then infinite. The variation of dp_1/dx_1 with x/λ is illustrated in Fig. 2.9.

In order to remove the singularity, Moulic and Yao (1989a,b) use the following new set of parabolic variables:

$$x_2 = x_1; \quad y_2 = \left(\frac{u_{ew}}{2x_1}\right)^{1/2} y_1;$$

$$u_2 = \frac{u_1}{u_{ew}}; \quad v_2 = \left(\frac{2x_1}{u_{ew}}\right)^{1/2} v_1. \tag{2.31}$$

In terms of these variables, the appropriate boundary layer equations are written as:

$$(2x_2)\frac{\partial u_2}{\partial x_2} + \frac{x_2}{u_{ew}}\frac{du_{ew}}{dx_2}u_2 - y_2\frac{\partial u_2}{\partial y_2} + \frac{\partial v_2}{\partial y_2} = 0 \tag{2.32}$$

$$(2x_2)u_2\frac{\partial u_2}{\partial x_2} + (v_2 - y_2u_2)\frac{\partial u_2}{\partial y_2} + \left(\frac{2x_2}{u_{ew}}\frac{du_{ew}}{dx_2} + \frac{2x_2\sigma_{x_2}\sigma_{x_2x_2}}{1+\sigma_{x_2}^2}\right)(u_2^2 - 1)$$

$$= \frac{Gr}{(u_{ew}Re)^2} + \frac{2x_2}{1+\sigma_{x_2}^2}\theta + (1+\sigma_{x_2}^2)\frac{\partial^2 u_2}{\partial y_2^2} \tag{2.33}$$

$$(2x_2)u_2\frac{\partial\theta}{\partial x_2} + (v_2 - y_2u_2)\frac{\partial\theta}{\partial y_2} = \frac{1}{Pr}(1+\sigma_{x_2}^2)\frac{\partial^2\theta}{\partial y_2^2} \tag{2.34}$$

subject to the following boundary conditions:

$$y_2 = 0: \quad u_2 = 0,\ v_2 = 0,\ \theta = 1$$
$$y_2 \to \infty: \quad u_2 = 1, \quad\quad\quad \theta = 0. \tag{2.35}$$

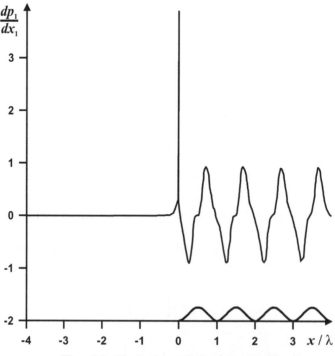

Fig. 2.9. Variation of dp_1/dx_1 with x/λ

A numerical solution of Eqs. (2.32)–(2.35) has been obtained (Moulic and Yao, 1989a,b) using a finite difference scheme (Blottner, 1970; Faghri et al., 1984; Blazek, 2001) and results have been presented for $\sigma_1 = (1/2)(1 - \cos x_2)$ with $a_1 = 0$ (smooth flat plate), $a_1 = 0.1$, and 0.3 for a value of $Pr = 1$. The required solution for u_{ew} is

$$u_{ew}(x_2) = 1 + a_2 \left[\frac{1}{2\pi} \int_0^\infty \frac{e^{-z_1 x_2}}{1 + z_1^2} dz_1 - \frac{1}{2} \cos x_2 \right] + O(a_2^2) \qquad (2.36)$$

It should be pointed out that there are also quite many published papers on this topic, such as, for example, Shankar and Sinha (1976); Vajravelu (1989); Chiu and Chou (1994); Yang et al. (1996); Hsu et al. (2000); Chu et al. (2002); Tashtoush and Al-Odat (2004); Wang and Chen (2005a,b); Molla and Hossain (2007); Molla et al. (2007); Molla and Gorla (2009); Nejad et al. (2015).

Chapter 3

Steady Natural Convection Flow in Fluid-Saturated Porous Media over Wavy Vertical Wall

A. Constant Temperature

Heat transfer by natural convection from various types of surfaces embedded in a porous medium is of fundamental importance in many practical applications, such as, insulation of buildings and equipment, energy storage and recovery systems, the storage of heat-generating materials such as grains and coal, geothermal reservoirs, nuclear waste disposal, and chemical reactor engineering (Sheffield and Metzner, 1976; Hickox et al., 1980; Wakao and Kaguei, 1982; Riley, 1988; Whitaker, 1989; Lemcoff et al., 1990; Farr et al., 1991; Lai, 2000; Vafai and Hadim, 2000; Sahimi, 2011; Sha, 2011; Narasimhan, 2013; Nield and Bejan, 2013; Singh et al., 2014; Umavathi and Shekar, 2015). Geophysical applications range from the flow of groundwater around hot intrusions to the stability of snow against avalanches. The above-mentioned applications involve Newtonian fluids that saturate porous media. However, in ceramic engineering, enhanced oil recovery, and filtration, there are instances where non-Newtonian fluid heat transfer studies in porous media assume importance. The significant research work on non-Newtonian fluid heat transfer in porous media has been well reviewed by Shenoy (1994) and can be referred to for details.

An excellent and extensive coverage of topics relating to convective Newtonian fluid flow in porous media can be found in the monographs by Kaviany (1991) as well as Nield and Bejan (2013). Thermal convection from a variety of surfaces embedded in porous media have been studied by a number of research workers as documented in the review articles (Cheng, 1985; Shenoy, 1994; Pop and Ingham, 2000); however, studies relating to embedded wavy surfaces have been missing. The first such studies are those of Rees and Pop (1994a,b), who considered the natural convection flow past constant temperature and constant heat flux wavy vertical surfaces embedded in Newtonian fluid-saturated porous media. In the following, the analysis and results of these works (Rees and Pop, 1994a,b) are briefly discussed.

The problem deals with natural convection Darcy flow along a wavy vertical wall placed in an isotropic porous medium saturated with a Newtonian fluid. The assumption that Darcy's law is obeyed implies that the seepage velocity is sufficiently small and in turn, the Reynolds/Grashof number based on a typical pore or particle diameter is of the order of unity or smaller. It is to be noted in passing that extensions of Darcy's law through inclusion of non-linear porous inertia term known as Forchheimer term (Nakayama and Pop, 1991; Shenoy, 1992; Nakayama and Shenoy, 1992a,b, 1993), and viscous effects term known as Brinkman term (Shenoy, 1993a,b), can be considered for additional investigations in this class of problems.

It is assumed that the wavy vertical wall (Fig. 1.3) is maintained at a constant temperature T_w and placed in an infinite expanse of Darcian fluid, which is at a constant temperature T_e, such that $T_w > T_e$. It is also assumed that all physical properties of the fluid (except the density in the buoyancy term) are constant and that the natural convection flow that results from this situation is at steady-state and free from effects of radiative heat transfer, viscous dissipation, and work done by pressure changes, as they are assumed to be negligible.

A non-dimensional form of Darcy's law and the energy equation using Boussinesq approximation can be written (Rees and Pop, 1994a) as follows:

$$\nabla^2 \psi_1 = Ra \frac{\partial \theta}{\partial y_1} \tag{3.1}$$

$$\nabla^2 \theta = \frac{\partial \psi_1}{\partial y_1} \frac{\partial \theta}{\partial x_1} - \frac{\partial \psi_1}{\partial x_1} \frac{\partial \theta}{\partial y_1} \tag{3.2}$$

where the non-dimensional variables are defined as follows:

$$x_1 = \frac{x}{\lambda}; \quad y_1 = \frac{y}{\lambda}; \quad \psi_1 = \frac{\psi}{\alpha}; \quad \theta = \frac{T - T_e}{T_w - T_e}. \tag{3.3}$$

Here, Ra is defined as follows:

$$Ra = Kg\beta(T_w - T_e)\lambda/(\alpha\nu). \tag{3.4}$$

It is designated as the modified Rayleigh number for a porous medium with K being the permeability of the medium. The above equations are to be solved under the following boundary condition:

$$\begin{aligned} y_1 = \sigma(x_1): \quad &\psi_1 = 0, \quad \theta = 1 \\ y_1 \to \infty: \quad &\frac{\partial \psi_1}{\partial y_1} = 0, \quad \theta = 0. \end{aligned} \tag{3.5}$$

It is now assumed that the Rayleigh number is large ($Ra \gg 1$) such that natural convection takes place within a boundary layer whose cross-stream width is substantially smaller than $O(1)$ amplitude of the surface waves. Accordingly, the new variables are then defined by subtracting out the effect of the surface waves as

$$x_2 = x_1; \quad y_2 = (y_1 - \sigma)Ra^{1/2}; \quad \psi_2 = \psi_1/Ra^{1/2}. \tag{3.6}$$

A straightforward substitution of Eq. (3.6) into Eqs. (3.1) and (3.2) yields the following:

$$\nabla_1^2 \psi_2 = \frac{\partial \theta}{\partial y_2} \tag{3.7}$$

$$\nabla_1^2 \theta = \frac{\partial \psi_2}{\partial y_2}\frac{\partial \theta}{\partial x_2} - \frac{\partial \psi_2}{\partial x_2}\frac{\partial \theta}{\partial y_2} \tag{3.8}$$

where the operator ∇_1^2 is defined as follows:

$$\nabla_1^2 = (1 + \sigma_{x_2}^2)\frac{\partial^2}{\partial y_2^2} - Ra^{-1/2}\left(2\sigma_{x_2}\frac{\partial^2}{\partial x_2 \partial y_2} + \sigma_{x_2 x_2}\frac{\partial}{\partial y_2}\right)$$
$$+ Ra^{-1}\frac{\partial^2}{\partial x_2^2}. \tag{3.9}$$

The boundary conditions of Eq. (3.5) now get transformed to

$$\begin{aligned} y_2 = 0: \quad &\psi_2 = 0, \quad \theta = 1 \\ y_2 \to \infty: \quad &\frac{\partial \psi_2}{\partial y_2} = 0, \quad \theta = 0. \end{aligned} \tag{3.10}$$

At large values of Rayleigh number ($Ra \rightarrow \infty$), thermal boundary layers are formed, and are described by the following boundary layer equations:

$$(1 + \sigma_{x_2}^2)\frac{\partial^2 \psi_2}{\partial y_2^2} = \frac{\partial \theta}{\partial y_2} \tag{3.11}$$

$$(1 + \sigma_{x_2}^2)\frac{\partial^2 \theta}{\partial y_2^2} = \frac{\partial \psi_2}{\partial y_2}\frac{\partial \theta}{\partial x_2} - \frac{\partial \psi_2}{\partial x_2}\frac{\partial \theta}{\partial y_2}. \tag{3.12}$$

A similarity solution of Eqs. (3.11) and (3.12) is possible. The dimensionless stream function and temperature take the following form:

$$\psi_2 = x_2^{1/2} f(\eta); \quad \theta = \theta(\eta) \tag{3.13}$$

where the similarity variable η is defined as

$$\eta = \frac{y_2}{(1 + \sigma_{x_2}^2)x_2^{1/2}}. \tag{3.14}$$

Using Eqs. (3.13) and (3.14) along with the fact that $\theta = f'(\eta)$, Eqs. (3.11) and (3.12) reduce to the following:

$$f''' + \frac{1}{2}ff'' = 0 \tag{3.15}$$

along with the boundary conditions

$$f(0) = 0; \quad f'(0) = 1; \quad f'(\infty) = 0. \tag{3.16}$$

The primes in the above Eqs. (3.15) and (3.16) denote differentiation with respect to η.

It is worth mentioning that Eq. (3.15) is precisely the same equation as that obtained by Cheng and Minkowycz (1977) in their study of thermal boundary layer flow induced by an isothermal flat vertical surface immersed in a porous medium. This indicates that flow from a wavy vertical surface placed in a porous medium depends only on the local slope of the surface wave. This is in contrast to an identical configuration in viscous fluids without porous medium considered by Yao (1983). As discussed in Chapter 2, the boundary layer equations for that case formed a set of parabolic partial differential equations whose solution has to be obtained numerically, thus implying that the flow is dependent on conditions upstream as well.

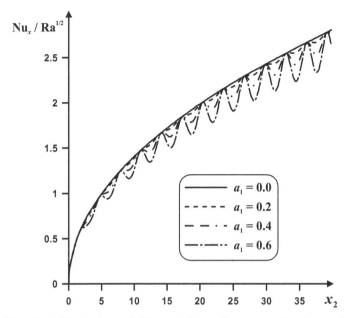

Fig. 3.1. Variation of Nu_x with x_2 for several values of a_1

In the present case of flow through porous media, the matter is much more simplified and the local Nusselt number can be expressed as

$$Nu/Ra^{1/2} = \frac{-\theta'(0)x_2^{1/2}}{(1+\sigma_{x_2}^2)^{1/2}} = \frac{0.44375x_2^{1/2}}{(1+\sigma_{x_2}^2)^{1/2}} \qquad (3.17)$$

where the numerical value of $-\theta'(0) = 0.44375$ has been taken from Rees and Bassom (1991).

The variation of Nu_x with x for various values of the dimensionless wave amplitude a is shown in Fig. 3.1. We can see that the values of Nu_x are less than or equal to that corresponding to a flat plate embedded in a porous medium, which can be explained in following way. When the heated surface is not vertical the component of the buoyancy force along the surface is reduced from its maximum value as in the case when the surface is vertical. Consequently, the boundary layer thickness is locally thicker and therefore the local rates of heat transfer at the surface are reduced.

B. Constant Heat Flux

In this section we consider the effects of transverse surface waves on the free convective boundary layer induced by a uniform heat flux vertical surface embedded in a porous medium. Since the transformed boundary layer equations are nonsimilar, they are solved numerically using the Keller box method (see Keller and Cebeci, 1971; Keller, 1978). The distribution of the wall temperature along the wavy surface is presented and the accuracy is verified by comparing with previous results for a vertical flat plate embedded in a porous medium obtained by Kumar et al. (1997); Kumar (2000); Kumar and Gupta (2005).

Consider a vertical surface with transverse waves embedded in a porous medium with constant ambient temperature, T_∞, as shown in Fig. 1.3. In particular, we assume that the surface profile is given by Eq. (1.7). The local heat flux rate normal to the surface is maintained at a constant value, q_w. The flow is considered to be steady and two-dimensional. In terms of dimensionless variables, the Darcy and energy equations can be written as

$$\frac{\partial^2 \psi}{\partial x^2} + \frac{\partial^2 \psi}{\partial y^2} = Ra \frac{\partial \theta}{\partial y}, \tag{3.18}$$

$$\frac{\partial^2 \theta}{\partial x^2} + \frac{\partial^2 \theta}{\partial y^2} = \frac{\partial \psi}{\partial y} \frac{\partial \theta}{\partial x} - \frac{\partial \psi}{\partial x} \frac{\partial \theta}{\partial y}. \tag{3.19}$$

The boundary conditions that apply are the following:

$$\psi = 0, \ \frac{\partial \theta}{\partial y} - a \cos(\pi x) \frac{\partial \theta}{\partial x} = -\sigma \text{ on } y = s_t(x) = \frac{a}{\pi} \sin(\pi x), \tag{3.20}$$

$$\frac{\partial \psi}{\partial y} \to 0, \ \theta \to 0 \text{ as } y \to \infty, \tag{3.21}$$

where σ, a function associated with the geometry of the surface, is given by

$$\sigma = \left(1 + \left(\frac{ds_t}{dx}\right)^2\right)^{1/2} = (1 + a^2 \cos^2(\pi x))^{1/2}. \tag{3.22}$$

The effect of the wavy surface boundary conditions can be transferred to the governing equations by means of the transformation

$$\hat{x} = x, \ \hat{y} = y - \frac{a}{\pi} \sin(\pi x). \tag{3.23}$$

Equations (3.18) and (3.19) become

$$\nabla_1^2 \psi = Ra \frac{\partial \theta}{\partial \hat{y}}, \tag{3.24}$$

$$\nabla_1^2 \theta = \frac{\partial \psi}{\partial \widehat{y}} \frac{\partial \theta}{\partial \widehat{x}} - \frac{\partial \psi}{\partial \widehat{x}} \frac{\partial \theta}{\partial \widehat{y}}, \tag{3.25}$$

and the boundary conditions are now

$$\psi = 0, \quad \frac{\partial \theta}{\partial \widehat{y}} - a \cos(\pi \widehat{x}) \frac{\partial \theta}{\partial \widehat{x}} = -\sigma \text{ on } \widehat{y} = 0, \tag{3.26}$$

$$\frac{\partial \psi}{\partial \widehat{y}} \rightarrow 0, \quad \theta \rightarrow 0 \text{ as } \widehat{y} \rightarrow \infty, \tag{3.27}$$

where

$$\nabla_1^2 = \frac{\partial^2}{\partial \widehat{x}^2} + \sigma^2 \frac{\partial^2}{\partial \widehat{y}^2} - 2a \cos(\pi \widehat{x}) \frac{\partial^2}{\partial \widehat{x} \partial \widehat{y}} + a\pi \sin(\pi \widehat{x}) \frac{\partial}{\partial \widehat{y}}. \tag{3.28}$$

Next, we introduce the boundary layer variables

$$\psi = Ra^{1/3} \widehat{x}^{2/3} f(\widehat{x}, \eta), \quad \theta = Ra^{-1/3} \widehat{x}^{1/3} h(\widehat{x}, \eta), \quad \eta = Ra^{1/3} \widehat{y}/\widehat{x}^{1/3}. \tag{3.29}$$

Thus, on introducing Eq. (3.29) into Eqs. (3.24) and (3.25), and formally letting $Ra \rightarrow \infty$, we obtain the following boundary layer equations at leading order:

$$\sigma^2 f_\eta = h, \tag{3.30}$$

$$\sigma^2 h_{\eta\eta} + \frac{2}{3} f h_\eta - \frac{1}{3} f_\eta h = \widehat{x}(f_\eta h_{\widehat{x}} - h_\eta f_{\widehat{x}}) \tag{3.31}$$

and the corresponding boundary conditions are

$$f = 0, \quad h_\eta = -1/\sigma \text{ on } \eta = 0, \quad h \rightarrow 0 \text{ as } \eta \rightarrow \infty. \tag{3.32}$$

It is also convenient to introduce a third transformation to render the boundary conditions independent of $\widehat{x}$:

$$f = \sigma^{1/3} \widehat{f}(\widehat{x}, \widehat{\eta}), \quad h = \sigma^{2/3} \widehat{h}(\widehat{x}, \widehat{\eta}), \quad \eta = \sigma^{5/3} \widehat{\eta}. \tag{3.33}$$

Equations (3.30) and (3.31) become

$$\widehat{f}_{\widehat{\eta}} = \widehat{h}, \tag{3.34}$$

$$\widehat{h}_{\widehat{\eta}\widehat{\eta}} + \frac{1}{3}(2\widehat{f}\widehat{h}_{\widehat{\eta}} - \widehat{f}_{\widehat{\eta}}\widehat{h}) = \widehat{x}(\widehat{f}_{\widehat{\eta}}\widehat{h}_{\widehat{x}} - \widehat{h}_{\widehat{\eta}}\widehat{f}_{\widehat{x}}) + \frac{\sigma'}{3\sigma}\widehat{x}(2\widehat{f}_{\widehat{\eta}}\widehat{h} - \widehat{f}\widehat{h}_{\widehat{\eta}}), \tag{3.35}$$

and the boundary conditions transform to

$$\widehat{f} = 0, \quad \widehat{h}_{\widehat{\eta}} = -1 \text{ on } \widehat{\eta} = 0, \quad \widehat{h} \rightarrow 0 \text{ as } \widehat{\eta} \rightarrow \infty, \tag{3.36}$$

where the prime denotes differentiation with respect to $\widehat{x}$.

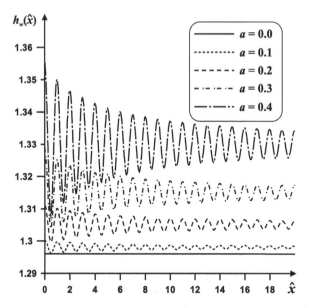

Fig. 3.2. Variation of the surface temperature, $h_w(\widehat{x})$,
for different values of a

Both the systems of Eqs. (3.30), (3.31), and (3.34), (3.35) have been solved numerically for different values of the wave amplitude, a, in the paper by Rees and Pop (1995a,b).

Figure 3.2 shows the distribution of the surface temperature, $h_w = h(\widehat{x}, 0)$ for different values of a. The main effect of the presence of surface undulations is to raise the temperature of the bounding surface above the value 1.29612, corresponding to that of a plane surface. This is a consequence of the fact that the component of the buoyancy force parallel to the wavy surface is less than or equal to that of a plane surface, as can be seen by the factor $1/\sigma^2$ in Eq. (3.30).

The wall temperature distribution, $h_w(\widehat{x})$, as a function of both $\widehat{x}$ and a, is displayed in Fig. 3.3. Although the temperature is found to increase (decrease) whenever the slope of σ increases (decreases), the presence of the boundary layer flow ensures that the positions of maximum temperature are shifted downstream of the positions of maximum slope. The two maxima in the surface temperature are caused by σ having two maxima. The effect of having such a nonuniformity on the surface temperature can be seen to persist a substantial distance downstream of the hump.

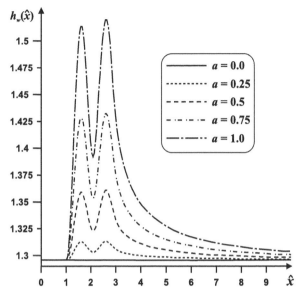

Fig. 3.3. Variation of the surface temperature, $h_w(\widehat{x})$, for a hump of length 2, centered at $\widehat{x}_c = 2$, for different values of a

It is worth mentioning, to this end, that there are also several published papers on this topic, such as, for example, Rees and Pop (1994a,b); Pop and Na (1995); Hossain et al. (2001); Neagu (2011).

Chapter 4

Natural Convective Flow of a Viscous Fluid in a Wavy Vertical Channel

Steady Natural Convection Flow

The rather complex problem of understanding natural convective heat transfer patterns in a viscous fluid confined between two vertical walls, one or both of which are wavy, has been the subject of a few analytical (Vajravelu and Sastri, 1978, 1980; Rao et al., 1983) and numerical (Watson and Poots, 1971) studies. The present section focusses on these studies, which may be of interest in certain special situations, with the idea of throwing some light on the flow and heat transfer characteristics.

The presentation is restricted to steady, incompressible, and laminar flow of constant property Newtonian fluids only. The effects of natural convection and heat source or sink are included at appropriate places. However, the effects of viscous dissipation (i.e., internal friction), pressure stress (i.e., work done by pressure forces), and mass transfer are omitted.

There are three types of channel flows considered, namely,

1) flow between a long vertical wall and a parallel wall,
2) flow between two symmetrically configured wavy walls, and
3) flow between two asymmetrically configured wavy walls.

1. Flow between a Long Vertical Wall and a Parallel Wall

Natural convection flow of a viscous incompressible Newtonian fluid is considered to take place in a channel, which is formed by a long wavy wall with $y = a\cos(bx)$ maintained at a constant temperature of T_0 and a flat vertical wall of uniform temperature T_1 located at a distance of $y = d$ from the axis of the wavy wall, as shown in Fig. 1.5. It is assumed that all physical properties of the fluid (except the density in the buoyancy term) are constant and that the natural convection flow which results from this situation is at steady-state, with the presence of heat source or sink term but without heat generation and viscous dissipation. It is further assumed that Boussinesq approximation can be invoked and that the wavelength of the wavy wall, which is proportional to $1/b$, is large.

The governing Eqs. (1.1)–(1.4) can then be written (Vajravelu and Sastri, 1978) in the non-dimensional form as follows:

$$\frac{\partial u_1}{\partial x_1} + \frac{\partial v_1}{\partial y_1} = 0 \tag{4.1}$$

$$u_1\frac{\partial u_1}{\partial x_1} + v_1\frac{\partial u_1}{\partial y_1} = -\frac{\partial P_1}{\partial x_1} + \nabla^2 u_1 + Gr\,\theta \tag{4.2}$$

$$u_1\frac{\partial v_1}{\partial x_1} + v_1\frac{\partial v_1}{\partial y_1} = -\frac{\partial P_1}{\partial y_1} + \nabla^2 v_1 \tag{4.3}$$

$$u_1\frac{\partial \theta}{\partial x_1} + v_1\frac{\partial \theta}{\partial y_1} = \frac{1}{Pr}(\nabla^2\theta + \alpha^*). \tag{4.4}$$

The above equations are to be solved under the following boundary conditions:

$$\begin{aligned} y_1 = \sigma(x_1) = a_1\cos(b_1 x_1): \quad & u_1 = 0,\ v_1 = 0,\ \theta = 1 \\ y_1 = 1: \quad\quad\quad\quad\quad\quad & u_1 = 0,\ v_1 = 0,\ \theta = \theta_1. \end{aligned} \tag{4.5}$$

The non-dimensional variables are defined as follows:

$$\begin{aligned} & x_1 = \frac{x}{d};\ y_1 = \frac{y}{d};\ a_1 = \frac{a}{d};\ b_1 = bd; \\ & P_1 = \frac{P}{\rho u_c^2};\ \theta = \frac{T - T_s}{T_0 - T_s};\ \theta_1 = \frac{T_1 - T_s}{T_0 - T_s}; \\ & u_1 = \frac{u}{u_c};\ v_1 = \frac{v}{u_c};\ u_c = (\nu/d) \end{aligned} \tag{4.6}$$

where the index s signifies the condition in the static fluid. In Eqs. (4.2) and (4.4), the Grashof number Gr and the heat source ($\alpha^* > 0$) or heat sink ($\alpha^* < 0$) parameter are defined as:

$$Gr = g\beta(T_0 - T_s)d^3/\nu^2; \quad \alpha^* = qd^2/[k(T_0 - T_s)]. \tag{4.7}$$

Vajravelu and Sastri (1978) based their analysis on the linearized forms of Eqs. (4.1)–(4.5) and an approximation to u_1, v_1, P_1, and θ, which are assumed to be of the following form:

$$
\begin{aligned}
u_1 &= u_{10}(y_1) + u_{11}(x_1, y_1); \quad v_1 = v_{11}(x_1, y_1) \\
P_1 &= P_{11}(x_1, y_1); \qquad\qquad \theta = \theta_0(y_1) + \theta_1(x_1, y_1)
\end{aligned}
\tag{4.8}
$$

where u_{11}, v_{11}, P_{11}, and θ_1 are small compared with the mean quantities u_{10} and θ_0, which are independent of x_1 since the flow has been assumed to be fully developed. It should be noted that the perturbed part of the solution is based on the long wave approximation, and this is, therefore, the contribution from the gradual waviness of the wavy wall.

Using (4.8), Eqs. (4.1)–(4.4) can be rewritten both under static fluid conditions and otherwise as

$$u''_{10} + Gr\,\theta_0 = 0, \quad \theta''_0 = -\alpha^* \tag{4.9}$$

$$\frac{\partial u_{11}}{\partial x_1} + \frac{\partial v_{11}}{\partial y_1} = 0 \tag{4.10}$$

$$u_{10}\frac{\partial u_{11}}{\partial x_1} + u'_{10}v_{11} = -\frac{\partial P_{11}}{\partial x_1} + \nabla^2 u_{11} + Gr\,\theta_1 \tag{4.11}$$

$$u_{10}\frac{\partial v_{11}}{\partial x_1} = -\frac{\partial P_{11}}{\partial y_1} + \nabla^2 v_{11} \tag{4.12}$$

$$u_{10}\frac{\partial \theta_1}{\partial x_1} + \theta'_0 v_{11} = \frac{1}{Pr}(\nabla^2 \theta_1) \tag{4.13}$$

where primes denote differentiation with respect to y_1. With the help of (4.8), the boundary conditions (4.5) can be easily simplified to

$$
\begin{aligned}
y_1 = 0: \quad u_{10} = 0, \quad \theta_0 = 1 \\
y_1 = 1: \quad u_{10} = 0, \quad \theta_0 = \theta_i
\end{aligned}
\tag{4.14}
$$

and

$$
\begin{aligned}
y_1 = 0: \quad u_{11} = -u_{10}, \quad v_{11} = 0, \quad \theta_1 = -\theta'_0 \\
y_1 = 1: \quad u_{11} = 0, \qquad\quad v_{11} = 0, \quad \theta_1 = 0.
\end{aligned}
\tag{4.15}
$$

The analytical solution of Eq. (4.9) along with the boundary conditions (4.14) is

$$
\begin{aligned}
u_{10} &= \frac{Gr}{12}\alpha^* \left(\frac{1}{2}y_1^3 - y_1^2 + \frac{1}{2}\right)y_1 + \frac{Gr}{6}(1 - \theta_i)(y_1^2 - 1)y_1 \\
&\quad + \frac{Gr}{2}(1 - y_1)y_1
\end{aligned}
\tag{4.16}
$$

$$
\theta_0 = \frac{1}{2}\alpha^*(1 - y_1)y_1 + (\theta_i - 1)y_1 + 1.
$$

The problem defined by Eqs. (4.10)–(4.13) subject to (4.15) is now considered. The stream function ψ_1 is defined as follows:

$$
u_{11} = -\frac{\partial \psi_1}{\partial y_1}, \quad v_{11} = \frac{\partial \psi_1}{\partial x_1}.
\tag{4.17}
$$

Substituting (4.17) into Eqs. (4.11)–(4.13) and eliminating the pressure P_{11}, the following two equations are obtained in terms of ψ_1 and θ_1:

$$
u_{10}\left(\frac{\partial^3 \psi_1}{\partial x_1^3} + \frac{\partial^3 \psi_1}{\partial y_1^3}\right) - u_{10}''\frac{\partial \psi_1}{\partial x_1} = \nabla^3 \psi_1 - Gr\frac{\partial \theta_1}{\partial y_1}
\tag{4.18}
$$

$$
u_{10}\frac{\partial \theta_1}{\partial x_1} + \theta_0'\frac{\partial \psi_1}{\partial x_1} = \frac{1}{Pr}(\nabla^2 \theta_1).
\tag{4.19}
$$

The above Eqs. (4.18) and (4.19) are subject to the following boundary conditions:

$$
\begin{aligned}
y_1 = 0: \quad &\frac{\partial \psi_1}{\partial y_1} = -u_{10}', \quad \frac{\partial \psi_1}{\partial x_1} = 0, \quad \theta_1 = -\theta_0' \\
y_1 = 1: \quad &\frac{\partial \psi_1}{\partial y_1} = 0, \quad \frac{\partial \psi_1}{\partial x_1} = 0, \quad \theta_1 = 0.
\end{aligned}
\tag{4.20}
$$

A solution of Eqs. (4.18) and (4.19) is sought, assuming the following form:

$$
\begin{aligned}
\psi_1 &= a_1 \exp(ib_1 x_1)f(y_1) \\
\theta_1 &= a_1 \exp(ib_1 x_1)h(y_1)
\end{aligned}
\tag{4.21}
$$

where $i = \sqrt{-1}$.

Attention is only focussed on the real parts of the solutions for the perturbed quantities u_{11}, v_{11} and θ_1

$$
\begin{aligned}
u_{11} &= a_1[\sin(b_1 x_1)f_i' - \cos(b_1 x_1)f_r'] \\
v_{11} &= -a_1 b_1[\sin(b_1 x_1)f_r + \cos(b_1 x_1)f_i] \\
\theta_1 &= a_1[\cos(b_1 x_1)h_r - \sin(b_1 x_1)h_i]
\end{aligned}
\tag{4.22}
$$

where
$$f = f_r + if_i, \quad f' = f'_r + if'_i, \quad h = h_r + ih_i. \tag{4.23}$$

The functions f and h are obtained by substituting (4.21) into Eqs. (4.18) and (4.19), as given below:

$$f^{iv} - ib_1[u_{10}(f'' - b_1^2 f) - u''_{10}f] - b_1^2(2f'' - b_1^2 f) = Gr\, h' \tag{4.24a}$$
$$h'' - b_1^2 h = Pr\, ib_1(u_{10}h + \theta'_0 f) \tag{4.24b}$$

subject to the boundary conditions

$$\begin{aligned}
y_1 = 0: \quad & f = 0, \quad f' = -u'_{10}, \quad h = -\theta'_0 \\
y_1 = 1: \quad & f = 0, \quad f' = 0, \quad h = 0.
\end{aligned} \tag{4.25}$$

A solution of Eqs. (4.24)–(4.25) is sought by expanding f and h in series for small frequency parameter b_1 of the form

$$\begin{aligned}
f &= f_0(y_1) + b_1 f_1(y_1) + b_1^2 f_2(y_1) + \text{higher order terms;} \\
h &= h_0(y_1) + b_1 h_1(y_1) + b_1^2 h_2(y_1) + \text{higher order terms.}
\end{aligned} \tag{4.26}$$

Substituting (4.26) into Eqs. (4.24)–(4.25) yields three sets of ordinary differential equations along with the corresponding boundary conditions for f_j and h_j ($j = 0, 1, 2$). These sets of equations have been solved analytically by Vajravelu and Sastri (1978) but are not presented here, for the sake of brevity.

It is worth mentioning that Eq. (4.9) along with the boundary conditions (4.14) describe natural convection flow in a channel both of whose walls are flat (Ostrach, 1952). The behavior of the mean velocity u_{10} and the mean temperature θ_0 profiles are discussed here to aid comparison of the present results with those for a flat walled channel. Typical profiles of u_{10} and θ_0 for $\theta_i = -1$ (i.e., the average temperatures of the walls are equal to that of the static fluid) and $\theta_i = 2$ (i.e., the temperatures of the walls are unequal), and for various values of Gr and α^* are presented in Figs. 4.1 and 4.2.

It is seen from Fig. 4.1 that, in the presence of heat sources ($\alpha^* > 0$, curves III and VI), u_{10} increases across the channel width whenever Gr increases. This behavior is reversed in the case of heat sinks ($\alpha^* < 0$, curves I and IV).

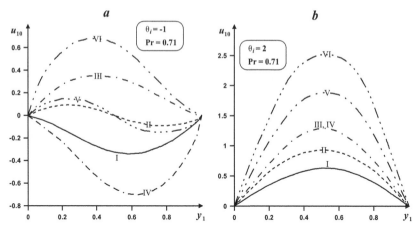

Fig. 4.1. Velocity profiles u_{10} for different values of θ_i

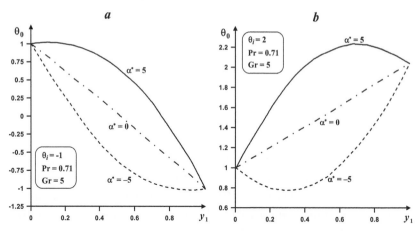

Fig. 4.2. Temperature profiles θ_0 for different values of θ_i

However, u_{10} is enhanced for $\theta_i = 2$ (Fig. 4.1b) by an increase in Gr for all values of α^*. A closer examination of Figs. 4.1a and 4.1b reveals that the fluid velocity u_{10} can reverse its direction in the case $\theta_i = -1$, while there is no such possibility when $\theta_i = 2$. This can be ascribed to the fact that for $\theta_i > 0$ both temperatures T_0 and T_1 are higher or lower than the temperature of the static fluid T_s, while for $\theta_i < 0$, only T_0 or T_1 can exceed T_s.

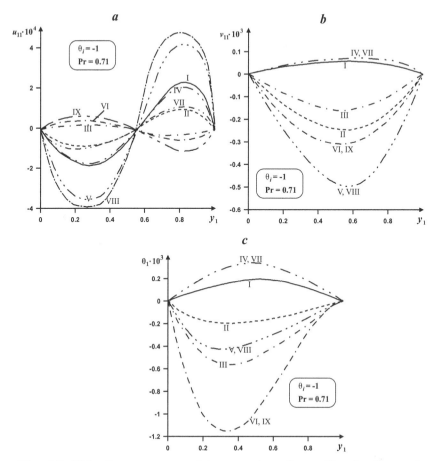

Fig. 4.3. Velocity u_{11}, v_{11} and temperature θ_1 profiles for $\theta_i = -1$

Figures 4.2a and 4.2b show that in the absence of heat sources or sinks ($\alpha^* = 0$), θ_0 is a linearly decreasing function of y_1 when $\theta_i = -1$, while on the other hand, it is a linearly increasing function of y_1 when $\theta_i = 2$. The presence of heat sources or sinks ($\alpha^* \neq 0$) lead to a parabolic nature of θ_0.

The perturbed quantities u_{11}, v_{11}, and θ_1 are plotted in Fig. 4.3 for $\theta_i = -1$, $Pr = 0.71$, and some values of Gr, b_1, and α^*.

It can be seen that for $\alpha^* \neq 0$, u_{11} increases steadily for a fixed value of y_1 up to approximately $y_1 = 0.55$, i.e., in the first half of the channel, while in the other half, u_{11} is a decreasing function of y_1. It is also noticed that, when $\alpha^* > 0$ (curves III and VI of Fig. 4.3a), an increase in b_1 leads to a considerable increase in u_{11} in the first half of the channel and that this behavior is reversed when $\alpha^* \leq 0$ (curves I,

IV and II, VII). Further, Fig. 4.3b indicates that, as α^* is increased, v_{11} diminishes sharply. Also, the effect of b_1 is to reduce v_{11} when $\alpha^* \geq 0$ and to enhance it when $\alpha^* < 0$. When Fig. 4.3c for θ_1 is compared with Fig. 4.3b for v_{11}, it can be seen that the variation of θ_1 with each of the parameters Gr, b_1, and α^* is in all respects similar to that of v_{11}.

Figure 4.4 describes the behavior of the total fluid velocity u ($= u_{10}+u_{11}$) for $\theta_i = -1$ (Fig. 4.4a) and $\theta_i = 2$ (Fig. 4.4b) when $Pr = 0.71$ and various values of Gr, b_1, and α^* are prescribed.

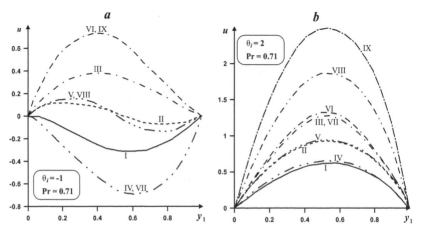

Fig. 4.4. Total fluid velocity profiles u for different values of θ_i

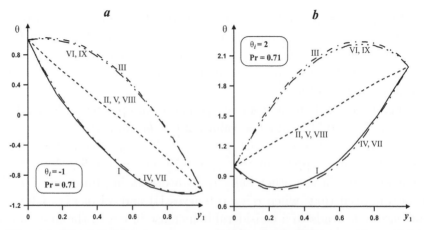

Fig. 4.5. Total fluid temperature profiles θ for different values of θ_i

It is seen that the effects of these parameters are quite different in both cases of $\theta_i = -1$ and $\theta_i = 2$, respectively. u is a decreasing function of both Gr and b_1 when $\alpha^* < 0$, while it is an increasing function when $\alpha^* \geq 0$. However, when $\theta_i = 2$, it is seen that u is always an increasing function of both Gr and b_1 for all values of α^*.

The total fluid temperature θ ($= \theta_0 + \theta_1$) is displayed in Fig. 4.5 for $\theta_i = -1$ (Fig. 4.5a) and $\theta_i = 2$ (Fig. 4.5b). These figures show that when $\theta_i = -1$, the effect of an increase in Gr is to increase θ for $\alpha^* < 0$, while the reverse is the case for $\theta_i = 2$.

2. Flow between Two Symmetrically Configured Wavy Walls

The problem presented in this section is that studied by Watson and Poots (1971), which discusses the convective motion occurring in a fluid confined between two wavy vertical walls whose surfaces are maintained at constant, but different, temperatures T_0 and T_1, respectively. Specifically, the focus of attention is on sinusoidal walls defined by Eqs. (1.10) and (1.11) as shown in Fig. 1.6a,b. It is assumed that:

(i) the convective motion is two-dimensional and laminar;

(ii) the cavity between the walls is of sufficient extent for the flow to be fully developed and periodic in the x-direction;

(iii) Boussinesq approximation is applicable, namely

$$(\rho - \rho_m)/\rho_m = -(T - T_m)/T_m,$$

where $T_m = (T_0 + T_1)/2$;

(iv) the heat source or sink is neglected ($\alpha^* = 0$) in the energy equation.

Under the above-mentioned assumptions, the governing equations are identical to those given by Eqs. (4.1)–(4.4), except for $\alpha^* = 0$ in Eq. (4.4). However, the non-dimensional temperature θ and the Grashof number Gr are defined differently for the present problem as follows:

$$\theta = 2(T - T_m)/(T_1 - T_0),$$

$$Gr = (gd_m^3/\nu^2)(T_1 - T_0)/(T_1 + T_0). \tag{4.27}$$

On introducing the dimensionless stream function ψ_1 in the usual manner as $u_1 = (\partial\psi_1/\partial y_1)$ and $v_1 = -(\partial\psi_1/\partial x_1)$, the governing Eqs. (4.2)–(4.4) after elimination of P_1 can be rewritten as

$$\nabla^4\psi_1 + \frac{\partial(\psi_1, \nabla^2\psi_1)}{\partial(x_1, y_1)} + Gr\frac{\partial\theta}{\partial y_1} = 0 \tag{4.28}$$

$$\nabla^2\theta + Pr\frac{\partial(\psi_1, \theta)}{\partial(x_1, y_1)} = 0. \tag{4.29}$$

The above Eqs. (4.28) and (4.29) are subject to the following boundary conditions:

$$y_1 = \pm\sigma(x_1) = \pm[1 \pm a_1 \sin(b_1 x_1)] :$$

$$\frac{\partial\psi_1}{\partial x_1} = \frac{\partial\psi_1}{\partial y_1} = 0, \ \theta = \mp 1. \tag{4.30}$$

A further transformation is made as follows:

$$x_2 = x_1, \ y_2 = \frac{y_1}{\sigma(x_1)}. \tag{4.31}$$

This transforms the space between the wavy walls into a region bounded by two smooth flat walls, wherein $y_2 = \pm 1$ for all x_2.

Equations (4.28) and (4.29) have a regular expansion for small values of a_1 of the form:

$$\begin{aligned}
\psi_1 &= \psi_{1,0}(y_2) + a_1\psi_{1,1}(x_2, y_2) \\
&\quad + a_1^2\psi_{1,2}(x_2, y_2) + \text{higher order terms} \\
\theta &= \theta_0(y_2) + a_1\theta_1(x_2, y_2) \\
&\quad + a_1^2\theta_2(x_2, y_2) + \text{higher order terms}
\end{aligned} \tag{4.32}$$

where $\psi_{1,0}$ and θ_0 are independent of x_2 because the flow is assumed to be fully developed. Substituting expansions (4.32) into Eqs. (4.28) and (4.29) along with the use of (4.31) gives three sets of differential equations for the functions $\psi_{1,i}$ and θ_i ($j = 0, 1, 2$) when like powers of a_1 are equated.

When the zeroth-order equations are considered, the leading order terms in Eq. (4.32) satisfy the following ordinary differential equations:

$$\psi_{1,0}^{IV} + Gr\,\theta_0^1 = 0 \tag{4.33a}$$

$$\theta_0^{II} = 0 \tag{4.33b}$$

subject to the boundary conditions

$$y_2 = \pm 1: \quad \psi_{1,0} = 0, \ \psi_{1,0}^I = 0, \ \theta_0 = \mp 1 \tag{4.34}$$

where the primes denote differentiation with respect to y_2. The solution of Eqs. (4.33) is

$$\psi_{1,0} = \frac{1}{24} Gr(y_2^2 - 1)^2 \qquad (4.35a)$$

$$\theta_0 = -y_2. \qquad (4.35b)$$

When first-order equations are considered, using the above Eq. (4.35), the equations for the terms of the order of a_1 can be written as

$$\nabla^4 \psi_{1,1} + Gr\, y_2 \frac{\partial \psi_{1,1}}{\partial x_2} - \frac{1}{6} Gr\, y_2 (y_2^2 - 1) \frac{\partial (\nabla^2 \psi_{1,1})}{\partial x_2} + Gr \frac{\partial \theta_1}{\partial x_2}$$

$$= \frac{1}{6} Gr \sin(b_1 x_2)[b_1^4 y_2^2 (y_2^2 - 1) - 4b_1^2 (6b_2^2 - 1) + 18] \qquad (4.36)$$

$$+ \frac{1}{36} b_1 Gr^2 \cos(b_1 x_2) y_2 (y_2^2 - 1)[b_1^2 y_2^2 (y_2^2 - 1) - 6y_2^2 + 2]$$

$$\nabla^2 \theta_1 - Pr \left[\frac{\partial \psi_{1,1}}{\partial x_2} + \frac{1}{6} Gr\, y_2 (y_2^2 - 1) \frac{\partial \theta_1}{\partial x_2} \right] = b_1^2 y_2 \sin(b_1 x_2) \qquad (4.37)$$

subject to the following boundary conditions:

$$y_2 = \pm 1 : \quad \psi_{1,1} = 0, \quad \frac{\partial \psi_{1,1}}{\partial y_2} = 0, \ \theta_1 = 0. \qquad (4.38)$$

Now, on the assumption that the flow is periodic in the x-direction, the functions $\psi_{1,1}$ and θ_1 can be expressed as follows:

$$\psi_{1,1} = f_1(y_2) \sin(b_1 x_2) + f_2(y_2) \cos(b_1 x_2)$$
$$\theta_1 = g_1(y_2) \sin(b_1 x_2) + g_2(y_2) \cos(b_1 x_2) \qquad (4.39)$$

where f_i and g_i satisfy the following ordinary differential equations:

$$L_1(D, f_i, g_i) = \frac{1}{6} Gr[b_1^4 y_2^2 (y_2^2 - 1) - 4b_1^2 (6y_2^2 - 1) + 18]$$

$$L_2(D, f_i, g_i) = \frac{1}{36} b_1 Gr^2 y_2 (y_2^2 - 1)[b_1^2 y_2^2 (y_2^2 - 1) - 6y_2^2 + 2] \qquad (4.40)$$

$$M_1(D, f_i, g_i) = b_1^2 y_2$$

$$M_2(D, f_i, g_i) = 0$$

and the following imposed boundary conditions:

$$f_i(\pm 1) = Df_i(\pm 1) = g_i(\pm 1) = 0, \qquad (4.41)$$

where $D = d/dy_2$. Equations (4.40) and (4.41) constitute a system of twelfth-order boundary value problems since the homogeneous parts of the equations are given by

$$
\begin{aligned}
L_1 \quad &= (D^2 - b_1^2)^2 f_1 - b_1 Gr\, y_2 f_2 + Gr\, Dg_1 \\
&+ \frac{1}{6} b_1 Gr\, y_2 (y_2^2 - 1)(D^2 - b_1^2) f_2 \\
L_2 \quad &= (D^2 - b_1^2)^2 f_2 - b_1 Gr\, y_2 f_1 + Gr\, Dg_2 \\
&+ \frac{1}{6} b_1 Gr\, y_2 (y_2^2 - 1)(D^2 - b_1^2) f_1 \\
M_1 \quad &= (D^2 - b_1^2) g_1 + b_1 Pr \left[f_2 + \frac{1}{6} Gr\, y_2 (y_2^2 - 1) g_2 \right] \\
M_2 \quad &= (D^2 - b_1^2) g_2 - b_1 Pr \left[f_1 + \frac{1}{6} Gr\, y_2 (y_2^2 - 1) g_1 \right].
\end{aligned}
\tag{4.42}
$$

It is important to mention that the functions f_i and g_i will be indeterminate if there exist eigensolutions of the coupled system

$$
L_i = M_i = 0,
\tag{4.43}
$$

which satisfies the boundary conditions (4.41). The existence of eigensolutions of (4.43) has been verified numerically (Watson and Poots, 1971) and these indicate a lowest critical $(b_1 - Gr)$ relation of the form

$$
F(b_1, Gr) = 0.
\tag{4.44}
$$

In practice, the critical values would also be expected to depend on the amplitude and frequency of waviness of the wavy wall. Such behavior is most likely to occur when $a_1 \sim O(1)$ and may be assumed negligible for sufficiently small a_1. For values of b_1 and Gr near those given by Eq. (4.44), the first-order corrections $\psi_{1,1}$ and θ_1 predict large increases in velocity and heat transfer, and this situation is unacceptable from a physical viewpoint.

The second-order equations can be obtained by substituting Eq. (4.32) into Eqs. (4.28) and (4.29), and equating the powers of a_1^2. These equations are not presented here but can be found in the work of Watson and Poots (1971). It is shown that the second-order perturbations, $\psi_{1,2}$ and θ_2, can be expressed as

$$
\begin{aligned}
\psi_{1,2} &= f_3(y_2) + f_4(y_2) \sin(2b_1 x_2) + f_5(y_2) \cos(2b_1 x_2) \\
\theta_2 &= g_3(y_2) + g_4(y_2) \sin(2b_1 x_2) + g_5(y_2) \cos(2b_1 x_2)
\end{aligned}
\tag{4.45}
$$

a   *b*

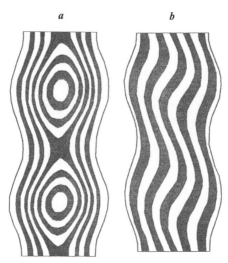

Fig. 4.6. Streamlines (a) and isotherms (b) for $Gr = 100$
[reprinted from Watson and Poots (1971) with permission from
Cambridge Journals]

where f_i and g_i satisfy an eighteenth-order system of inhomogeneous
ordinary differential equations. The singular behavior of these equations
for b_1 and Gr is now given by

$$F(2b_1, Gr) = 0. \tag{4.46}$$

The equations for the functions f_i and g_i $(j = 1, 2, \ldots, 5)$ are solved
numerically by Watson and Poots (1971) for $Pr = 0.72$ (gas), $0.5 \leq b_1 \leq 2.0$, and $Gr < 1000$. It is found that the minimum critical Grashof
number for natural convection flow between two wavy vertical walls of
small amplitude occurs at $Gr = 502.1$ with $b_1 = 1.4$, which corresponds
to the critical value for marginal stability between smooth flat vertical
walls.

Flow streamline and isotherms and temperature fields are presented
in Figs. 4.6–4.8 for $a_1 = 0.15$, $b_1 = 1.5$, and three different values of
$Gr = 100$, 200, and 300.

As is expected, the temperature and velocity fields show a strong dependence on the Gr value. For small Gr, the isotherms are on contours
parallel to the wavy walls with a slow flow wrapping itself smoothly
around the walls. In this case, the mechanism governing the flow and
heat transfer is, as in the case of flat walls, controlled by a balance
of the viscous and body forces, and heat is transferred by conduction
alone.

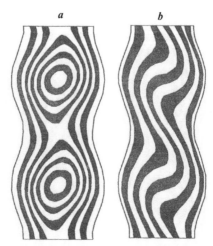

Fig. 4.7. Streamlines (a) and isotherms (b) for $Gr = 200$
[reprinted from Watson and Poots (1971) with permission from
Cambridge Journals]

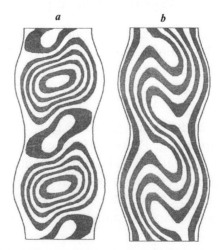

Fig. 4.8. Streamlines (a) and isotherms (b) for $Gr = 300$
[reprinted from Watson and Poots (1971) with permission from
Cambridge Journals]

On increasing the value of Gr to 100, the first effects of the waviness
are seen in the appearance of vortices symmetrically placed in the cavity
(Fig. 4.6a).

At this value of Gr, heat transfer by convection is still unimportant
(Fig. 4.6b). As Gr is further increased to a value of 200, the higher
fluid velocities near the walls makes the vortex motion more dominant

(Fig. 4.7a) and this leads to an increase in the role played by convection in the heat transfer process. The isotherms thus begin to twist (Fig. 4.7b), with hot isotherms being displaced towards the cold wall and vice-versa. Figures 4.8a,b for $Gr = 300$ clearly show that the effects of the curvature of the walls in the vicinity of the neck is to cause the fluid to break away. It thus appears feasible that for $Gr > 300$, further distortion of the isotherms will occur and then the convergence of the series (4.32) is doubtful.

The effects of a_1, b_1, and Gr on the wall heat flux can also be examined. The integrated heat flux over an area defined by unit width of the wall and over one wavelength $2\pi/b_1$ is given by

$$q_w(a_1, b_1) = -k \int_S (\overrightarrow{n} \cdot \nabla T)_w dS \qquad (4.47)$$

where the following hold when Eq. (1.10) is used:

$$\overrightarrow{n} = \left(-\frac{ab\cos(bx)}{[1 + a^2b^2\cos^2(bx)]^{1/2}}, \frac{1}{[1 + a^2b^2\cos^2(bx)]^{1/2}} \right)$$

$$\nabla T = \left(\frac{\partial T}{\partial x} - ab\cos(bx)\frac{\partial T}{\partial y}, \frac{\partial T}{\partial y} \right) \qquad (4.48)$$

$$dS = [1 + a^2b^2\cos^2(bx)]^{1/2}dx.$$

In terms of non-dimensional variables and on using (4.31), we have

$$\frac{q_w(a_1, b_1)}{k(T_1 - T_0)} = -\int_0^{2\pi/b_1} \frac{[1 + a_1^2 b_1^2 \cos^2(b_1 x_2)]}{[1 + a_1 \sin(b_1 x_2)]} \left(\frac{\partial \theta}{\partial y_2} \right)_{y_2=1} dx_2 \qquad (4.49)$$

where $\partial \theta / \partial x_2 = 0$ on the wall. Substituting θ as given by Eqs. (4.32), (4.39), and (4.45), and expanding in the powers of small a_1, gives

$$\frac{q_w(a_1, b_1)}{k(T_1 - T_0)} = \frac{2\pi}{b_1} \left\{ 1 + \frac{a_1^2}{2}[1 + b_1^2 + g_1^I(1) - 2g_3^I(1)] + O(a_1^4) \right\}. \qquad (4.50)$$

The dimensionless number Na_1 defined by

$$Na_1 = \frac{q_w(a_1, b_1) - q_w(0, b_1)}{q_w(0, b_1)} \qquad (4.51)$$

is used for comparing the heat flux with that for the reference case $a_1 = 0$ (flat walls). Thus, neglecting terms of $O(a_1^4)$ gives

$$Na_1 = \frac{1}{2}a_1^2[1 + b_1^2 + g_1^I(1) - 2g_3^I(1)]. \qquad (4.52)$$

In Fig. 4.9, a plot of Na_1 against b_1 is shown for $a_1 = 0.15$ and different values of Gr.

It can be seen from the figure that there are considerable changes in the percentage with the increase being much more marked for values of Gr near the critical value. For any given increase in the heat transfer rate, the process of heat transfer between the walls will become most efficient at the lowest possible values of Gr. The optimum values of (b_1, Gr) are thus given by the locus of the minimum of the (b_1, Gr) versus Na_1 curves. This locus has been indicated in Fig. 4.9 by the dashed curve.

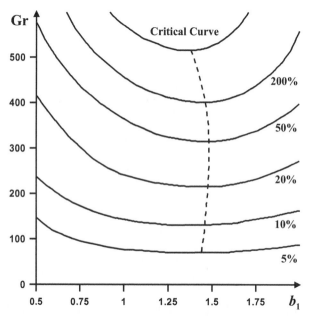

Fig. 4.9. Variation of Na_1 with b_1 for different values of Gr

Table 4.1. Fraction Increase in Heat Flux: Values of Na_1/a_1^2 [reprinted from Watson and Poots (1971) with permission from Cambridge Journals]

Gr \ b_1	0.50	0.75	1.00	1.25	1.50	1.75	2.00
50	0.878	1.200	1.480	1.666	1.751	1.758	1.725
100	1.741	2.611	3.269	3.638	3.698	3.491	3.127
150	2.842	4.220	5.290	5.967	6.091	5.612	4.732
200	3.978	5.875	7.640	9.065	9.524	8.616	6.772
250	5.095	7.748	10.92	14.15	15.59	13.70	9.760
300	6.235	10.17	16.26	23.98	28.11	23.35	14.40

Table 4.1 lists typical values of the fraction increase in heat flux, Na_1/a_1^2, for various values of b_1 and Gr. It can be seen that the heat transfer increases substantially as both b_1 and Gr increase.

3. Flow between Two Asymmetrically Configured Wavy Walls

Natural convection flow between two asymmetrically configured wavy vertical walls described by Eqs. (2.11) and (2.13), with ϕ equal to 0, $\pi/2$, π and $3\pi/2$ to denote changes in the orientation of the channel, has been considered theoretically by Vajravelu and Sastri (1980). A model for the flow together with the coordinate system is shown in Fig. 1.7. The walls are maintained at constant temperatures T_0 and T_1, respectively.

Assuming the flow to be laminar and two-dimensional, the governing equations for this case are exactly the same as Eqs. (4.1)–(4.4). However, the imposed boundary conditions are different and are given as

$$y_1 = \sigma_R(x_1) = a_1 \cos(b_1 x_1): \qquad u_1 = 0, \; v_1 = 0, \; \theta = 1$$
$$y_1 = \sigma_L(x_1) = -1 + a_1 \cos(b_1 x_1 + \phi): \quad u_1 = 0, \; v_1 = 0, \; \theta = \theta_1 \tag{4.53}$$

where $\sigma_R(x_1)$ and $\sigma_L(x_1)$ denote the right and left wavy walls, respectively.

Using the method of perturbations (Aziz and Na, 1984), under the boundary conditions (4.53), a solution of Eqs. (4.1)–(4.4) is sought in the form given by Eq. (4.22). Again, the mean part of the solution (u_{10}, θ_0) corresponds to the flow between two smooth flat vertical walls, while the perturbed part of the solution $(u_{11}, v_{11}, \theta_1)$ is the contribution of the waviness of the walls. The analytical solution of (u_{10}, θ_0) coincides with (4.16) after modifications resulting from different choices of the origin of the coordinates in Fig. 1.7. The solution $(u_{11}, v_{11}, \theta_1)$, on the other hand, can be obtained following the same method as used for Eqs. (4.10)–(4.13) subject to the corresponding boundary conditions resulting from Eq. (4.53). This solution is, however, not presented here because it is discussed in Vajravelu and Sastri (1980).

Figures 4.10–4.12 show the profiles for $(u_{11}, v_{11}, \theta_1)$ for $\theta_i = -1$, $Pr = 0.71$, $b_1 = 0.01$ and different values of Gr, α^*, and ϕ when $b_1 x_1 = 0$ and $\pi/2$, respectively (see Table 4.2). It can be seen from these figures that quantities $(u_{11}, v_{11}, \theta_1)$ are significantly affected by the parameter ϕ indicating that they are quite different in the several channels under consideration. It can also be seen that the perturbed velocity and temperature profiles $(u_{11}, v_{11}, \theta_1)$ are oscillatory in nature and this behavioral pattern is more prominent when the parameters a_1, Gr, and α^* take higher values.

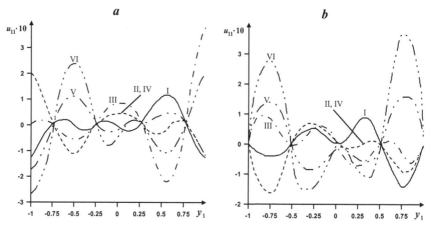

Fig. 4.10. Velocity profiles u_{11} for different values of Gr, α^*, and ϕ (see Table 4.2)

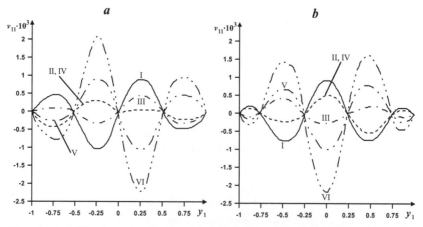

Fig. 4.11. Velocity profiles v_{11} for different values of Gr, α^*, and ϕ (see Table 4.2)

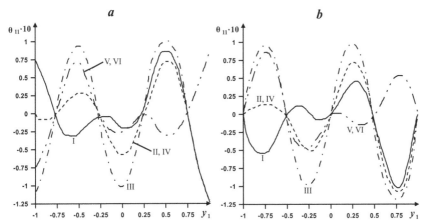

Fig. 4.12. Temperature profiles θ_{11} for different values of Gr, α^*, and ϕ (see Table 4.2)

Table 4.2. It refers to the curves I-VI in Figs. 4.10, 4.11 and 4.12.

	I	II	III	IV	V	VI
Gr	5	5	5	5	5	10
α^*	-5	-5	-5	-5	5	5
ϕ	0	$\pi/2$	π	$3\pi/2$	0	0

Finally, it is worth mentioning that this problem has been extended by Vajravelu (1989) by considering the effects of a constant magnetic field applied in the horizontal direction. Using the long wave approximation, the governing equations of the problem are solved by the perturbation technique, for both hydrodynamic and hydromagnetic cases. Numerical results are presented for stream function, temperature, flow, and heat transfer characteristics. At each stage, a comparison is made between the hydrodynamic and the hydromagnetic cases. The values of the stream function in the hydrodynamic case are found to be larger than those in the hydromagnetic case. This result is as per the expectations that a magnetic field exerts a retarding force on the free convection flow.

Chapter 5

Forced Convective Flow in a Wavy Horizontal Channel

A. Forced Convection Flow in Viscous Fluids

The published work relating to fluid flow confined between two horizontal walls has been concentrated on predominantly forced convection flow (Burns and Parkes, 1967; Sorenson and Stewart, 1974; Sobey, 1980; Stephanoff et al., 1980; O'Brien and Sparrow, 1982; Tsangaris and Leiter, 1984; Nishimura et al., 1984, 1985, 1986, 1990; Amano, 1985; Amano et al. 1987; Asako and Faghri, 1987; Asako et al., 1988; Garg and Maji, 1988; Rubin and Himansu, 1989; Ali and Ramadhyani, 1992; Saniei and Dini, 1993; Taneda and Vittori, 1996; Yang et al., 1997; Metwally and Manglik, 2000; Muley et al., 2002, 2006; Manglik et al., 2005) of constant property Newtonian fluids only. Two types of channels have been considered, each having frequent changes in cross-section — 1) symmetric and 2) asymmetric. In the following subsections, an outlook on the existing research work in these flow situations is presented.

1. Symmetrically Configured Wavy Walls

Viscous flows in sinusoidal converging–diverging channels have been treated by Burns and Parkes (1967) under the assumption of a very small Reynolds number (Stokes approximation). The stream function is obtained as a Fourier cosine series. Tsangaris and Leiter (1984) replaced the subsequent approximation satisfying the no-slip boundary condition at the walls by expressing the stream function as a Fourier series, not in the physical plane but in the transformed one, thereby converting the wavy boundary into a smooth one. Sobey (1980) has numerically treated the steady and unsteady flows through furrowed channels. His focus has been on the Reynolds number effect for separated flows. Thus, he has established a relation between the geometric parameters and Reynolds number for which separation occurs.

When steady, laminar flow of an incompressible viscous fluid through a wavy symmetrically configured horizontal channel as shown in Fig. 1.6c is considered, Navier–Stokes equations of motion on eliminating pressure p are reduced to the following and written in non-dimensional form as

$$Re\left(\frac{\partial \psi_1}{\partial y_1}\frac{\partial}{\partial x_1} - \frac{\partial \psi_1}{\partial x_1}\frac{\partial}{\partial y_1}\right)\nabla^2\psi_1 = \nabla^4\psi_1. \tag{5.1}$$

Here ∇^2 is the Laplace operator, ∇^4 is the biharmonic operator, and the stream function ψ is defined in the usual way as

$$u_1 = \partial\psi_1/\partial y_1, \quad v_1 = -\partial\psi_1/\partial x_1.$$

In the above Eq. (5.1), the non-dimensional variables are defined as follows:

$$x_1 = x/d_m, \quad y_1 = y/d_m, \quad \psi_1 = \psi/(U_0 d_m) \tag{5.2}$$

and $Re = U_0 d_m/\nu$ is the Reynolds number.

Equation (5.1) is to be solved subject to the following boundary conditions

$$y_1 = \pm\sigma(x_1): \quad \frac{\partial\psi_1}{\partial x_1} = 0, \quad \frac{\partial\psi_1}{\partial y_1} = 0 \tag{5.3}$$

and the condition of a symmetric flow with respect to the x_1-axis

$$\psi_1(x_1, +y_1) = -\psi_1(x_1, -y_1). \tag{5.4}$$

Equation (5.1) is now transformed by the introduction of the following variables

$$x_1 = x_1, \quad y_2 = y_1/\sigma(x_1), \tag{5.5}$$

which transforms the wavy channel to a channel confined by parallel walls such that

$$y_2 = \pm 1. \tag{5.6}$$

Thus, Eq. (5.1) is rewritten as

$$Re \left(\frac{\partial \psi_1}{\partial y_2} \frac{\partial}{\partial x_2} - \frac{\partial \psi_1}{\partial x_2} \frac{\partial}{\partial y_2} \right) \nabla_1^2 \psi_1 = \sigma \nabla_1^4 \psi_1 \tag{5.7}$$

where

$$\nabla_1^2 = \frac{\partial^2}{\partial x_2^2} + \frac{1}{\sigma^2} (1 + y_2^2 \sigma_{x_2}^2) \frac{\partial^2}{\partial y_2^2} - 2 y_2 \frac{\sigma_{x_2}}{\sigma} \frac{\partial^2}{\partial x_2 \partial y_2}$$

$$+ \frac{y_2}{\sigma^2} (2 \sigma_{x_2}^2 - \sigma \sigma_{x_2 x_2}) \frac{\partial}{\partial y_2}. \tag{5.8}$$

The appropriate boundary conditions for Eq. (5.7) are

$$y_2 = \pm 1 : \quad \frac{\partial \psi_1}{\partial x_2} = 0, \quad \frac{\partial \psi_1}{\partial y_2} = 0 \tag{5.9}$$

along with the symmetric flow condition

$$\psi_1(x_2, +y_2) = -\psi_1(x_2, -y_2). \tag{5.10}$$

Weakly furrowed channels are now considered on the basis that

$$\sigma(x_2) = 1 + a_1 f(x_2) \tag{5.11}$$

where the amplitude a_1 of the wavy walls is assumed to be small. This implies a search for the solution of Eq. (5.7) through an expansion of the following form

$$\psi_1 = \psi_{1,0}(y_2) + a_1 \psi_{1,1}(x_2, y_2) + \text{higher order terms.} \tag{5.12}$$

When the zeroth-order equations are considered, the leading order terms in (5.7) satisfy the following differential equation:

$$\psi_{1,0}^{IV} = 0 \tag{5.13}$$

subject to the boundary conditions

$$y_2 = \pm 1 : \quad \psi_{1,0} = 0, \quad \psi_{1,0}^1 = 0 \tag{5.14}$$

where the primes denote differentiation with respect to y_2. The solution of Eqs. (5.13)–(5.14) can be obtained in a straightforward manner as

$$\psi_{1,0} = y_2 - \frac{1}{3} y_2^3 \tag{5.15}$$

which describes the well-known Poiseuille flow. When first-order equations are considered, using the above Eq. (5.7), the equations for the terms of the order of a_1 can be written as

$$\frac{\partial^4 \psi_{1,1}}{\partial x_2^4} + 2\frac{\partial^4 \psi_{1,1}}{\partial x_2^2 \partial y_2^2} + \frac{\partial^4 \psi_{1,1}}{\partial y_2^4} + 12y_2 f_{x_2 x_2} - y_2(1-y_2^2)f_{x_2 x_2 x_2}$$

$$= Re\left\{ (1-y_2^2)\left[\frac{\partial^3 \psi_{1,1}}{\partial x_2^3} + \frac{\partial^3 \psi_{1,1}}{\partial x_2 \partial y_2^2} \right.\right.$$

$$\left.\left. +4y_2 f_{x_2} - y_2(1-y_2^2)f_{x_2 x_2 x_2} \right] + 2\frac{\partial \psi_{1,1}}{\partial x_2} \right\} \qquad (5.16)$$

subject to the boundary conditions

$$y_2 = \pm 1 : \ \psi_{1,1} = 0, \ \frac{\partial \psi_{1,1}}{\partial y_2} = 0. \qquad (5.17)$$

In order to obtain an analytical solution of Eqs. (5.16)–(5.17), Tsangaris and Leiter (1984) have considered a channel with sinusoidal walls such that

$$f(x_2) = -(1 + \cos x_2) \qquad (5.18)$$

which allows for a solution of $\psi_{1,1}$ of the following form

$$\psi_{1,1} = [f_1(y_2) - y_2(1-y_2^2)]\cos x_2 + f_2(y_2)\sin x_2. \qquad (5.19)$$

Substituting (5.19) into Eqs. (5.16)–(5.17) gives

$$f_1^{IV} - 2f_1^{II} + f_1 = Re[(1-y_2^2)(f_2^{II} - f_2) + 2f_2] \qquad (5.20)$$

$$f_2^{IV} - 2f_2^{II} + f_2 = Re[(1-y_2^2)(f_1^{II} - f_1) + 2f_1] \qquad (5.21)$$

subject to the following conditions

$$f_1(\pm 1) = 0, \ f_1^{I}(\pm 1) = -2, \ f_2(\pm 1) = 0, \ f_2^{I}(\pm 1) = 0. \qquad (5.22)$$

The numerical solution of these equations has been obtained (Tsangaris and Leiter, 1984) for $a_1 = 0.2$ and 0.3 over a wide range of Re values. The longitudinal u_1 and transversal v_1 velocity profiles are plotted across sections of the channel at $x_2 = 0, 1, 2, 3, 4, 5,$ and 6 in Figs. 5.1 and 5.2 for $Re = 1, 10, 15, 200,$ and 400. It can be seen that for small Reynolds numbers, for example, $Re = 1$ (Fig. 5.1a) and $Re = 10$ (Fig. 5.1b), the viscous effects dominate the inertial ones. However, for larger Reynolds numbers, for example, $Re = 75$ (Fig. 5.1c), there is an obvious influence of the fluid inertia on the velocity profile.

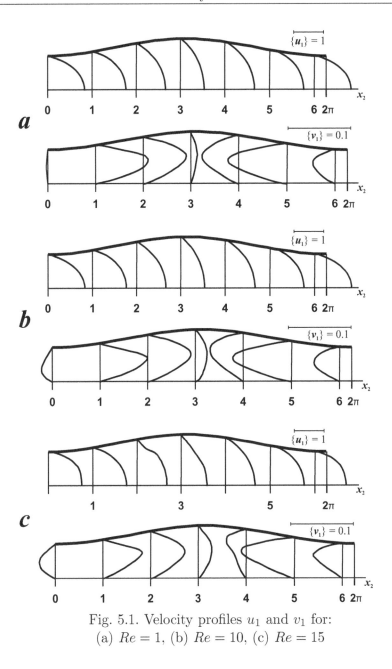

Fig. 5.1. Velocity profiles u_1 and v_1 for:
(a) $Re = 1$, (b) $Re = 10$, (c) $Re = 15$

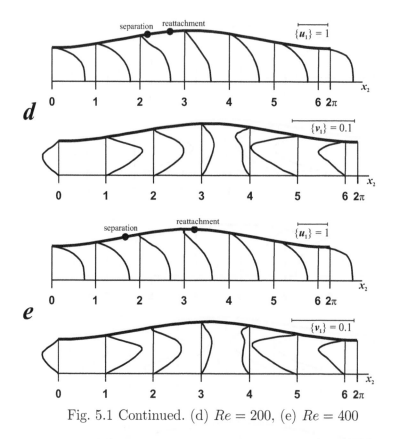

Fig. 5.1 Continued. (d) $Re = 200$, (e) $Re = 400$

In the diverging part of the channel, the profiles of u_1 are flattened at the centerline. The flow becomes more asymmetric in symmetrically lying cross-sections of the converging and diverging portions of the channel, with regard to the distribution of the u_1 and v_1 profiles. The negative and positive signs of v_1 for $x_2 = 0$ and $x_2 = \pi$ give a clear demonstration of the effect of inertia. Further increase in Re to a value of 200 (Fig. 5.1d) leads to a separation flow from the walls. The profiles of u_1 approach tangentially to y_2-planes near the wall and backflow is registered in some cross-sections of the channel. A separation bubble is recognized as the backflow becomes stronger and also more extensive when $Re = 400$ (Fig. 5.1e). In Fig. 5.2, the velocity field in the domain of the bubble vortex for $a_1 = 0.3$ and $Re = 400$ has been sketched.

The center of the bubble vortex, namely, the point where the velocity vanishes, can be determined as in Fig. 5.3. It is worth noting from this figure that the closed streamlines have an absolute value larger than 0.6666, which is the value of ψ_1 corresponding to the wavy boundary.

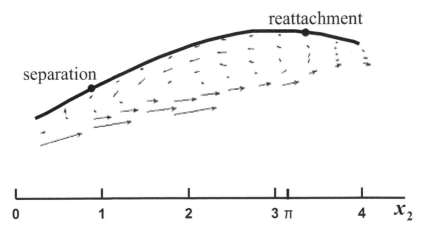

Fig. 5.2. Velocity field for $a_1 = 0.3$ and $Re = 400$ [reprinted from Tsangaris and Leiter (1984) with permission from Springer]

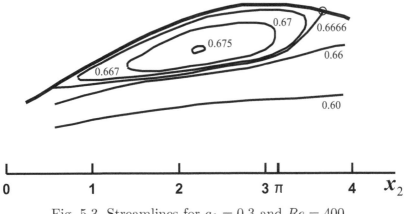

Fig. 5.3. Streamlines for $a_1 = 0.3$ and $Re = 400$

It is customary, from a pragmatic viewpoint, to present the skin friction coefficient at the walls of the channel. It can be shown, after some straightforward algebra, that the skin friction coefficient for the upper wall $y_2 = +1$ is given (Tsangaris and Leiter, 1984) as follows

$$\tau_w = \frac{(1 - a_1^2 \sin^2 x_2)^2 \{-2 + a_1[(f_1^{II}(+1) + 6) \cos x_2 + f_2^{II} \sin x_2]\}}{[1 - a_1^2(1 + \cos x_2)]^2(1 + a_1^2 \sin^2 x_2)}.$$

$$(5.23)$$

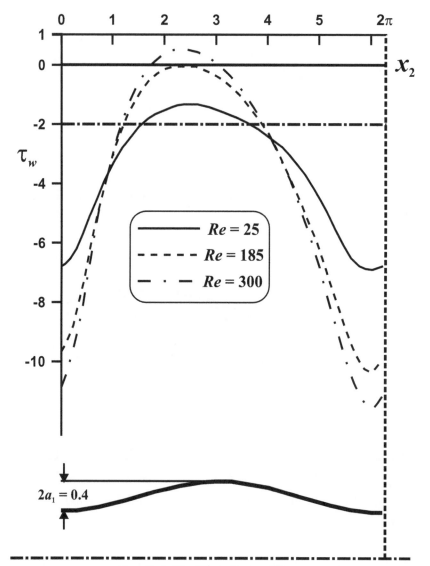

Fig. 5.4. Profiles of τ_w with x_2

It is pertinent to note that the term -2 within the brackets in the above Eq. (5.23) is the Poiseuille term slope in the (x_2, y_2)-plane. Figure 5.4 depicts the variation of wall shear stress against x_2 for $Re = 25$, 185, and 300, respectively, when $a_1 = 0.2$.

It can be seen that τ_w varies periodically with x_2 and has a phase difference in line with the variation of the shape of the wall. Its maximum value lies upstream of the maximum channel cross-section.

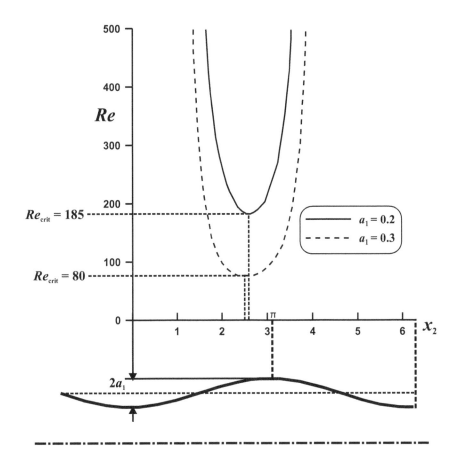

Fig. 5.5. Position of the points where the flow separates and reattaches

In Eq. (5.23), $\tau_w = 0$ locates the position of the points where the flow separates and reattaches, and this is shown in Fig. 5.5.

It can be easily shown that these points are defined by

$$x_{2S} = 2\arctan\frac{B - \sqrt{A^2 + B^2 - C^2}}{A + C} + \pi$$

$$x_{2R} = 2\arctan\frac{B + \sqrt{A^2 + B^2 - C^2}}{A + C} + \pi$$

(5.24)

under the condition that $A^2 + B^2 \geq C^2$ and $Re > Re_{crit}$, where the critical Reynolds number is obtained from $A^2 + B^2 - C^2 = 0$, with A, B, C given by

$$A = f_1^{II}(+1) + 6, \quad B = f_2^{II}(+1), \quad C = \frac{2}{a_1}.$$

(5.25)

It can be seen from Fig. 5.5 that the separation point lies slightly upstream of the maximum width of the wavy wall of the channel and the separation regions for $Re > Re_{crit}$ are shifted downstream.

Garg and Maji (1988) have also solved the same problem (Tsangaris and Leiter, 1984) numerically by a finite difference method. They have treated the heat transfer aspect in addition to the flow. Further, their solution is valid for any value of wall-amplitude-to-wavelength ratio, while that of the earlier investigation (Tsangaris and Leiter, 1984) has a severe restriction and is applicable only to very small values of amplitude-to-wavelength ratio. In the following, the work of Garg and Maji (1988) is briefly described and discussed.

The physical domain considered is shown in Fig. 1.6c (considering only the first cycle). The channel is assumed to be isothermal and the walls are defined by Eq. (1.5), which is taken to be of the following form

$$\overline{\sigma}(x) = d - (a/2)[1 + \cos(2\pi x/\lambda)]. \qquad (5.26)$$

Constant physical properties are assumed and viscous dissipation is neglected in the energy equation. Using these assumptions, the governing equations of continuity, momentum, and energy in terms of fluid enthalpy can be written in the non-dimensional form as follows:

$$\frac{\partial u_1}{\partial x_1} + \frac{\partial v_1}{\partial y_1} = 0 \qquad (5.27)$$

$$u_1 \frac{\partial u_1}{\partial x_1} + v_1 \frac{\partial u_1}{\partial y_1} = -\frac{\partial p_1}{\partial x_1} + \frac{1}{Re} \nabla^2 u_1 \qquad (5.28)$$

$$u_1 \frac{\partial v_1}{\partial x_1} + v_1 \frac{\partial v_1}{\partial y_1} = -\frac{\partial p_1}{\partial y_1} + \frac{1}{Re} \nabla^2 v_1 \qquad (5.29)$$

$$u_1 \frac{\partial h_1}{\partial x_1} + v_1 \frac{\partial h_1}{\partial y_1} = \frac{1}{Re} \nabla^2 h_1. \qquad (5.30)$$

The non-dimensional variables are defined by

$$\begin{gathered} x_1 = 2x/\lambda; \ y_1 = 2y/\lambda; \ u_1 = u/U_0; \ v_1 = v/U_0; \\ p_1 = (p - p_0)/(\rho U_0^2); \ h_1 = (h - h_w)/(h_0 - h_w) \end{gathered} \qquad (5.31)$$

and further, the following variable is also defined as

$$\sigma(x_1) = \overline{\sigma}(x)/\lambda. \qquad (5.32)$$

Note that U_0, p_0, and h_0 are the uniform velocity, pressure, and enthalpy at the entrance of the channel and h_w is the uniform enthalpy at the wall. Due to symmetry about the channel centerline, only one half of the channel needs to be considered in the analysis. Thus, the boundary conditions imposed on Eqs. (5.27)–(5.30) are

$$x_1 = 0 : u_1 = 1, \ p_1 = 0, \ h_1 = 1$$

$$y_1 = 0 : \frac{\partial u_2}{\partial y_1} = 0, \ v_1 = 0, \ \frac{\partial h_1}{\partial y_1} = 1 \qquad (5.33)$$

$$y_1 = \frac{\sigma}{2} : u_1 = 0, \ v_1 = 0, \ h_1 = 0$$

and

$$u_1(x_1^*, y_1) = u_1[(x_1^* + 2), y_1],$$
$$v_1(x_1^*, y_1) = v_1[(x_1^* + 2), y_1] \qquad (5.34)$$
$$h_1(x_1^*, y_1)/h_1[(x_1^* + 2), y_1] = h_1[(x_1^* + 2), y_1]/h_1[(x_1^* + 4), y_1]$$

where x_1^* is an arbitrary location in the fully developed region.

It should be mentioned here that the boundary conditions (5.34) imply the periodic nature of the fully developed flow. For the isothermal wall condition, the cross-sectional shape of the enthalpy difference $[h_1(x_1, y_1) - h_{1w}]$ repeats itself periodically in the fully developed region, but the level decreases in the streamwise direction (Prata and Sparrow, 1984). In fact, the ratio h_1/h_{1b}, where h_{1b} is the dimensionless bulk enthalpy defined later in Eq. (5.41), is constant in the fully developed region. This conclusion leads to the last boundary condition in Eqs. (5.34).

Next, the variables introduced in Eq. (5.5) are used in order to transform the wavy physical domain into a rectangular computational domain such that $y_2 = \pm 1/2$ at all points on the curved boundaries.

It can be easily shown that the following relationship between velocity components (u_1, v_1) and the wavy surface equation $\sigma(x_2)$ is applicable

$$u_2 = (1 + F^2)^{1/2} u_1; \quad v_2 = v_1 - F u_1 \qquad (5.35)$$

where

$$F = y_2 \sigma_{x^2}. \qquad (5.36)$$

Substituting (5.35)–(5.36) into (5.27)–(5.30) leaves Eqs. (5.27), (5.28), and (5.30) unchanged, but transforms Eq. (5.29) to

$$
\begin{aligned}
u_1 \frac{\partial v_2}{\partial x_1} + v_2 \frac{\partial v_2}{\partial y_1} = {} & F \frac{\partial p_1}{\partial x_1} - \frac{\partial p_1}{\partial y_1} + \frac{1}{Re} \nabla^2 v_2 \\
& - u_1 \left(u_1 \frac{\partial F}{\partial x_1} + v_2 \frac{\partial F}{\partial y_1} \right) + \frac{u_1}{Re} \nabla^2 F \\
& + \frac{2}{Re} \left(\frac{\partial u_1}{\partial x_1} \frac{\partial F}{\partial x_1} + \frac{\partial u_1}{\partial y_1} \frac{\partial F}{\partial y_1} \right).
\end{aligned} \tag{5.37}
$$

Thus, the velocity components solved through the above are u_1 and v_2. Nevertheless, the velocity component u_2 can be obtained through Eqs. (5.35)–(5.36). The boundary conditions (5.33)–(5.34) for Eqs. (5.27), (5.28), (5.30), and (5.37) now become

$$
x_2 = 0 : \; u_1 = 1, \; p_1 = 0, \; h_1 = 1
$$

$$
y_2 = 0 : \; \frac{\partial u_1}{\partial y_2} = 0, \; v_2 = 0, \; \frac{\partial h_1}{\partial y_2} = 1 \tag{5.38}
$$

$$
y_2 = \frac{1}{2} : \; u_1 = 0, \; v_2 = 0, \; h_1 = 0
$$

and

$$
\begin{aligned}
& u_1(x_2^*, y_2) = u_1[(x_2^* + 2), y_2], \\
& v_2(x_2^*, y_2) = v_2[(x_2^* + 2), y_2], \\
& h_1(x_2^*, y_2)/h_1[(x_2^* + 2), y_2] = h_1[(x_2^* + 2), y_2]/h_1[(x_2^* + 4), y_2]
\end{aligned} \tag{5.39}
$$

where $x_2^* = x_1^*$.

The local Nusselt number, based on twice the mean width, $4(d-a)$, of the wavy channel, is given by

$$
Nu = - \frac{4(d-a)}{h_b - h_w} \left(\frac{\partial h}{\partial n} \right) \Big|_{y = \bar{\sigma}(x)} \tag{5.40}
$$

where the bulk enthalpy h_b is defined as

$$
h_b = \int_0^{\bar{\sigma}(x)} uh\,dy \Big/ \int_0^{\bar{\sigma}(x)} u\,dy \tag{5.41}
$$

or, in non-dimensional form, by

$$
h_{1b} = \int_0^{\sigma(x_1)/2} u_1 h_1\,dy_1 \Big/ \int_0^{\sigma(x_1)/2} u_1\,dy_1. \tag{5.42}
$$

Using (5.5), the following is obtained at any location of x_2

$$h_{1b}(x_2) = \int_0^{1/2} \sigma(x_2)u_1h_1dy_2 / \int_0^{1/2} \sigma(x_2)u_1dy_2$$

$$= 2\frac{\sigma(x_2)}{\sigma(0)} \int_0^{1/2} u_1h_1dy_2. \tag{5.43}$$

At axial locations where $F = 0$, $\partial/\partial n = \partial/\partial y$, and at such locations, $x_2 = 0, 1, 2, \ldots$ At these conditions, the Nusselt number can be calculated by

$$Nu = -\frac{4[(d/\lambda) - a_1]}{h_{1b}\sigma(x_2)} \left(\frac{\partial h_1}{\partial y_2}\right)\Big|_{y_2=1/2}. \tag{5.44}$$

In order to obtain the discretization form of Eqs. (5.27), (5.28), (5.30), and (5.37), they are first integrated (Garg and Maji, 1988) over a control volume bounded by lines of constant x_2 and constant y_2, as shown in Fig. 5.6. The resulting equations are then discretized using the method developed by Patankar (1980), and these are not given here, but are available in Garg and Maji (1988).

Some representative results taken from Garg and Maji (1988) are presented in Figs. 5.6 and 5.7 for $a_1 = 0.1$ and 0.25 at Reynolds numbers of 100 and 500, respectively. The Prandtl number is fixed at 0.72 and d/λ at 0.5. Due to the symmetry about the channel centerline, the figures show the various profiles in only half of the domain.

Figures 5.6 and 5.7 show the behavior of the velocity components (u_1, v_1) when $a_1 = 0.1$, $Re = 500$ and $a_1 = 0.25$, $Re = 100$ at various x_1 locations. It can be seen from Fig. 5.6 that in the diverging part of the channel the u_1-profiles are flattened at the centerline. Further, it can be also seen from Figs. 5.6 and 5.7 that the flow becomes more symmetric in the symmetrically lying cross-sections of the converging and diverging parts of the channel.

Next, it is seen from Fig. 5.6a that a small backflow exists near the wall of the channel for $a_1 = 0.1$ and $Re = 500$, which is not observed for $a_1 = 0.25$ and $Re = 100$. Garg and Maji (1988) concluded that when Re is increased to 400, the flow begins to separate. The corresponding separation points (S) and reattachment points (R) are also shown in Fig. 5.6a. It is stated (Garg and Maji, 1988) that as Re increases, the (S) point moves upstream and the (R) point moves downstream.

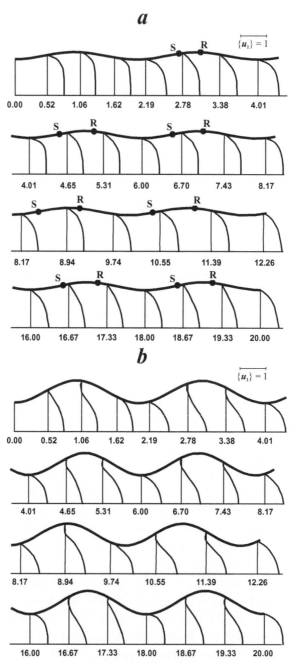

Fig. 5.6. Velocity profiles u_1 for: $a_1 = 0.1$, $Re = 500$ (a);
$a_1 = 0.25$, $Re = 100$ (b)

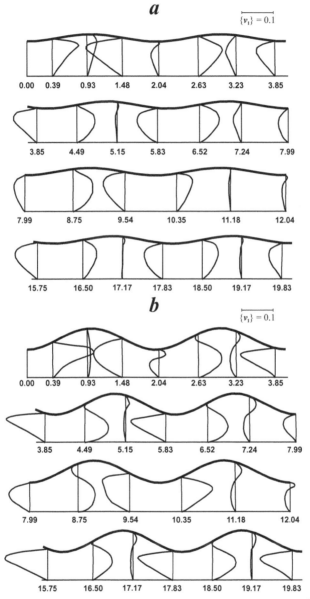

Fig. 5.7. Velocity profiles v_1 for: $a_1 = 0.1$, $Re = 500$ (a);
$a_1 = 0.25$, $Re = 100$ (b)

The separated flow region grows not only with Re but also with the amplitude a_1. Moreover, for higher a_1 and Re, separated flow occurs in the converging portion of the channels as well.

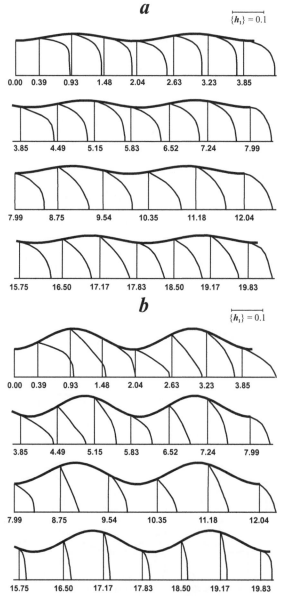

Fig. 5.8. Profiles of the fluid enthalpy for:
$a_1 = 0.1$, $Re = 500$ (a); $a_1 = 0.25$, $Re = 100$ (b)

The behavior of the fluid enthalpy for the same combination of a_1, Re and $Pr = 0.72$ is illustrated in Fig. 5.8. It is seen that as a_1 increases, the difference between the fluid enthalpy at the centerline and that at the wall decreases substantially.

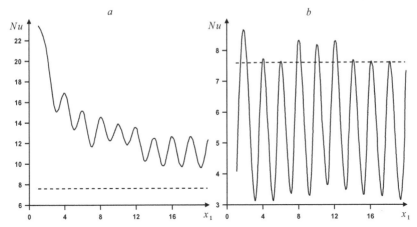

Fig. 5.9. Variation of the local Nusselt number for: $a_1 = 0.1$, $Re = 500$ (a); $a_1 = 0.25$, $Re = 100$ (b)

A similar effect can also be observed for a decrease in Re. Further, it can be seen that a point of inflection appears in the enthalpy profiles in the separated region, whereas in the non-separated region the enthalpy profile is parabolic.

Figure 5.9 presents the variation of the local Nusselt number versus x_1. The dashed line corresponds to the constant value of $Nu(= 7.54)$ for fully developed flow in a flat channel with isothermal walls.

It is observed that Nu is larger and varies almost sinusoidally in a wavy channel, and also that increases with Re. As expected, Nu is larger at the minimum cross-section and smaller at the maximum cross-section.

2. Asymmetrically Configured Wavy Walls

In this subsection, the work of Vajravelu (1980) on steady heat transfer in a viscous fluid flowing in a horizontal wavy channel is presented. The geometry under consideration is that which is shown in Fig. 1.8a,b,c wherein the wavy walls are described in the non-dimensional form as

$$\sigma_U(x_1) = 1 + a_1 \cos(b_1 x_1)$$
$$\sigma_L(x_1) = -1 + a_1 \cos(b_1 x_1 + \phi). \tag{5.45}$$

Here $\sigma_U(x_1)$ and $\sigma_L(x_1)$ denote the shape functions of upper and lower wavy walls, respectively, with ϕ equal to 0, $\pi/2$, and $3\pi/2$ to denote changes in the orientation of the channel. It is assumed that the walls are maintained at uniform temperatures of T_0 and T_1, respectively. It is further assumed that the wavelength of the wavy wall is large and that the volumetric heat source or sink term in the energy equation is constant.

Under these assumptions, the equations that govern the steady two-dimensional flow and heat transfer in a viscous incompressible fluid occupying the channel can be written (Vajravelu, 1980) in terms of the non-dimensional variables as follows:

$$\frac{\partial u_1}{\partial x_1} + \frac{\partial v_1}{\partial y_1} = 0 \tag{5.46}$$

$$u_1 \frac{\partial u_1}{\partial x_1} + v_1 \frac{\partial u_1}{\partial y_1} = -\frac{\partial p_1}{\partial x_1} + \nabla^2 u_1 \tag{5.47}$$

$$u_1 \frac{\partial v_1}{\partial x_1} + v_1 \frac{\partial v_1}{\partial y_1} = -\frac{\partial p_1}{\partial y_1} + \nabla^2 v_1 \tag{5.48}$$

$$u_1 \frac{\partial \theta}{\partial x_1} + v_1 \frac{\partial \theta}{\partial y_1} = \frac{1}{Pr}(\nabla^2 \theta + \alpha^*). \tag{5.49}$$

The above equations are to be solved under the following boundary conditions:

$$y_1 = \sigma_U(x_1): \ u_1 = 0, \ v_1 = 0, \ \theta = 1$$
$$y_1 = \sigma_L(x_1): \ u_1 = 0, \ v_1 = 0, \ \theta = \theta_i. \tag{5.50}$$

The non-dimensional variables are the same as those defined by Eq. (5.8), except that d is now the half width of the wavy channel.

The solution of the above-stated problem is sought in the form

$$u_1 = u_{10}(y_1) + u_{11}(x_1, y_1); \ v_1 = v_{11}(x_1, y_1)$$
$$p_1 = p_{10}(x_1) + p_{11}(x_1, y_1); \ \theta = \theta_0(y_1) + \theta_1(x_1, y_1) \tag{5.51}$$

where u_{11}, v_{11}, p_{11}, and θ_1 are small compared with the mean quantities u_{10}, p_{10}, and θ_0, which are independent of x_1 since the flow has been assumed to be fully developed.

Substituting (5.51) into Eqs. (5.46)–(5.49), and making use of (5.45) and (5.50), yields

$$u_{10}'' = \frac{dp_{10}}{dx_1}, \quad \theta_0'' = -\alpha^* \tag{5.52}$$

$$y_1 = 1 : \ u_{10} = 0, \ \theta_0 = 1$$
$$y_1 = -1 : \ u_{10} = 0, \ \theta_0 = \theta_i \tag{5.53}$$

and

$$\frac{\partial u_{11}}{\partial x_1} + \frac{\partial v_{11}}{\partial y_1} = 0$$

$$u_{10}\frac{\partial u_{11}}{\partial x_1} + u'_{10}v_{11} = -\frac{\partial p_{11}}{\partial x_1} + \nabla^2 u_{11}$$

$$u_{10}\frac{\partial v_{11}}{\partial x_1} = -\frac{\partial p_{11}}{\partial y_1} + \nabla^2 v_{11} \tag{5.54}$$

$$u_{10}\frac{\partial \theta_1}{\partial x_1} + \theta'_0 v_{11} = \frac{1}{Pr}(\nabla^2 \theta_1)$$

$$y_1 = 1 : \quad u_{11} = -u'_{10}, \ v_{11} = 0, \ \theta_1 = -\theta'_0$$
$$y_1 = -1 : \quad u_{11} = -u'_{10}\cos\phi, \ v_{11} = 0, \ \theta_1 = -\theta'_0\cos\phi \tag{5.55}$$

where primes denote differentiation with respect to y_1.

The analytical solution of Eqs. (5.52)–(5.53) describes the plane Poiseuille flow and is shown to be

$$u_{10} = \frac{1}{2}(1 - y_1^2)$$

$$\theta_0 = -\frac{1}{2}\alpha^*(1 - y_1^2) + \frac{1 - \theta_1}{2}y_1 + \frac{1 - \theta_i}{2} \tag{5.56}$$

where $\partial p_{10}/\partial x_1$ is taken equal to -1.

From the second and third Eqs. of (5.54)–(5.55), p_{11} is eliminated to give

$$u_{10}\left(\frac{\partial^3 \psi_1}{\partial x_1^3} + \frac{\partial^3 \psi_1}{\partial y_1^3}\right) - u''_{10}\frac{\partial \psi_1}{\partial x_1} = \nabla^4 \psi_1 \tag{5.57}$$

$$u_{10}\frac{\partial \theta_1}{\partial x_1} + \theta'_0\frac{\partial \psi_1}{\partial x_1} = \frac{1}{Pr}(\nabla^2 \theta_1) \tag{5.58}$$

where the stream function ψ_1 is defined as follows

$$u_{11} = -\frac{\partial \psi_1}{\partial y_1}, \quad v_{11} = \frac{\partial \psi_1}{\partial x_1}. \tag{5.59}$$

A solution is now sought assuming the following

$$\psi_1 = a_1 \exp(ib_1 x_1) f(y_1)$$
$$\theta_1 = a_1 \exp(ib_1 x_1) h(y_1) \tag{5.60}$$

where $i = \sqrt{-1}$. Eqs. (5.57) and (5.58) then become

$$f^{IV} - ib_1[u_{10}(f'' - b_1^2 f) - u_{10}'' f] - b_1^2(2f'' - b_1^2 f) = 0 \qquad (5.61)$$

$$h'' - b_1^2 h = Pr\, ib_1(u_{10}h + \theta_0' f). \qquad (5.62)$$

The boundary conditions (5.55) for u_{11} and v_{11} can be written in terms of ψ_1 as

$$
\begin{aligned}
y_1 = 1: \quad & \frac{\partial \psi_1}{\partial y_1} = -u_{10}', \quad \frac{\partial \psi_1}{\partial x_1} = 0 \\
y_1 = -1: \quad & \frac{\partial \psi_1}{\partial y_1} = u_{10}' \cos \phi, \quad \frac{\partial \psi_1}{\partial x_1} = 0.
\end{aligned}
\qquad (5.63)
$$

Vajravelu (1980) assumed that for small $b_1 (\ll 1)$, the functions f and h can be expanded as

$$
\begin{aligned}
f &= f_0(y_1) + b_1 f_1(y_1) + b_1^2 f_2(y_1) + \text{higher order terms} \\
h &= h_0(y_1) + b_1 h_1(y_1) + b_1^2 h_2(y_1) + \text{higher order terms}
\end{aligned}
\qquad (5.64)
$$

Substituting (5.64) into Eqs. (5.61)–(5.62) gives the following, after some manipulation:

$$
\begin{aligned}
& f_0^{IV} = 0, \quad h_0^{II} = 0 \\
& y_1 = 0: \quad f_0 = 0, \quad f_0' = -u_{10}', \quad h_0 = -\theta_0' \\
& y_1 = 1: \quad f_0 = 0, \quad f_0' = u_{10}' \cos \phi, \quad h_0 = -\theta_0' \cos \phi
\end{aligned}
\qquad (5.65)
$$

and

$$
\begin{aligned}
& f_1^{IV} - i(u_{10}f_0'' - u_{10}'' f_0) = 0 \\
& h_1'' = Pr\, i(u_{10}h_0 + \theta_0' f_0) \\
& f_2^{IV} - i(u_{10}f_1'' - u_{10}'' f_1) - 2f_0'' = 0 \\
& h_2'' = Pr\, i(u_{10}h_1 + \theta_0' f_1).
\end{aligned}
\qquad (5.66)
$$

Figures 5.10 and 5.11 depict the behavior of the total velocity component $u_1(= u_{10} + u_{11})$ and of the total temperature $\theta(= \theta_0 + \theta_1)$ (see Tables 5.1 and 5.2).

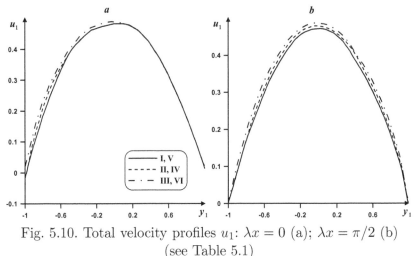

Fig. 5.10. Total velocity profiles u_1: $\lambda x = 0$ (a); $\lambda x = \pi/2$ (b)
(see Table 5.1)

Table 5.1. Referring to the curves I-VI in Figs. 5.10.

	I	II	III	IV	V	VI
λ	0.01	0.01	0.01	0.01	0.02	0.02
ϕ	0	$\pi/2$	π	$3\pi/2$	0	π

It can be seen again that u_1 and θ are significantly affected by the change in the shape of the channel wall, that is, by the parameter ϕ. Moreover, by increasing the value of α^*, a greater θ-profile can be expected.

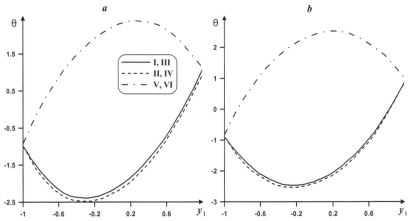

Fig. 5.11. Total temperature profiles θ for $\lambda = 0.01$,
$m = -1$: $\lambda x = 0$ (a); $\lambda x = \pi/2$ (b) (see Table 5.2)

Table 5.2. Referring to the curves I-VI in Figs. 5.11.

	I	II	III	IV	V	VI
Pr	0.71	0.71	0.71	0.71	0.71	7
α	-5	-5	-5	-5	5	5
ϕ	0	$\pi/2$	π	$3\pi/2$	0	0

B. Convective Flow in Fluid-Saturated Porous Media

Thermal convection in a saturated porous medium confined between two wavy walls of mean horizontal disposition, with the medium being heated from below, has been studied theoretically by Rees and Riley (1986). They solved the coupled full Darcy and energy equations both analytically and numerically. The main intention behind this study was to acquire a better understanding of convective flow within a saturated porous medium such as a folded rock stratum. The research interest on this subject was initiated first by Horton and Rogers (1945) and followed by Lapwood (1948), who investigated the onset of convection in a porous medium bounded between two smooth horizontal boundaries of infinite extent with constant, but unequal temperatures. Subsequent papers are due to Palm et al. (1972) and Straus (1974), who also dealt with smooth surfaces. These works are basically the porous-medium analog of the well-known Rayleigh–Benard problem. The effects of spatially periodic boundary conditions upon the Rayleigh–Benard stability problem, wherein the bounding surfaces have temperatures that vary periodically, or the surfaces themselves are wavy, has been studied by Kelly and Pal (1978). The amplitudes of the spatial non-uniformities are assumed to be small, and the wavelength is set equal to the critical wavelength for the onset of Rayleigh–Benard convection. The specific problem (Kelly and Pal, 1978) is referred to as involving resonant wavelength excitation. For values of the mean Rayleigh number below the classical critical value, the mean Nusselt number and the mean flow have been determined as a function of Rayleigh number, Prandtl number, and the modulation amplitude. For values of the Rayleigh number close to the classical critical value, the effects of non-uniformities are greatly amplified. The response of convective flow to spatially periodic forcing at a period different from that at the critical wavelength has been investigated experimentally by Lowe et al. (1986). They used an electrodynamic instability in a thin layer of nematic liquid for reasons of experimental convenience. Their experiments reveal a wide variety of flows with multiple periodicities in the presence of period forcing. It is believed (Lowe et al., 1986) that most of these phenomena would be exhibited by ordinary Rayleigh–Benard convection, though they used electrohydrodynamic instability. However, a number of issues have not been addressed by their experiments (Lowe et al., 1986), and hence, they have not been able to truly test the predictions of Kelly and Pal (1978). The work of Kelly and Pal (1978) treats viscous fluids without the presence of porous media. When it comes to the case of wavy surfaces in identical situations but in the presence of porous media, the work of Rees and Riley (1986) becomes important. In the following, the work of Rees and Riley (1986) is briefly outlined and discussed.

A porous medium of mean vertical depth $2d$, which is confined between two wavy walls of infinite horizontal extent and saturated with a Newtonian fluid, is considered. The coordinate system for the case under study is shown in Fig. 1.8d. The upper and lower walls are maintained at constant temperatures T_0 and T_1, where $T_0 < T_1$.

The non-dimensional form of Darcy's law and the energy equation using Boussinesq approximation can be written (Rees and Riley, 1986) in the form

$$\nabla^2 \psi_1 = Ra \frac{\partial \theta}{\partial x_1} \tag{5.67}$$

$$\nabla^2 \theta = \frac{\partial \psi_1}{\partial x_1} \frac{\partial \theta}{\partial y_1} - \frac{\partial \psi_1}{\partial y_1} \frac{\partial \theta}{\partial x_1} + \frac{\partial \theta}{\partial t_1} \tag{5.68}$$

where the non-dimensional variables are defined as follows:

$$x_1 = x/d, \ y_1 = y/d, \ \psi_1 = (\rho_f C_f / \alpha)\psi$$
$$t_1 = (\alpha/d^2 \rho_p C_p)t, \ \theta = [T - 0.5(T_1 + T_0)]/T_m \tag{5.69}$$

where $T_m = 0.5(T_1 - T_0)$, and the Rayleigh number is defined as

$$Ra = g\beta K T_m d/(\alpha\nu).$$

Equations (5.67) and (5.68) are to be solved under the following boundary conditions

$$y_1 = 1 + \delta\sigma_U(x_1): \ \psi_1 = 0, \ \theta = -1$$
$$y_1 = -1 - \delta\sigma_L(x_1): \ \psi_1 = 0, \ \theta = 1 \tag{5.70}$$

with $\delta \ll 1$. Here $\sigma_U(x_1)$ and $\sigma_L(x_1)$ are taken to be of the following forms:

$$\sigma_U(x_1) = a_U \cos(bx_1 - \phi)$$
$$\sigma_L(x_1) = a_L \cos(bx_1 + \phi). \tag{5.71}$$

The following variables are now introduced as done by Rees and Riley (1986) such that

$$x_2 = x_1, \quad y_2 = \frac{2y_1 + \delta[\sigma_L(x_1) - \sigma_U(x_1)]}{2 + \delta[\sigma_L(x_1) + \sigma_U(x_1)]}, \tag{5.72}$$

which transforms the upper and lower wavy walls to parallel smooth walls such that $y_2 = \pm 1$. Equations (5.67) and (5.68) then become

$$\nabla_1^2 \psi_1 = Ra \left(S_1^2 \frac{\partial \theta}{\partial x_2} + \delta S_1 S_2 \frac{\partial \theta}{\partial y_2} \right) \tag{5.73}$$

$$\nabla_1^2 \theta = 2S_1 \left(\frac{\partial \psi_1}{\partial x_2} \frac{\partial \theta}{\partial y_2} - \frac{\partial \psi_1}{\partial y_2} \frac{\partial \theta}{\partial x_2} \right) + S_1^2 \frac{\partial \theta}{\partial t_1} \tag{5.74}$$

where

$$\nabla_1^2 = S_1^2 \frac{\partial^2}{\partial x_2^2} + (4 + \delta^2 S_1^2) \frac{\partial^2}{\partial y_2^2} + 2\delta S_1 S_2 \frac{\partial^2}{\partial x_2 \partial y_2}$$

$$+ \{ \delta^2 S_1 [(1 - y_2)\sigma_{Lx_2x_2} - (1 + y_2)\sigma_{Ux_2x_2}]$$

$$- 2\delta^2 S_1 (\sigma_{Lx_2} + \sigma_{Ux_2}) \} \frac{\partial}{\partial y_2} \tag{5.75}$$

with

$$S_1 = 2 + \delta(\sigma_L + \sigma_U)$$
$$S_2 = ((1 - y_2)\sigma_{Lx_2} - (1 + y_2)\sigma_{Ux_2}. \tag{5.76}$$

Attention is now focussed on the steady-state case and it is assumed that Ra is sufficiently below Ra_c, which denotes the critical value of Ra. Thus, ψ_1 and θ are expanded as follows:

$$\psi_1 = \delta\psi_{1,1}(x_2, y_2) + \delta^2\psi_{1,2}(x_2, y_2) + \text{higher order terms}$$
$$\theta = -y_2 + \delta\theta_1(x_2, y_2) + \delta^2\theta_2(x_2, y_2) + \text{higher order terms}. \tag{5.77}$$

Insertion of (5.77) into Eqs. (5.73) and (5.74), and equating like powers of δ, gives the following equations for $O(\delta)$ functions:

$$\nabla^2 \psi_{1,1} - Ra \frac{\partial \theta_1}{\partial x_2} \tag{5.78}$$

$$= \frac{1}{2} Ra\, b[(1 - y_2)a_L \sin(bx_2 + \phi) - (1 + y_2)a_U \sin(bx_2 - \phi)]$$

$$\nabla^2 \theta_1 + \frac{\partial \psi_{1,1}}{\partial x_2} \tag{5.79}$$

$$= -\frac{1}{2}b^2[(1 - y_2)a_L \cos(bx_2 + \phi) + (1 + y_2)a_U \cos(bx_2 - \phi)]$$

subject to the following boundary conditions

$$y_2 = \pm 1 : \quad \psi_{1,1} = 0, \ \theta_1 = 0. \tag{5.80}$$

The solution of these equations are sought of the form

$$\psi_{1,1} = f_1(y_2)a_L \sin(bx_2 + \phi) + f_2(y_2)a_U \sin(bx_2 - \phi)$$
$$\theta_1 = g_1(y_2)a_L \sin(bx_2 + \phi) + g_2(y_2)a_U \sin(bx_2 - \phi) \tag{5.81}$$

where the analytical expressions for f_1, f_2, g_1, and g_2 are available in Rees and Riley (1986). It has been shown (Rees and Riley, 1986) that expansions (5.77) become singular when $Ra \to Ra_c$, where Ra_c is given by

$$Ra_c^{1/2} = \left(b^2 + \frac{1}{4}\pi^2\right)/b, \tag{5.82}$$

which is precisely the expression for the neutral stability curve in the Lapwood (1948) problem. However, when the porous layer is varicose, that is, for $a_L = a_U$ and $\theta = 0$, the first order solution $\psi_{1,1}$ and θ_1 remains bounded near the neutral stability curve.

In Fig. 5.12, results of the Rees and Riley (1986) solution for $\psi_{1,1}$ are graphically presented for $Ra = 5$, $a_L = a_U$ and at various values of wall phase ϕ. It can be seen that for the varicose configuration (Fig. 5.12), the flow consists of a set of four counter-rotating cells per wall wave number b, as expected from symmetry considerations.

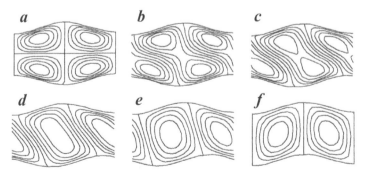

Fig. 5.12. Streamlines at $Ra = 5$, $a_L = a_U$ for: $\phi = 0°$ (a); $\phi = 5°$ (b); $\phi = 10°$ (c); $\phi = 20°$ (d); $\phi = 45°$ (e); $\phi = 90°$ (f) [reprinted from Rees and Riley (1986) with permission from Cambridge Journals]

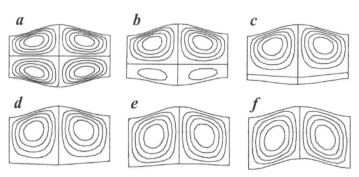

Fig. 5.13. Streamlines at $Ra = 5$, $\beta = 0$ for various values of the wall amplitude: $a_L = a_U$ (a); $a_L = 0.75a_U$ (b); $a_L = 0.5a_U$ (c); $a_L = 0.25a_U$ (d); $a_L = -0.25a_U$ (e); $a_L = -a_U$ (f) [reprinted from Rees and Riley (1986) with permission from Cambridge Journals]

But this pattern is quickly distorted by small deviations of ϕ from zero, with pairs of cells coalescing as $|\phi|$ increases. A symmetrical two-cell pattern, when $\phi = 0$ and $a_L = -a_U$, i.e., for the sinuous case, is also as expected (Fig. 5.12). This is shown in Fig. 5.13 along with the effect of varying a_L, while holding a_U fixed with $\phi = 0$ and $Ra = 5$.

The average Nusselt number from the wavy walls can be calculated using Eq. (5.59). For the lower wall this is given by

$$\overline{Nu} = \frac{b}{2\pi} \int_0^{2\pi/b} \frac{1 + \delta^2 (\sigma_{Lx_2})^2}{1 + (\delta/2)(\sigma_L + \sigma_U)} \left(\frac{\partial \theta}{\partial y_2}\right)\Bigg|_{y_2=-1} dx_2 \qquad (5.83)$$

Using Eqs. (5.77) and (5.81), the following can be derived

$$\overline{Nu} = -1 + \delta^2 [(a_L^2 + a_U^2 + 2a_L a_U \cos 2\phi)\overline{Nu}_V$$
$$+ (a_L^2 + a_U^2 - 2a_L a_U \cos 2\phi)\overline{Nu}_S] + O(\delta^4) \qquad (5.84)$$

where $\overline{Nu}_V$ and $\overline{Nu}_S$ corresponding to varicose ($a_L = a_U$, $\phi = 0$) and sinuous ($a_L = -a_U$, $\phi = 0$) contributions, respectively, are given by

$$\overline{Nu}_V = \frac{b}{64} Ra^{1/2} \left(\frac{\coth \gamma_1}{\gamma_1} - \frac{\coth \gamma_2}{\gamma_2} + \coth^2 \gamma_2 - \coth^2 \gamma_1\right)$$
$$- \frac{1}{16}(\gamma_1 \coth \gamma_1 + \gamma_2 \coth \gamma_2) \qquad (5.85)$$

$$\overline{Nu}_S = \frac{b}{64} Ra^{1/2} \left(\frac{\tanh \gamma_1}{\gamma_1} - \frac{\tanh \gamma_2}{\gamma_2} + \tanh^2 \gamma_2 - \tanh^2 \gamma_1\right)$$
$$- \frac{1}{16}(\gamma_1 \tanh \gamma_1 + \gamma_2 \tanh \gamma_2).$$

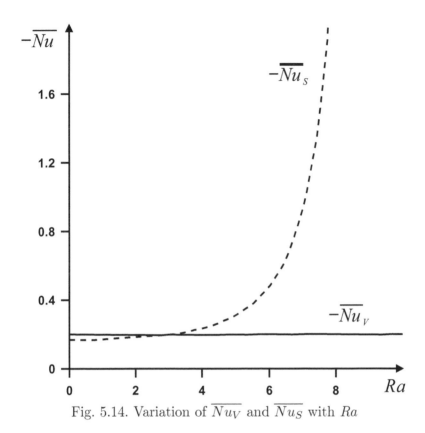

Fig. 5.14. Variation of $\overline{Nu}_V$ and $\overline{Nu}_S$ with Ra

In Fig. 5.14, $\overline{Nu}_V$ and $\overline{Nu}_S$ are shown as a function of Ra. It can be seen that both $\overline{Nu}_V$ and $\overline{Nu}_S$ are always negative. Since the coefficients in Eq. (5.84) are non-negative, it is concluded that the induced convection enhances the boundary heat transfer.

It should again be mentioned that the asymptotic solution (5.77) is valid for small wall amplitudes δ and for Rayleigh numbers varying from zero to near critical value Ra_c. In order to obtain two-dimensional solutions for finite wall amplitudes and for larger Rayleigh numbers, Rees and Riley (1986) have integrated Eqs. (5.73) and (5.74) numerically with $\delta = 1$ and $a_L = a_U = a_1$. The solution domain consists of one critical wavelength of symmetry (varicose), or, equivalently, one 4×2 rectangle in the transformed (x_2, y_2)-plane. The Dufort–Frankel numerical method as described in Roache (1972) in combination with the second-order finite difference scheme proposed by Arakawa (1966) has been used (Rees and Riley, 1986).

Chapter 6

Convective Flow in a Wavy Tube

The fluid flow path through packed beds, porous media, blood vessels, etc., can be modelled closely by wavy tubes due to their convergent–divergent characteristics. Such tubes are commonly termed periodically constricted tubes (PCT), though there are other names given to this flow geometry as indicated in the Introduction. The research interest in PCT was initiated by the early work of Petersen (1958), who modelled a porous medium as a set of PCT in parallel in order to predict porous diffusion rates. A year later, Michaels (1959) showed how a simplified treatment of diffusion in a pore could be done using the PCT model. At the same time, Houpeurt (1959) also used the PCT model to study flow of gases through pores. After this initial work in the late 1950s, there was renewed interest in PCT only in the early 1970s. There have been various studies treating different aspects of flow through PCT in order to elucidate the effects of geometry on the flow characteristics. The various chosen geometries of the PCT can be categorized into the following five groups: square wave, conical, corrugated, parabolic, and sinusoidal.

A. Square Wave

Periodically constricted tubes that have a square wave form have been studied by Dullien and Azzam (1973) and Azzam and Dullien (1977). Square wave tubes were formed by joining together short, alternating segments of capillaries of two different diameters. Axial cross-sections of the tubes thus have the shape of a square wave on each side of the symmetry axis as shown in Fig. 1.9. Experiments were performed using four different hole diameters and five different combinations of diameters, thus bringing the total number of test capillaries to fifteen. With the main intention of simulating flow in porous media, Dullien and Azzam (1973) performed flow rate–pressure gradient measurements in periodically non-uniform capillary tubes with a square wave type. They covered a wide range of Reynolds numbers, thereby needing the use of the Forchheimer modified Darcy's equation for data analysis. Since the flow geometry in actual porous media is not exactly of the square wave type, results of Dullien and Azzam (1973) only have qualitative implications for flow through porous media. In fact, they have themselves indicated that their results may be applied to interpretation of flow phenomena in actual porous media only with great caution. The dimensionless permeability, which is the ratio of the measured-to-Poiseuille permeability, was found to be a function of the small-to-large capillary diameter ratio and the capillary diameter-to-wavelength ratio.

Azzam and Dullien (1977) theoretically studied the effect of tube geometry on pressure drop, flow patterns, and velocity profiles. They solved the complete Navier–Stokes equation, with the retention of the nonlinear inertia terms, numerically using the upwind finite difference scheme. For the purpose of numerical solution, only one period of the square wave was sufficient. As shown in Fig. 1.9, four linear dimensions define the geometry of the domain of the solution completely. These are the length L_s and radius r_s of the small diameter segment and the length L_L and radius r_L of the large diameter segment. The total length of the period is thus defined as

$$\lambda = L_s + L_L. \tag{6.1}$$

The distance from the wall to the axis of symmetry can then be written as

$$r_w(x) = r_s, \quad 0 \le x \le L_s$$
$$r_w(x) = r_L, \quad L_s \le x \le L_L \tag{6.2}$$

In the following, the numerical work of Azzam and Dullien (1977) is outlined and their results are presented and discussed.

It is assumed that an incompressible Newtonian fluid is flowing through the wavy tube shown in Fig. 1.9 under steady-state, isothermal conditions. The non-dimensional components of the Navier–Stokes equation in terms of stream function and vorticity in cylindrical coordinates can then be written as

$$\frac{1}{r_1}\frac{\partial\overline{\psi}}{\partial x_1}\frac{\partial\overline{\omega}}{\partial r_1} - \frac{1}{r_1}\frac{\partial\overline{\psi}}{\partial r_1}\frac{\partial\overline{\omega}}{\partial x_1} - \frac{\overline{\omega}}{r_1^2}\frac{\partial\overline{\psi}}{\partial x_1}$$

$$= \frac{1}{Re}\left[\frac{\overline{\omega}}{r_1^2} - \frac{\partial^2\overline{\omega}}{\partial x_1^2} - \frac{\partial^2\overline{\omega}}{\partial r_1^2} - \frac{1}{r_1}\frac{\partial\overline{\omega}}{\partial r_1}\right] \tag{6.3}$$

$$\overline{\omega} = -\left[\frac{\partial}{\partial x_1}\left(\frac{1}{r_1}\frac{\partial\overline{\psi}}{\partial x_1}\right) + \frac{\partial}{\partial r_1}\left(\frac{1}{r_1}\frac{\partial\overline{\psi}}{\partial r_1}\right)\right] \tag{6.4}$$

where dimensionless stream function $\overline{\psi}$ and dimensionless non-vanishing component of vorticity $\overline{\omega}$ are defined as follows:

$$u_1 = \frac{1}{r_1}\frac{\partial\overline{\psi}}{\partial r_1}, \quad v_1 = -\frac{1}{r_1}\frac{\partial\overline{\psi}}{\partial x_1}, \quad \overline{\omega} = \frac{\partial v_1}{\partial x_1} - \frac{\partial u_1}{\partial r_1} \tag{6.5}$$

and the non-dimensional variables are given as:

$$x_1 = x/\lambda, \ r_1 = r/\lambda, \ u_1 = u/U_e, \ v_1 = v/U_e, \ Re = U_e\lambda/\nu. \tag{6.6}$$

The wavelength λ is chosen as the characteristic length while the average velocity $U_e = Q/\pi r_s^2$ is chosen to be the characteristic velocity. It is to be noted that in the $(u_1 - v_1 - p_1)$ system of equations, the dimensionless pressure is given by the following:

$$p_1 = p/\rho U_e^2. \tag{6.7}$$

For the purpose of numerical calculations, Eqs. (6.3) and (6.4) are written in the following general form (Gosman et al., 1969; Shih, 1984):

$$A_1\left[\frac{\partial}{\partial x_1}\left(\phi_p\frac{\partial\overline{\psi}}{\partial r_1}\right) - \frac{\partial}{\partial r_1}\left(\phi_p\frac{\partial\overline{\psi}}{\partial x_1}\right)\right] - \frac{\partial}{\partial x_1}\left[A_2 r_1\frac{\partial}{\partial x_1}(A_3\phi_p)\right]$$

$$- \left[A_2 r_1\frac{\partial}{\partial r_1}(A_3\phi_p)\right] + A_4 r_1 = 0. \tag{6.8}$$

In the above Eq. (6.8), ϕ_p represents the property that is being transported. Thus, it is equal to $(\overline{\omega}/r_1)$ for vorticity and $\overline{\psi}$ for stream function. The variables A_1 to A_4 can be evaluated as follows:

For $\phi_p = (\overline{\omega}/r_1)$, $\quad A_1 = r_1^2, \ A_2 = r_1^2, \ A_3 = 1/Re, \ A_4 = 0. \tag{6.9}$

For $\phi_p = \overline{\psi}$, $A_1 = 0$, $A_2 = 1/r_1^2$, $A_3 = 1$, $A_4 = -(\overline{\omega}/r_1)$.   (6.10)

It should be noted that the values of the dependent variables $(\overline{\omega}/r_1)$ and $\overline{\psi}$ or their gradients must be specified at all points along the boundaries since the equations to be solved are of the elliptical type. Assuming no slip condition at the wall, the following chosen (Payatakis et al., 1973) constant value for $\overline{\psi}$ satisfies the condition

$$\overline{\psi}(x_1, r_{1w}) = (1/2)r_{1s}^2.$$ (6.11)

The boundary condition for the vorticity at the wall is written using the approach in Gosman et al. (1969) as

$$\left(\frac{\overline{\omega}}{r_1}\right)_w = -\left[\frac{3(\overline{\psi}_m - \overline{\psi}_w)}{r_{1w}^2 m^2} + \frac{1}{2}\left(\frac{\overline{\omega}}{r_1}\right)_m \frac{r_{1m}}{r_{1w}}\right].$$ (6.12)

The solution domain is shown in Fig. 6.1. The iteration process is initialized by assuming profiles of a fully developed flow in a uniform tube of radius r_{1s} along the entrance and the exit. Thus,

$$\overline{\psi}(0, r_1) = \overline{\psi}(1, J) = r_1^2\left[1 - \left(\frac{1}{2}\right)\left(\frac{r_1}{r_{1s}}\right)^2\right]$$ (6.13)

$$\frac{\overline{\omega}}{r_1}(0, r_1) = \frac{\overline{\omega}}{r_1}(1, J) = 4/r_{1s}^2.$$ (6.14)

The values of $\overline{\psi}$ and $(\overline{\omega}/r_1)$ on the nodal points along I-lines 2 through IN shown in Fig. 6.1 are evaluated using the finite difference analogue of Eq. (6.8). For the next iteration cycle, the values along the $(IN+1)$st line are calculated using the values along the (IN)th line and the (2)nd line. The values calculated along the $(IN+1)$st line are then substituted for values along the (1)st line. This iteration process is continued until the condition along the entrance and exit become equal.

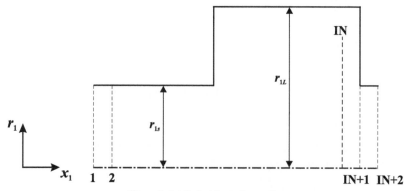

Fig. 6.1. Solution domain

a)

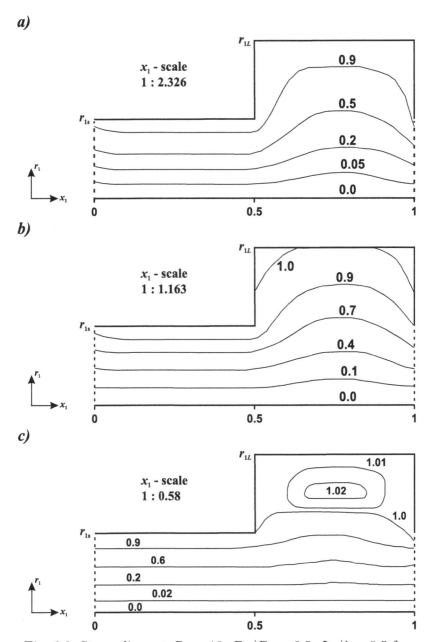

b)

c)

Fig. 6.2. Streamlines at $Re = 10$, $D_1/D_2 = 0.5$, $L_1/\lambda = 0.5$ for:
$D_1/\lambda = 0.215$ (a); $D_1/\lambda = 0.43$ (b); $D_1/\lambda = 0.86$ (c)

The boundary conditions along the axis of symmetry are written as follows:

$$\overline{\psi}(x_1, 0) = 0 \tag{6.15}$$

$$\left(\frac{\overline{\omega}}{r_1}\right)_0 = 8\frac{(\overline{\psi}_{mm} - \overline{\psi}_0)/r_{1mm}^2 - (\overline{\psi}_m - \overline{\psi}_0)/r_{1m}^2}{r_{1m}^2 - r_{1mm}^2}. \tag{6.16}$$

Choosing a non-uniform grid, Azzam and Dullien (1977) employed the upwind finite difference numerical scheme (Gosman et al., 1969; Shih, 1984).

Convergence criteria used for $\overline{\psi}$ and $(\overline{\omega}/r_1)$ for all cases of numerical experiments, as follows:

$$|\overline{\psi}_k - \overline{\psi}_{k+1}| \leq 1 \cdot 10^{-7} \tag{6.17}$$

$$\left[\left(\frac{\overline{\omega}}{r_1}\right)_k - \left(\frac{\overline{\omega}}{r_1}\right)_{k+1}\right] \leq 1 \cdot 10^{-6}. \tag{6.18}$$

Good convergence is achieved because the finite difference equations are written in such a way as to satisfy the Scarborough criterion (Scarborough, 1958). Stream function and vorticity distribution are first obtained and then the velocity distributions are calculated from the finite difference analogue of Eq. (6.5). From this information, the pressure distribution is calculated by integrating the Navier–Stokes equation written in terms of u_1, v_1, p_1.

Figures 6.2 show calculated flow patterns, namely, lines of constant stream function using a normalized stream function.

B. Conical

Wavy tubes with alternating converging–diverging conical sections placed end to end have been studied by Sparrow and Prata (1983). The main intention behind this study has been to acquire a better understanding of the use of such wavy tubes in enhanced heat transfer devices. Numerical solutions and experiments have been performed for laminar flow with isothermal boundary conditions at the wall. Comparisons have been made with the numerically and experimentally determined cycle-averaged Nusselt numbers. Though numerical solutions have been restricted to the fully developed regime, the experiments encompassed both entrance and fully developed regions. In the following, the work of Sparrow and Prata (1983) is presented.

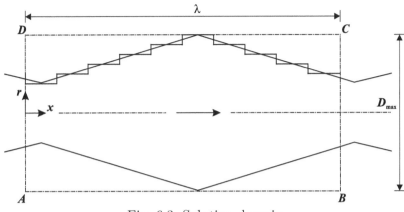

Fig. 6.3. Solution domain

A typical portion of a periodically converging–diverging tube with conical sections is shown in Fig. 1.10. Four variables have been shown in the figure, namely, the maximum and minimum diameters $D_{\max}$ and $D_{\min}$, the cyclelength λ, and the half taper angle γ. Only three of these variables are independent, and therefore, $D_{\max}$, λ and γ are chosen to define the geometry completely. Thus, the two dimensionless groups that are specified for numerical calculations are $\lambda/D_{\max}$ and γ. The solution domain used for the numerical work is shown in Fig. 6.3. The domain is bounded by envelope $ABCDA$ whose axial length is equal to the cyclelength λ and whose cross-section is a circle with diameter $D_{\max}$.

The inlet face AD is displaced just upstream of the initial cross-section of the diverging portion of the cycle, and the exit face BC is similarly displaced from the final cross-section of the converging portion of the cycle. This displacement is equal to $\Delta x/2$, where Δx is the axial length of the control volumes used in the finite difference representation. The sloping wall of the flow passage is approximated by a succession of 20 steps of the same height and width. It can be seen that the solution domain includes the flow passage itself plus an exterior region shown by the area enclosed by the dashed line in Fig. 6.3, thus rendering the solution domain as a cylinder of uniform diameter $D_{\max}$. Consequently, a general purpose computer program for cylindrical coordinates can be employed for solving the problem. In the appended exterior region, thermal conductivity as well as viscosity of the fluid are assumed to be large enough to be approximated as infinity. This assumption along with the specification of zero velocity and uniform temperature on the cylindrical boundary of the solution domain, AB and CD, ensure that the stepped boundary of the flow passage is also a surface of zero velocity and uniform temperature.

The velocity field repeats itself at axial stations separated by the cyclelength λ, and this can be expressed as follows:

$$u(x,r) = u(x + \lambda, r) \tag{6.19}$$

$$v(x,r) = v(x + \lambda, r) \tag{6.20}$$

where x represents any arbitrary station in the fully developed region. Cross-sectional pressure distributions at x and at $x + \lambda$ are identical in shape, but the pressure level is necessarily lower farther downstream as

$$p(x,r) - p(x + \lambda, r) = p(x + \lambda, r) - p(x + 2\lambda, r) = \ldots \tag{6.21}$$

This periodic condition is expressed as

$$p(x,r) = -\beta' x + p'(x,r) \tag{6.22}$$

$$p'(x,r) = p'(x + \lambda, r) \tag{6.23}$$

where β' is a constant. Pressure gradients, based on constant property and axisymmetry in x, r coordinates, are written as:

$$-\frac{\partial p}{\partial x} = \beta' - \frac{\partial p'}{\partial x}, \quad \frac{\partial p}{\partial r} = \frac{\partial p'}{\partial r}. \tag{6.24}$$

When isothermal wall conditions are considered, namely, when T_w is constant, the cross-sectional shape of the temperature difference $T(x,r) - T_w$ repeats itself periodically in the fully developed regime, though the level decreases in the streamwise direction. Thus,

$$\theta(x,r) = \theta(x + \lambda, r) \tag{6.25}$$

where

$$\theta(x,r) = \frac{T(x,r) - T_w}{T_b(x) - T_w} \tag{6.26}$$

and the local wall-to-bulk temperature difference over the flow cross-section area A is defined as:

$$T_b(x) - T_w = \frac{\displaystyle\int_A u(T - T_w)dA}{\displaystyle\int_A udA}. \tag{6.27}$$

The dimensionless variables and parameters are defined as follows:

$$x_1 = x/D^*, \quad r_1 = r/D^*, \quad u_1 = uD^*/\nu,$$
$$v_1 = vD^*/\nu, \quad p_1 = p'/\rho(\nu/D^*)^2, \quad \beta_1 = \beta'/\rho(\nu/D^*)^2 \tag{6.28}$$

where the characteristic length scale D^* is the equivalent diameter for the periodic tube and is defined such that its heat surface area (for a given axial length) is equal to that for a straight tube of diameter D^*. Thus,

$$D^* = \frac{A_{cyc}}{\pi \lambda} \tag{6.29}$$

where the per-cycle heat transfer A_{cyc} is defined as follows:

$$A_{cyc} = \frac{\pi(D_{max}^2 - D_{min}^2)}{2 \sin \gamma}. \tag{6.30}$$

Using the above dimensionless variables, the non-dimensional form of the governing equations are:

$$\frac{\partial(r_1 u_1)}{\partial x_1} + \frac{\partial(r_1 v_1)}{\partial r_1} = 0 \tag{6.31}$$

$$u_1 \frac{\partial u_1}{\partial x_1} + v_1 \frac{\partial u_1}{\partial r_1} = \beta' - \frac{\partial p_1}{\partial x_1} + \nabla^2 u_1 \tag{6.32}$$

$$u_1 \frac{\partial v_1}{\partial x_1} + v_1 \frac{\partial v_1}{\partial r_1} = -\frac{\partial p_1}{\partial r_1} + \nabla^2 v_1 - \frac{v_1}{r_1^2} \tag{6.33}$$

$$u_1 \frac{\partial \theta}{\partial x_1} + v_1 \frac{\partial \theta}{\partial r_1} - \frac{\nabla^2 \theta}{Pr} = \frac{\xi}{Pr} \tag{6.34}$$

where

$$\xi = \left(2\frac{\partial \theta}{\partial x_1} - Pr\, u_1 \theta\right) + \theta\left(\xi^2 + \frac{\partial \xi}{\partial x_1}\right) \tag{6.35}$$

and

$$\xi = \frac{d(T_b - T_w)/dx_1}{T_b - T_w}. \tag{6.36}$$

The boundary conditions imposed on the above equations are as follows:
at $r_1 = D_{max}/2D^*$

$$u_1 = v_1 = \theta = 0 \tag{6.37}$$

At the inlet and outlet faces of the solution domain AD and BC in Fig. 6.3, periodicity conditions are imposed instead of the boundary conditions as

$$\phi_p(0, r_1) = \phi_p(\lambda/D^*, r_1) \text{ for } \phi_p = u_1, v_1, p_1, \theta \qquad (6.38)$$

and the periodicity of the temperature difference parameter gives

$$\xi(0) = \xi(\lambda/D^*). \qquad (6.39)$$

The governing Eqs. (6.31)–(6.34) are solved using the numerical solution scheme presented by Patankar et al. (1977). The heat transfer results are obtained in the form of cycle-averaged Nusselt number $Nu = hD^*/k$ with $h = q/A_{cyc}/\Delta T_{wb}$, where q is the rate of heat transfer from the wall to the fluid per cycle and ΔT_{wb} is the arithmetic mean temperature difference corresponding to station 1 (the inlet face of the cycle) and station 2 (the outlet face of the cycle). The expression for Nu is written as follows:

$$Nu = 2\frac{D^*}{\lambda}\frac{1-\kappa}{1+\kappa}\left[\frac{Re\,Pr}{4} - \frac{1}{\pi}\int_1\left(\frac{\partial\theta}{\partial x_1} + \xi\theta\right)dA\right] \qquad (6.40)$$

where the Prandtl number is defined as $\mu C_p/k$, the Reynolds number as $4\rho Q/\mu\pi D^*$, and

$$\kappa = \frac{(T_w - T_b)_2}{(T_w - T_b)_1}. \qquad (6.41)$$

The subscripts 1 and 2 denote the corresponding stations 1 and 2. The integral term on the right-hand side of Eq. (6.40) and κ are evaluated numerically.

In order to validate the numerical results of heat transfer, experiments (Sparrow and Prata, 1983) have been done by conducting mass transfer studies using a naphthalene sublimation technique. The inside of the metallic tube is coated with a layer of solid naphthalene applied by a casting process and tapered at different angles by the use of a tapered center body during the casting. Various half-taper angles γ equal to 0 (straight tube), 3.4, 6.3, and $10°$ are used in the experiments, but the cycle aspect ratio is maintained constant at 1.52 in all cases.

The apparatus is operated in the open-circuit mode with air being drawn into the inlet of the test section from a temperature-controlled laboratory room. Mass of naphthalene sublimated from each module during every data run is measured. The naphthalene wall temperature, the flow rate, and the ambient are monitored periodically. Boundary condition of the naphthalene sublimation experiments is maintained in such a way that it corresponds to a uniform wall-temperature boundary condition in an analogous heat transfer experiment. Based on the heat/mass transfer analogy, the Sherwood number during experiments at a Schmidt number of 2.5 can be considered to be equivalent to Nusselt number at Prandtl number of 2.5.

Figure 6.4 shows a typical set of experimental data for entrance region and fully developed flow at $\gamma = 10°$, $Re = 100$, $Pr = 2.5$, and $\lambda/D_{\max} = 1.52$. The local per-module and per-cycle Nusselt numbers are normalized by the fully developed per-cycle values. It can be seen that the Nusselt numbers for the converging modules exceed those of the diverging modules.

Fully developed heat transfer is achieved at axial distances corresponding to those beyond the tenth module. Comparison between experiments and numerical calculations in Table 6.1 shows good agreement and suggests that the heat transfer deviations between PCT and straight tubes are rather small.

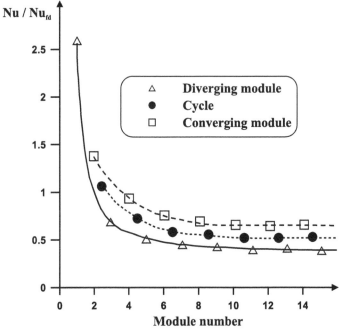

Fig. 6.4. Experimental data

Table 6.1. Comparison of Experimentally and Numerically Determined Fully Developed Nusselt Numbers for $Pr = 2.5$ and $\lambda/D_{\max} = 1.52$ [reprinted from Sparrow and Prata (1983) with permission from Taylor & Francis]

γ	Re	Nu_{PCT}/Nu_{ST} Experiment	Nu_{PCT}/Nu_{ST} Numerical
3.4°	85.4	0.991	0.995
	194.2	0.991	0.997
6.3°	97.4	1.014	0.989
	152.8	0.991	0.995
10.0°	100.0	0.968	0.981
	184.0	0.959	0.989

Table 6.2. Numerical Results for Fully Developed Nusselt Numbers [reprinted from Sparrow and Prata (1983) with permission from Taylor & Francis]

Re	$\lambda/D_{\max} = 1.52$ $Pr = 0.7$	$Pr = 2.5$	$Pr = 5.0$	$\lambda/D_{\max} = 2.29$ $Pr = 0.7$	$Pr = 2.5$	$Pr = 5.0$
(a) $\gamma = 2°$						
100	3.62	3.65	3.66	3.55	3.64	3.66
200	3.63	3.65	3.68	3.61	3.65	3.69
400	3.64	3.66	3.70	3.63	3.66	3.73
600	3.65	3.67	3.73	3.63	3.67	3.78
800	3.65	3.68	3.75	3.63	3.69	3.82
1000	3.65	3.68	3.78	3.64	3.70	3.87
(a) $\gamma = 3.4°$						
100	3.59	3.64	3.59	3.53	3.63	3.68
200	3.62	3.65	3.61	3.58	3.65	3.74
400	3.63	3.67	3.78	3.60	3.68	3.85
600	3.63	3.69	3.85	3.60	3.71	3.96
800	3.63	3.71	3.93	3.60	3.74	4.06
1000	3.63	3.73	4.00	3.61	3.78	4.16
(a) $\gamma = 6.3°$						
100	3.45	3.62	3.75	3.45	3.60	3.74
200	3.49	3.65	3.89	3.48	3.64	3.88
400	3.49	3.71	4.15	3.49	3.70	4.12
600	3.49	3.76	4.33	3.49	3.76	4.32
800	3.48	3.80	4.47	3.48	3.81	4.49
1000	3.48	3.84	4.59	3.48	3.86	4.64
(a) $\gamma = 10.0°$						
100	3.46	3.59	3.78	3.27	3.53	3.83
200	3.46	3.63	3.95	3.27	3.57	4.00
400	3.44	3.70	4.19	3.27	3.65	4.25
600	3.44	3.76	4.39	3.28	3.72	4.41
800	3.44	3.81	4.52	3.28	3.78	4.53
1000	3.43	3.85	4.63	3.29	3.81	4.64
(a) $\gamma = 14.8°$						
100	3.27	3.50	3.85	2.92	3.51	4.18
200	3.25	3.56	4.02	2.97	3.63	4.37
400	3.25	3.65	4.24	3.05	3.80	4.66
600	3.27	3.72	4.37	3.10	3.89	4.81
800	3.29	3.75	4.47	3.12	3.95	4.92
1000	3.30	3.82	4.55	3.14	4.00	5.03

The numerical results for fully developed per-cycle Nusselt numbers are presented in Table 6.2 in more detail. For the straight tube, the fully developed Nusselt number is known to have a value of 3.66 independent of Reynolds number and Prandtl number.

It can be seen that for PCT, Nusselt numbers depend on the Prandtl number value. For $Pr = 0.7$, Nusselt numbers for the PCT generally fall below those for the straight tube; for $Pr = 2.5$, their values are equivalent in value to those for the straight tube; while for $Pr = 5$, their values are larger than those for the straight tube. However, the moderate enhancements in heat transfer are accompanied by a substantial increase in pressure drop.

Computer-generated streamline maps are presented in Figs. 6.5. Although these figures are for $\lambda/D_{\max} = 2.29$, the qualitative trends are applicable to $\lambda/D_{\max} = 1.52$ as well.

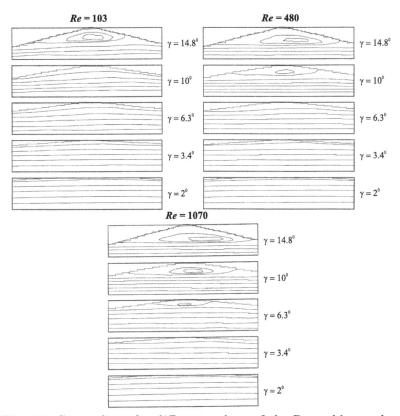

Fig. 6.5. Streamlines for different values of the Reynolds number [reprinted from Sparrow and Prata (1983) with permission from Taylor & Francis]

In all the figures, bowing of streamlines begins at small γ. However, it can be seen from Fig. 6.5a that the full-fledged recirculation zone appears at $\gamma = 14.8°$ for $Re = 103$; but, it appears at smaller values of γ with increasing Reynolds number, as is evident from Figs. 6.5b and 6.5c. Furthermore, the eye of the recirculation pattern in Fig. 6.5c gets displaced toward the downstream portion of the cycle. When these flow fields are considered in conjunction with the Nusselt numbers in Table 6.2, it is seen that the bowing of the streamlines at small γ and the recirculation at larger γ have opposite heat transfer effects at lower and higher Prandtl numbers. It is, thus, quite likely that for PCT, the transverse heat conduction is a more effective transport mechanism than transverse fluid motions at lower Prandtl numbers, with an opposite relationship at higher Prandtl numbers.

C. Corrugated

Wavy tubes with wall corrugations of the type shown in Fig. 1.11 have been studied by Savvides and Gerrard (1984). The interest in this study has been due to the practical problems concerned with blood flow in arterial prostheses. Steady and oscillating axisymmetric laminar flows in corrugated tubes are determined by a finite difference solution of the vorticity and continuity equations for an incompressible Newtonian fluid. In steady flow, the onset of flow separation and the growth of the separated region of flow is determined, and in unsteady flow, the effect of amplitude ratio, particle paths, and Reynolds number on mean velocity and flow resistance is studied. Comparisons have been made with other computational and experimental data wherever possible. In the medical application of this work (Savvides and Gerrard, 1984), the main concerns are whether sustained stagnant regions occur in the corrugations and whether there is a large change in the resistance value when compared with the straight cylindrical tube. In order to address this issue, investigation is made with an arterial waveform that contains six harmonics. But it is found that effect of the corrugations is minimal for stagnation or increased resistance. In the following, the work of Savvides and Gerrard (1984) is presented.

The velocity field repeats itself at axial stations separated by the cyclelength λ of the corrugation, and this can be written as:

$$u = u_m + u_a \cos \frac{2\pi t}{t_p} \tag{6.42}$$

where t is the time, t_p is the period of oscillation, u_m is the mean axial velocity at the section of diameter $D_{\max}$, and u_a is the mean velocity amplitude at that section. It is to be noted that lower-case letters are used to denote velocities that are functions of radial distance. Capital letters are used for cross-sectional mean velocities. Thus, U_a is defined as the cross-sectional mean amplitude and U_m the cross-sectional mean velocity at the section of $D_{\max}$.

It is assumed that an incompressible Newtonian fluid is flowing through the wavy tube shown in Fig. 1.11 under laminar, axisymmetric conditions. The non-dimensional components of the Navier–Stokes equation in terms of stream function and vorticity in cylindrical coordinates can then be written (Savvides and Gerrard, 1984) as

$$\frac{\partial \overline{\omega}}{\partial t} + \frac{1}{r_1} \frac{\partial \overline{\psi}}{\partial x_1} \frac{\partial \overline{\omega}}{\partial r_1} - \frac{1}{r_1} \frac{\partial \overline{\psi}}{\partial r_1} \frac{\partial \overline{\omega}}{\partial x_1} - \frac{\overline{\omega}}{r_1^2} \frac{\partial \overline{\psi}}{\partial x_1}$$

$$= \frac{1}{Re} \left[\frac{\partial^2 \overline{\omega}}{\partial r_1^2} + \frac{1}{r_1} \frac{\partial \overline{\omega}}{\partial r_1} + \frac{\partial^2 \overline{\omega}}{\partial x_1^2} - \frac{\overline{\omega}}{r_1^2} \right] \tag{6.43}$$

$$\overline{\omega} = \frac{1}{r_1} \left[\frac{\partial^2 \overline{\psi}}{\partial r_1^2} - \frac{1}{r_1} \frac{\partial \overline{\psi}}{\partial r_1} + \frac{\partial^2 \overline{\psi}}{\partial x_1^2} \right] \tag{6.44}$$

where dimensionless stream function $\overline{\psi}$ and dimensionless non-vanishing component of vorticity $\overline{\omega}$ are defined as follows:

$$u_1 = -\frac{1}{r_1} \frac{\partial \overline{\psi}}{\partial r_1}, \ v_1 = \frac{1}{r_1} \frac{\partial \overline{\psi}}{\partial x_1}, \ \overline{\omega} = \frac{\partial v_1}{\partial x_1} - \frac{\partial u_1}{\partial r_1} \tag{6.45}$$

and the non-dimensional variables are given as:

$$x_1 = x/D_{\max}, \ r_1 = r/D_{\max},$$

$$u_1 = u/U_a, \ v_1 = v/U_a, \ Re = U_c D_{\max}/\nu. \tag{6.46}$$

The maximum diameter $D_{\max}$ is chosen as the characteristic length while the cross-sectional velocity amplitude U_a is chosen to be the characteristic velocity. In steady flow, where $U_a = 0$, U_m is used as the characteristic velocity. In the definition of Reynolds number in Eq. (6.46), U_c is U_m in steady flow and U_a in oscillating flow.

A central finite difference scheme is used with a two-time-level Dufort–Frankel substitution for time-dependent terms, as described by Roache (1972) and used by other researchers (Pearson, 1965; Macagno and Hung, 1967; Williams, 1969; Gerrard, 1971; Gillani and Swanson,

1976; Butler, 1979; Sobey, 1980) for solving a variety of different problems. The finite difference forms of Eqs. (6.43) and (6.44) are used for getting the updated values of vorticity and stream functions as follows:

$$\overline{\omega}_{i,j}^{k+1}\left(1 + \frac{\Delta t}{Re\, r_1^2} + \frac{2\Delta t(h_1^2 + h_2^2)}{Re\, h_1^2 h_2^2}\right)/2\Delta t$$

$$= \overline{\omega}_{i,j}^{k-1}\left(1 + \frac{\Delta t}{Re\, r_1^2} + \frac{2\Delta t(h_1^2 + h_2^2)}{Re\, h_1^2 h_2^2}\right)/2\Delta t$$

$$+ \left\{\frac{Re\, r_1}{4h_1 h_2}[(\overline{\psi}_{i+1,j} - \overline{\psi}_{i-1,j})(\overline{\omega}_{i,j-1} - \overline{\omega}_{i,j+1})\right.$$

$$+(\overline{\psi}_{i,j+1} - \overline{\psi}_{i,j-1})(\overline{\omega}_{i+1,j} - \overline{\omega}_{i-1,j})] + \frac{Re}{2h_1}\overline{\omega}_{i-1,j}(\overline{\psi}_{i+1,j} - \overline{\psi}_{i-1,j})$$

$$+\frac{r_1^2}{h_2^2}(\overline{\omega}_{i,j+1} + \overline{\omega}_{i,j-1}) + \frac{r_1}{2h_2}(\overline{\omega}_{i,j+1} - \overline{\omega}_{i,j-1})$$

$$\left. +\frac{r_1^2}{h_1^2}(\overline{\omega}_{i+1,j} - \overline{\omega}_{i-1,j})\right\} / Re\, r_1^2 \qquad (6.47)$$

$$\overline{\psi}_{i,j}^{k+1} = \left\{h_1^2(\overline{\psi}_{i+1,j} + \overline{\psi}_{i-1,j}) + h_2^2(\overline{\psi}_{i,j+1} - \overline{\psi}_{i,j-1})\right.$$

$$-\frac{h_1 h_2^2}{2r_1}(\overline{\psi}_{i,j+1} - \overline{\psi}_{i,j-1}) - h_1^2 h_2^2 r_1 \overline{\omega}_{i,j}^{k+1}\right\}/2(h_1^2 + h_2^2). \qquad (6.48)$$

The dimensionless mesh lengths in the $r-$ and $x-$ directions are h_1 and h_2, respectively. The subscripts i, j refer to mesh points in the $x-$ and $r-$ directions. The superscripts refer to the number of time steps. Note that wherever the superscript has been omitted, it is implied to be k. The vorticity equation thus gives $\overline{\omega}_{ij}$ at time $(k+1)\Delta t$, where Δt is the time step, in terms of $\overline{\omega}_{ij}$ at time $(k-1)\Delta t$, and $\overline{\psi}$ and $\overline{\omega}$ at adjacent mesh points at time $k\Delta t$.

The boundary conditions imposed on Eqs. (6.47) and (6.48) are as follows:

At the tube axis

$$\overline{\psi} = 0 \qquad (6.49)$$

$$\overline{\omega} = 0. \qquad (6.50)$$

On the boundary walls

$$\overline{\psi} = \overline{\psi}_B = \begin{cases} -0.125 & \text{for steady flow} \\[2mm] -0.125\left(\dfrac{U_m}{U_a} + \cos\dfrac{2\pi t}{t_p}\right) & \text{for oscillatory flow.} \end{cases} \qquad (6.51)$$

Values of $\bar{\omega}$ at the boundary wall are determined by extrapolation from known values in the flow. This is done by expanding $\bar{\psi}$ and $\bar{\omega}$ at the wall in Taylor series and applying the conditions of no slip and zero normal velocity at the wall. Thus, if the point on the sloping wall at which the value of $\bar{\omega}$ is required is B, then $B+1$ is the point where the normal to the wall at B crosses the next gridline. The distance from B to $B+1$ is $h_3 = h_1 \cos\gamma$. The expression for $\bar{\omega}_B$ is then obtained as:

$$\bar{\omega}_B = \frac{\bar{\psi}_B - \bar{\psi}_{B+1} + \dfrac{h_3^2}{6}\bar{\omega}_{B+1}r_{1B} - \dfrac{h_3^2}{24}\dfrac{2+\cos^2\gamma}{\cos\gamma}\bar{\omega}_{B+1}}{\dfrac{h_3^2}{3}\left[\dfrac{h_3}{2}\dfrac{1+\cos^2\gamma}{\cos\gamma} - r_{1B}\right] - \dfrac{h_3^2}{24}\dfrac{2+\cos^2\gamma}{\cos\gamma}}. \qquad (6.52)$$

Choice of time step and mesh length is governed by the considerations of accuracy, computational stability, and compatibility with the boundary data. The relationship between mesh lengths and time step is found by trial and error. For steady flow, the smaller of the values from the following two limiting expressions is used:

$$\frac{1}{15}\frac{h_1^2 h_2^2 Re}{h_1^2 + h_2^2} < \Delta t < \frac{1}{30}h_2^2 Re. \qquad (6.53)$$

For oscillatory flow, the following relationship is used

$$\Delta t < \frac{1}{30}\frac{D_{\max}}{\nu}h_2^2. \qquad (6.54)$$

The dimensionless mesh lengths are computed from the following expressions:

$$h_1 = \frac{2a}{ND_{\max}} = \frac{1}{2(M+N)}, \quad h_2 = \frac{\lambda}{2ND_{\max}} \qquad (6.55)$$

where M and N are the number of grid points. There are $2N$ grid points in the axial direction and $M+I$ grid points in the radial direction, with $I = 0, 1, 2, \ldots, N$.

Results of the numerical calculations for steady flow are shown in Fig. 6.6. The lines along which the stream function has a boundary value of -0.125 are indicated by the $+$ symbols in Fig. 6.6. These are the separation–reattachment streamlines.

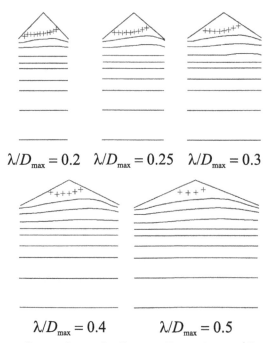

$\lambda/D_{\max} = 0.2 \quad \lambda/D_{\max} = 0.25 \quad \lambda/D_{\max} = 0.3$

$\lambda/D_{\max} = 0.4 \qquad \lambda/D_{\max} = 0.5$

Fig. 6.6. Streamlines of steady flow at $Re = 250$, $a/D_{\max} = 0.05$ for different values of $\lambda/D_{\max}$. Values of stream function are 0, 0.02, 0.05, 0.07, 0.09, 0.10, 0.11, 0.12, 0.124, 0.125 [reprinted from Savvides and Gerrard (1984) with permission from Cambridge Journals]

These + symbols are separated by one mesh in the axial direction. It can be seen that at large-enough values of $\lambda/D_{\max}$ the flow does not separate and the streamlines follow the wall shape. As $\lambda/D_{\max}$ decreases below some critical value, the flow separates at the apex of the corrugation. After its inception at the apex, the separation region then grows as $\lambda/D_{\max}$ decreases, eventually filling most of the corrugation. The critical value of $\lambda/D_{\max}$ for the flow to separate increases as Re and $a/D_{\max}$ increase.

The effect of Reynolds number on the streamlines is shown in Fig. 6.7 for $\lambda/D_{\max} = 0.5$ and $a/D_{\max} = 0.0835$. It can be seen that separation first occurs at a small Reynolds number.

The separated region grows with increasing Reynolds number, and the vortex that is formed spreads and shifts its center downstream. This shift is more pronounced at larger values of $a/D_{\max}$. This flow behavior pattern can be expected from the possible absence of eddies and thus the existence of separation at sharp corners (Moffatt, 1964).

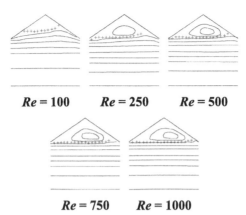

Fig. 6.7. Streamlines of steady flow at $\lambda/D_{\max} = 0.5$, $a/D_{\max} = 0.0835$ for different values of the Reynolds number. Values of stream function extra to Fig. 6.6 are 0.126 and 0.127 [reprinted from Savvides and Gerrard (1984) with permission from Cambridge Journals]

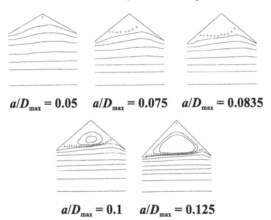

Fig. 6.8. Streamlines of steady flow at $\lambda/D_{\max} = 0.5$, $Re = 100$ for different values of $a/D_{\max}$. Values of stream function as in Fig. 6.7 [reprinted from Savvides and Gerrard (1984) with permission from Cambridge Journals]

Figure 6.8 shows the effect of varying $a/D_{\max}$ at fixed values of Re and $\lambda/D_{\max}$. Same as in Fig. 6.7, extra streamlines are included to show the circulation within the separated region.

An attempt to represent the vortex strength is done in Fig. 6.9 by plotting $a/D_{\max}$ versus S_1. It can be concluded that vortex strength increases rapidly with $a/D_{\max}$ for all Re; it also increases at fixed $a/D_{\max}$ up to Re approaching 1000, where the variation is much diminished.

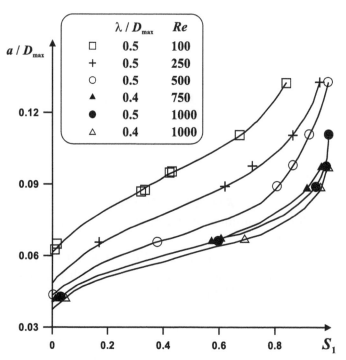

Fig. 6.9. Ratio of separated area to corrugation area
as a function of $a/D_{\max}$

Savvides and Gerrard (1984) have compared their computed results
with those of Deiber and Schowalter (1979) for the PCT. All the computations and experiments of these authors are for much larger values of
$\lambda/D_{\max}$ than those treated by Savvides and Gerrard (1984). However,
a comparison has been attempted for a selected value of $\lambda/D_{\max} = 2.41$
and $a/D_{\max} = 0.231$.

Steady flow discussed in the above paragraphs is, undoubtedly, of
fundamental importance, but its practical significance is far less than
that of the more relevant case of oscillating flow, which mimics blood
flow through arteries rather closely. Oscillating flow is characterized by
five non-dimensional parameters: $\lambda/D_{\max}$, $a/D_{\max}$, Reynolds number
Re, Stokes number St_k, and velocity ratio $U_R = U_m/U_a$. Particle paths
strongly depend upon U_R and Strouhal number $St_H = D_{\max}/U_a t_p =
2St_k^2/\pi Re$. Since the number of independent non-dimensional parame-
ters are large, Savvides and Gerrard (1984) have restricted their computations to one particular tube geometry corresponding to $\lambda/D_{\max} = 0.2$
and $a/D_{\max} = 0.05$, which closely simulate conditions involved in arterial prostheses.

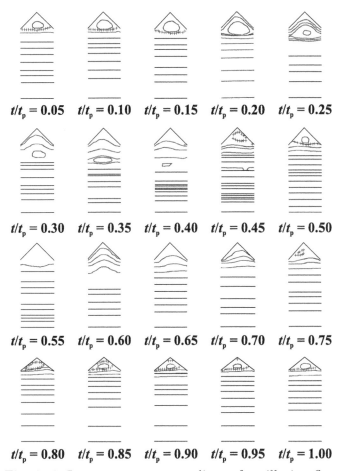

$t/t_p = 0.05$ $t/t_p = 0.10$ $t/t_p = 0.15$ $t/t_p = 0.20$ $t/t_p = 0.25$

$t/t_p = 0.30$ $t/t_p = 0.35$ $t/t_p = 0.40$ $t/t_p = 0.45$ $t/t_p = 0.50$

$t/t_p = 0.55$ $t/t_p = 0.60$ $t/t_p = 0.65$ $t/t_p = 0.70$ $t/t_p = 0.75$

$t/t_p = 0.80$ $t/t_p = 0.85$ $t/t_p = 0.90$ $t/t_p = 0.95$ $t/t_p = 1.00$

Fig. 6.10. Instantaneous streamlines of oscillating flow
[reprinted from Savvides and Gerrard (1984) with permission from
Cambridge Journals]

For the boundary value of the stream function as given by Eq. (6.51),
Fig. 6.10 shows the sequence of calculated streamlines at $U_R = 1$,
$St_k = 10$, and $Re = 300$.

It can be seen that at $t/t_p = 0.05$ or $\to 0$, maximum flow occurs and
the separation region has a similar shape to that in steady flow at the
same Reynolds number, but is larger in size. As volumetric flow rate
decreases, the recirculating region grows is size, the separated region
bulges to more than fill the corrugation, and the flow becomes asymmetrical. The center of the recirculating streamlines moves downstream up
to $t/t_p = 0.35$ as it moves out of the corrugation. As the volumetric flow
rate approaches zero at $t/t_p = 0.5$, this center moves back and before

disappearing eventually at about midradius at t/t_p between 0.55 and
0.6, recirculating streamlines cross the boundary into the upstream corrugation. The flow is most complicated at around $t/t_p = 0.5$. However,
at $t/t_p = 0.65$, the flow becomes unidirectional and has the appearance
of steady flow at low Reynolds number. The instantaneous stream lines
with the wall value of the stream function are marked on the figures
by plus signs. These represent separation streamlines when they spring
from the wall, and only when they are stationary or slowly changing.
During the time interval from $t/t_p = 0.75$ through 1.0 to 0.1, there is a
streamline leaving the wall close to the corner, the position of that is not
changing throughout the whole of this interval. At $t/t_p = .075$, there
is a closed loop of + points that is not attached to the wall, and this
represents a circulatory flow; at this time, the flow separates from the
corner and almost immediately reattaches. From $t/t_p = 0.9$ onwards,
it is seen that separation streamlines proceed across the corrugation to
reattach on the downstream wall.

The major difference in the streamlines between unsteady oscillating flow and steady flow is that in the former, separation takes place
from the vicinity of the corner while in the latter, it occurs at the apex
as the Reynolds number is increased. In unsteady flow, separation occurs when the flow near the wall is accelerating.

If particle path is followed, then it is observed that all particles
within a corrugation leave the corrugation between the period from
$t/t_p = 0$ to 1 and have a radial motion that is not very much less
than their axial motion. Particles outside the corrugation move almost
entirely in the axial direction, and so convection plays no part in the
radial motion of the vorticity in this region.

Figure 6.11 shows the effect of increasing Stokes number St_k, when
all other conditions are maintained the same as in Fig. 6.10. It can be
seen that differences in the flows in the corrugations are not very large
wherever the cycle is dominated by separation of the flow because the
Reynolds number in the two figures are the same.

The value of St_k in Fig. 6.11 is twice the value of Fig. 6.10 and
hence, diffusion of vorticity is half as great as is evident at times $t/t_p = 0.3 - 0.5$. When the flow rate is nearly zero at the time period after
$t/t_p = 0.5$, the streamline patterns appear very different in the two
figures.

It is be noted that when $U_R = 0$, the waveform corresponds to a
sinusoidal oscillation of the volume flow rate about a zero mean value.

When a comparison is made of the flow patterns with $U_R = 0$ and with $U_R = 1$, they are found to be identical at corresponding times. The flux ratio has only a small effect, except on the timescale, and thus, on the diffusion of the vorticity from the wall. As in the case of $U_R = 1$, the particles in the $U_R = 0$ case, too, are convected out of the corrugation, but are convected radially by only very small amounts when they are away from the immediate neighborhood.

The main objective of the above-discussed analysis (Savvides and Gerrard, 1984) is to apply the knowledge gained through the numerical calculations of flow through a corrugated tube to arterial prostheses.

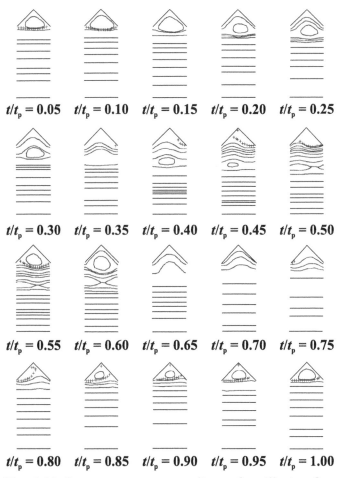

$t/t_p = 0.05$ $t/t_p = 0.10$ $t/t_p = 0.15$ $t/t_p = 0.20$ $t/t_p = 0.25$

$t/t_p = 0.30$ $t/t_p = 0.35$ $t/t_p = 0.40$ $t/t_p = 0.45$ $t/t_p = 0.50$

$t/t_p = 0.55$ $t/t_p = 0.60$ $t/t_p = 0.65$ $t/t_p = 0.70$ $t/t_p = 0.75$

$t/t_p = 0.80$ $t/t_p = 0.85$ $t/t_p = 0.90$ $t/t_p = 0.95$ $t/t_p = 1.00$

Fig. 6.11. Instantaneous streamlines of oscillating flow [reprinted from Savvides and Gerrard (1984) with permission from Cambridge Journals]

It is known that blood tends to clot when in contact with a foreign material such as a prosthesis. The likelihood of clotting is greater in the regions of stasis or regions of flow that are not washed out at each cycle of the heartbeat. In arterial prostheses of small diameter, corrugations are relatively large and if the effect of the corrugations is to cause local stasis, then this may promote thrombus formation. Savvides and Gerrard (1984) have found that the mixing in the wall region is seen to be enhanced by the presence of the corrugations and there are no stagnation regions in the corrugated tube in the normal arterial waveform cases studied by them. It is, however, well-known that the waveforms in atherosclerotic arteries are far from normal, even after a bypass graft operation. In all likelihood, therefore, the reason for the increased failure rate of corrugated prostheses over cylindrical tubes is the abnormal waveform.

Investigation on laminar flow of a suspension in corrugated tubes has been carried out by Gopalan and Ponnalagarsamy (1992). The interest in this study is mainly from the biofluid mechanics viewpoint, particularly for the design of medical appliances such as blood pumps and oxygenators. Treating blood as a suspension of erthyrocytes in plasma, the study of the flow characteristics of suspensions in corrugated tubes attains relevance. It is observed that the wall geometry and the presence of particles both have a combined effect on the flow patterns to a considerable extent. The mixing phenomenon due to secondary flow is enhanced appreciably with the flow of relatively large volume fraction of low diffusive particles in high inertial fluids. Friction loss in a corrugated tube is found to be larger than that in a smooth one. However, it is reduced considerably with the addition of sparingly diffusive particles in the flow.

Another significant work, which deals with flow through corrugated pipes, is that of Dodson et al. (1971). The study deals with Newtonian and non-Newtonian fluids and carries out a theoretical as well as an experimental investigation. The equation of state used for the fluid in theoretical analysis is based on the well-known Oldroyd model (Oldroyd, 1950, 1958). The theory predicts that, in the Newtonian case, the corrugated wall of the pipe causes a reduction in flow rate when compared with the value expected on the basis of a straight pipe of the same mean radius. This behavior is confirmed by experiment. In the case of an elastico-viscous non-Newtonian fluid, theoretical analysis indicates that, under certain conditions, the reduction in flow rate may be amplified or reduced by the non-Newtonian properties of the fluid. These predictions are also confirmed (Dodson et al., 1971) by experiment.

It is to be noted, however, that the word "corrugated" as used by Dodson et al. (1971) as well as Gopalan and Ponnalagarsamy (1992) differs from the definition that is used by Savvides and Gerrard (1984). The asymmetric and symmetric wavy wall studied (Gopalan and Ponnalagarsamy, 1992) are, in fact, sinusoidal in shape. The entire analysis (Gopalan and Ponnalagarsamy, 1992) has been done under the assumption of a small amplitude wall waviness of this sinusoidal tube. Similarly, the work of Dodson et al. (1971) actually considers a pipe of circular cross-section whose radius varies sinusoidally along the longitudinal axis. Thus, the results of Gopalan and Ponnalagarsamy (1992) as well as Dodson et al. (1971) are to be viewed in the context of analyzes and discussions presented in the subsection dealing with the sinusoidal profile.

D. Parabolic

The problem of creeping Newtonian fluid flow through periodically constricted tubes composed of symmetric segments, which are parabolas of revolution as shown in Fig. 1.12, has been studied by Neira and Payatakes (1978). This particular wall profile of a PCT was introduced by Petersen (1958), Houpert (1959), and Michaels (1959) in the late 1950s. The revived interest in this profile in the 1970s can be seen from the works of Payatakes et al. (1973) and Franzen (1977). The profile represents flow through a porous medium or packed bed made up of a systematic distribution of spheres, as is evident from Fig. 1.12. Neira and Payatakes (1978) have obtained a collocation solution of the creeping flow of Newtonian fluids through such PCT. The domain of solution as shown in Fig. 1.12 is half a segment, which is transformed into a rectangular one by proper transformations such that all the boundary conditions are appropriately satisfied. The axial and radial velocity components are obtained in analytical form and the pressure drop along a tube segment is calculated both from the volume integration of the viscous dissipation function and from the line integration of the equation of motion. In the following, the work of Neira and Payatakes (1978) is outlined and discussed.

The equation of motion for steady creeping flow is written as

$$\nabla^4 \overline{\psi} = 0 \tag{6.56}$$

where the operator ∇^2 is defined as follows:

$$\nabla^2 \equiv \frac{\partial^2}{\partial r_1^2} - \frac{1}{r_1}\frac{\partial}{\partial r_1} + \frac{\partial^2}{\partial x_1^2}. \tag{6.57}$$

The dimensionless stream function is defined as

$$u_1 = -\frac{1}{r_1}\frac{\partial \overline{\psi}}{\partial r_1}, \quad v_1 = \frac{1}{r_1}\frac{\partial \overline{\psi}}{\partial x_1} \tag{6.58}$$

where the non-dimensional variables are

$$x_1 = x/\lambda, \ r_1 = r/\lambda, \ u_1 = u/U_e, \ v_1 = v/U_e, \ Re = U_e\lambda/\nu. \tag{6.59}$$

The length λ of one segment is chosen as the characteristic length while the average velocity $U_e = Q/\pi r_s^2$ is chosen as the characteristic velocity. It is to be noted that in the $(u_1 - v_1 - P_1)$ system of equations, the dimensionless pressure is given by the following:

$$P_1 = P/\rho U_e^2 = (p - \rho g x)/\rho U_e^2. \tag{6.60}$$

The dimensionless radius of the tube wall is given by the following:

$$r_{1w} = \begin{cases} r_{1s} + 4(r_{1L} - r_{1s})x_1^2 & \text{for} \quad 0 \le x_1 \le 1/2 \\ r_{1s} + 4(r_{1L} - r_{1s})(1 - x_1)^2 & \text{for} \quad 1/2 \le x_1 \le 1. \end{cases} \tag{6.61}$$

Equation (6.56) is to be integrated under the following boundary conditions:

$$\overline{\psi} = \frac{\partial \overline{\psi}}{\partial r_1} = \frac{\partial \overline{\psi}}{\partial x_1} = 0 \text{ at } r_1 = r_{1w}(x_1) \tag{6.62}$$

$$\overline{\psi}(0, x_1) = r_{1s}^2/2 = \text{constant} \tag{6.63}$$

$$\frac{1}{r_1}\frac{\partial \overline{\psi}}{\partial x_1} = \frac{\partial}{\partial r_1}\left(\frac{1}{r_1}\frac{\partial \overline{\psi}}{\partial r_1}\right) = 0 \text{ at } r_1 = 0. \tag{6.64}$$

The periodicity condition of fully developed flow gives

$$\overline{\psi}(r_1, x_1) = \overline{\psi}(r_1, x_1 + i) \text{ where } i \text{ is any integer.} \tag{6.65}$$

Further, the symmetric nature of the creeping flow requires that

$$\overline{\psi}(r_1, x_1) = \overline{\psi}(r_1, -x_1). \tag{6.66}$$

Based on the periodicity and the symmetry of the solution, the region of integration can be reduced to the region between the planes $x_1 = 0$ and $x_1 = 1/2$, corresponding to the shaded portion in Fig. 1.12.

A new set of transformed coordinate systems (r_2, x_2) will now be chosen in order to facilitate a collocation solution. It is found that simultaneous selection of the coordinate transformation and the trial function is more convenient than choosing them sequentially. Thus, rather general expressions for r_2, x_2 and $\psi(r_2, x_2)$ are selected, and then, by imposing the required conditions of the problem, the free parameters included in those general expressions are determined. The chosen transformed coordinates and the trial function are as follows:

$$r_2 = \left(\frac{r_1}{r_{1w}}\right)^2 \left\{ \left[1 - \left(\frac{r_1}{r_{1w}}\right)^2\right] \chi + 1 \right\} \tag{6.67}$$

$$x_2 = x_1 \tag{6.68}$$

where

$$\chi = c - 2(c+1)\left(\frac{r_{1w}}{r_2}\right) + (c+3)\left(\frac{r_{1w}}{r_2}\right)^2. \tag{6.69}$$

Here c is an adjustable parameter. The new coordinate system based on r_2, x_2 for $c = 1$.

The trial function is chosen as follows:

$$\overline{\psi}_N(r_2, x_2) = \overline{\psi}_0(r_1) + \sum_{k=1}^{\overline{N}} C_k r_2 (1 - r_2)^{i+1} \cos(j-1)2\pi x_2 \tag{6.70}$$

where

$$\overline{\psi}_0(r_1) = (r_{1s}^2/2)(1 - r_1)^2$$
$$k = (j-1)n_r + i \tag{6.71}$$
$$i = 1, \ldots, n_r, \quad j = 1, \ldots, n_x, \quad \overline{N} = n_r n_x.$$

The coefficients C_k are determined by equating the residuals of the differential Eq. (6.56) to zero at $\overline{N}$ collocation points in the domain of interest of the independent variables. The residual equations form a set of $\overline{N}$ linear equations for the $\overline{N}$ unknown coefficients C_k.

Collocation points in the x_2 direction are chosen as the zeroes of the first omitted cosine function of the trial expansion, while in the r_2 direction, the collocation points are chosen as the roots of the Jacobi polynomials defined by the orthogonality property.

Axial and radial components of velocity are obtained in analytical form from the approximate solution of ψ_N using the following expressions:

$$u_1 = -\frac{1}{r_1}\left(\frac{\partial r_2}{\partial r_1}\right)_{x_1}\left(\frac{\partial \overline{\psi}}{\partial r_2}\right)_{x_2} \tag{6.72}$$

$$v_1 = \frac{1}{r_1}\left[\left(\frac{\partial r_2}{\partial x_1}\right)_{r_1}\left(\frac{\partial \overline{\psi}}{\partial r_2}\right)_{x_2} + \left(\frac{\partial \overline{\psi}}{\partial x_2}\right)_{r_2}\right]. \tag{6.73}$$

The pressure drop along one tube segment is calculated numerically from the approximate solution ψ_N in two ways:

1) by volume integration of the viscous dissipation function over half the tube segment, and

2) by line integration of the equation for dP_1 with expressions obtained from equation of motion.

The pressure drop calculated from viscous dissipation is denoted as ΔP_{1VD} and expressed as follows:

$$-\Delta P_{1VD} = \frac{2}{Re\pi r_1^2}\int_V \overline{\Phi}_v dV \tag{6.74}$$

where

$$\overline{\Phi}_v = 2\left[\left(\frac{\partial u_1}{\partial x_1}\right)^2 + \left(\frac{v_1}{r_1}\right)^2 + \left(\frac{\partial v_1}{\partial r_1}\right)^2\right] + \left(\frac{\partial v_1}{\partial x_1} + \frac{\partial u_1}{\partial r_1}\right)^2. \tag{6.75}$$

$\overline{\Phi}_v$ can easily be expressed in analytical form in terms of $\overline{\psi}_N$ and its derivatives.

The pressure drop calculated from line integration is written as:

$$dP_1 = \frac{\partial P_1}{\partial x_1}dx_1 + \frac{\partial P_1}{\partial r_1}dr_1 \tag{6.76}$$

where

$$\frac{\partial P_1}{\partial x_1} = \frac{1}{Re}\left\{\frac{1}{r_1}\frac{\partial}{\partial r_1}\left(r_1\frac{\partial u_1}{\partial r_1}\right) + \left(\frac{\partial^2 u_1}{\partial x_1^2}\right)\right\} \tag{6.77}$$

$$\frac{\partial P_1}{\partial r_1} = \frac{1}{Re}\left\{\frac{\partial}{\partial r_1}\left[\frac{1}{r_1}\frac{\partial}{\partial r_1}(r_1 v_1)\right] + \left(\frac{\partial^2 v_1}{\partial x_1^2}\right)\right\}. \tag{6.78}$$

The above equations can easily be expressed in terms of $\overline{\psi}_N$ and its derivatives, thus allowing the evaluation of the pressure gradient at any point.

The use and performance of the collocation method is illustrated (Neira and Payatakes, 1978) by determining the solution for a PCT with a geometry corresponding to $r_{1s} = 0.1760$ and $r_{1L} = 0.3975$, which is typical of a unit cell of packed bed of sand. The choice of the parameter c is based on an examination of the behavior of the solution obtained from the initial guess ψ_0 of the trial function. It is found that such an approximate solution is rather sensitive to the value of c. Hence, c is adjusted to a value such that the calculated rate of viscous dissipation based on ψ_0 becomes minimum. For most geometries of interest in modeling granular porous media, a minimum ΔP_{1VD} is obtained for $c = 1$. Based on the same criteria as used in the adjustment of parameter c, the convergence of the complete trial function is tested for an increasing number of collocation points with $c = 1$. It is found that for $n_r = n_x \geq 4$, solutions for the stream function, velocity components, and pressure drop remain almost unchanged. Figures 6.12 to 6.13 show the solutions obtained for the streamlines, and the axial and radial velocity profiles, respectively, for the specific geometry discussed above.

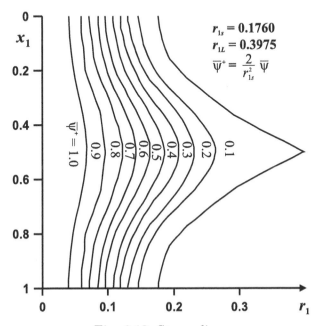

Fig. 6.12. Streamlines

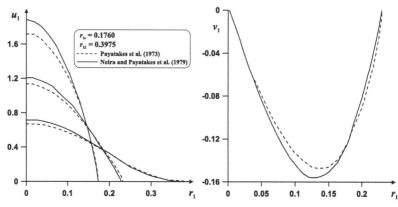

Fig. 6.13. Axial and radial velocity profiles

The figures also show the results of Payatakes et al. (1973) for the sake of comparison. The solution obtained by Payatakes et al. (1973) employed a finite difference network containing about 350 nodes in the region $0 \leq x_1 \leq 1/2$. It can be seen that the results of the two methods used for obtaining the solution differ by less than 7% at points of maximum deviation. This difference is most likely due to the discretization error in the finite difference solution (Payatakes et al., 1973; Neira and Payatakes, 1978).

E. Sinusoidal

Periodically constricted tubes with sinusoidal wave forms of wall profile have been the most popular (Forrester and Young, 1970a,b; Chow and Soda, 1972, 1973; Payatakes et al., 1973; Fedkiw and Newman, 1977, 1978, 1979, 1987; Neira and Payatakes, 1978, 1979; Deiber and Schowalter, 1979; Prata and Sparrow, 1984; Tilton and Payatakes, 1984; Lahbabi and Chang, 1986) among the wavy tubes studied by various researchers. Different methods of solutions have been used in order to carry out a theoretical investigation of this flow situation. In the following, some of the works of significance that treat flow through sinusoidal tubes are presented and discussed.

1. Analytical Solutions

Forrester and Young (1970a,b) developed equations to approximately describe the flow of an incompressible fluid through an axisymmetric PCT and conducted an experimental program to verify the theoretical findings.

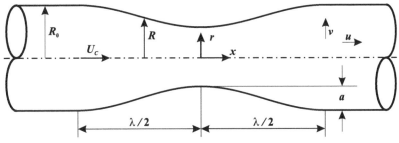

Fig. 6.14. Solution domain

Their interest in PCT was from a biomedical viewpoint. The presence of stenosis in arteries causes blood to flow with a converging–diverging character and implications of such flow in occlusive vascular disease are important to know.

In the theoretical analysis, Forrester and Young (1970a) used approximate techniques to obtain a solution for flow in a tube with a cosine-shaped constriction, whose validity is restricted to the amplitude a being small compared with the tube radius R_0 and wavelength λ. It is basically assumed that the tube has a circular cross-section at all locations and that this cross-section is constant everywhere except in one converging–diverging section. The specific boundary form of the stenosis used is described by the following equation:

$$R = R_0 - a[1 + \cos(2\pi x/\lambda)] \qquad (6.79)$$

for $-\lambda/2 \le x \le \lambda/2$ with a constant radius of R_0 outside this range, as shown in Fig. 6.14.

Although blood is actually a non-Newtonian suspension of cells in plasma, it is considered to flow as a homogeneous, Newtonian fluid in vessels greater than approximately 0.5 mm in diameter (McDonald, 1960). It has also been found (Merrill et al., 1966; Merrill and Pelletier, 1967) that the non-Newtonian characteristics of blood become apparent only at shear rates lower than 100 $\sec^{-1}$. It is thus expected that the non-Newtonian characteristics of blood would not be important in the study of Forrester and Young (1970a), except possibly in the immediate vicinity of the separation region where low shear rates may occur.

Governing equations for steady, laminar, axisymmetric flow of an incompressible Newtonian fluid can then be written as

$$\frac{\partial u}{\partial x} + \frac{1}{r}\frac{\partial(rv)}{\partial r} = 0 \qquad (6.80)$$

$$u\frac{\partial u}{\partial x} + v\frac{\partial u}{\partial r} = -\frac{1}{\rho}\frac{\partial p}{\partial x} + \nu\left(\frac{\partial^2 u}{\partial r^2} + \frac{1}{r}\frac{\partial u}{\partial r} + \frac{\partial^2 u}{\partial x^2}\right) \tag{6.81}$$

$$u\frac{\partial v}{\partial x} + v\frac{\partial v}{\partial r} = -\frac{1}{\rho}\frac{\partial p}{\partial r} + \nu\left(\frac{\partial^2 v}{\partial r^2} + \frac{1}{r}\frac{\partial v}{\partial r} - \frac{v}{r^2} + \frac{\partial^2 v}{\partial x^2}\right). \tag{6.82}$$

Performing an order of magnitude analysis in the usual manner, Eqs. (6.81) and (6.82) can be rewritten as follows:

$$u\frac{\partial u}{\partial x} + v\frac{\partial u}{\partial r} = -\frac{1}{\rho}\frac{\partial p}{\partial x} + \nu\left(\frac{\partial^2 u}{\partial r^2} + \frac{1}{r}\frac{\partial u}{\partial r}\right) \tag{6.83}$$

$$\frac{\partial p}{\partial r} = 0. \tag{6.84}$$

Equation (6.83) is now integrated across the tube, and boundary condition $u = v = 0$ at $r = R$ is applied to get

$$\frac{d}{dx}\int_0^R r u^2 dr = -\frac{1}{\rho}\frac{\partial p}{\partial x}\frac{R^2}{2} + \nu R\left(\frac{\partial u}{\partial r}\right)_R. \tag{6.85}$$

The integrated form of the continuity equation is now written as

$$Q = \pi R^2 U_c = \int_0^R 2\pi r u\, dr \tag{6.86}$$

where U_c is the mean velocity at any given cross-section with radius R, and Q is the volumetric flow rate. It is assumed that the radial dependence of the axial velocity can be expressed as a fourth-order polynomial and that the following boundary conditions are imposed

$$\text{at}\quad r = R: \quad u = 0; \quad \frac{dp}{dx} = \mu\left(\frac{\partial^2 u}{\partial r^2} + \frac{1}{r}\frac{\partial u}{\partial r}\right)$$

$$\text{at}\quad r = 0: \quad u = U_c; \quad \frac{\partial u}{\partial r} = 0; \quad \frac{\partial^2 u}{\partial r^2} = -\frac{2U_c}{R^2}. \tag{6.87}$$

The velocity profile is then obtained as

$$\frac{u}{U_c} = \left(\frac{-\Delta p_1 + 10}{7}\right) r_{1w} + \left(\frac{3\Delta p_1 + 5}{7}\right) r_{1w}^2$$

$$+ \left(\frac{-3\Delta p_1 - 12}{7}\right) r_{1w}^3 + \left(\frac{\Delta p_1 + 4}{7}\right) r_{1w}^4 \tag{6.88}$$

where

$$\Delta p_1 = \frac{R^2}{\mu U_c}\frac{dp}{dx}. \tag{6.89}$$

Further, Eq. (6.88) can be substituted into Eq. (6.86) and integrated
to give

$$U_c = \frac{210}{97\pi} \frac{1}{R^2} \left(Q + \frac{\pi}{105\mu} R^4 \frac{dp}{dx} \right). \tag{6.90}$$

It can be seen that the velocity profile in Eq. (6.88) is a function of the
dimensionless pressure gradient. The parameter Δp_1 can be determined
by using Eq. (6.88) in Eq. (6.85) and numerically solving the non-linear
ordinary differential equation. However, Forrester and Young (1970a)
resort to the use of an approximation at this stage in order to evolve
a simpler solution to the problem. The non-linear terms arise from the
evaluation of the integral in Eq. (6.85). These can be avoided through
the use (Forrester and Young, 1970a) of a parabolic velocity profile
given below, as a first approximation.

$$\frac{u}{U_c} = 2[1 - (r/R)^2] \tag{6.91}$$

For mild stenoses, the velocity profile is known (Young, 1968) to approach the parabolic distribution for low Reynolds numbers. Hence,
the validity of the assumption is expected to increase with decreasing
Reynolds number.

Substituting (6.91) into the integral of Eq. (6.85) gives

$$\frac{d}{dx} \left(\frac{2}{3} R^2 U_c^2 \right) = -\frac{1}{\rho} \frac{dp}{dx} \frac{R^2}{2} + \nu R \left(\frac{\partial u}{\partial r} \right)_R. \tag{6.92}$$

Substituting the expressions for U_c and $(\partial u/\partial r)_R$ from Eqs. (6.86) and
(6.88) along with the use of Eq. (6.90) gives

$$\frac{dp}{dx} = \frac{5432}{1575\pi^2} \frac{\rho Q^2}{R^5} \frac{dR}{dx} - \frac{8\mu Q}{\pi R^4} \tag{6.93}$$

and in dimensionless terms

$$\frac{R_0}{\rho U_{e0}^2} \frac{dp}{dx} = \frac{5432}{1575} \frac{R_0^5}{R^5} \frac{dR}{dx} - \frac{16}{Re_0} \frac{R_0^4}{R^4} \tag{6.94}$$

where

$$Re_0 = \frac{\rho(2R_0)U_{e0}}{\mu}. \tag{6.95}$$

In the converging portion of the tube, dR/dx is negative so that a
favorable pressure gradient exists; whereas, in the diverging portion,

dR/dx is positive so that an adverse pressure gradient $(dp/dx \geq 0)$ may develop for which the following condition must hold:

$$Re_0 \frac{R_0}{R} \frac{dR}{dx} \geq 4.64. \tag{6.96}$$

With R defined by Eq. (6.79), $[(1/R)(dR/dx)]$ will have a maximum value somewhere near the middle of the diverging section, and there will be a some minimum value of Re_0 that will be needed to satisfy Eq. (6.96).

Combining Eqs. (6.88)–(6.90) and (6.93) gives the following expression for the velocity profile:

$$\frac{u}{U_{e0}} = Re_0 \frac{R_0^3}{R^3} \frac{dR}{dx} \left[-\frac{308}{1575} r_{1w} + \frac{1204}{1575} r_{1w}^2 - \frac{4}{5} r_{1w}^3 + \frac{4}{15} r_{1w}^4 \right]$$
$$+ 2\frac{R_0^2}{R^2}(2r_{1w} - r_{1w}^2). \tag{6.97}$$

Two theoretical profiles, as predicted by Eq. (6.97) at various axial positions of $x/\lambda = -0.5$, -0.2175, 0, 0.2175, and 0.5 for Reynolds numbers of 100 and 150 are plotted to scale in Fig. 6.15 for the tube geometry of $\lambda = 8R_0 = 48a$ (which is the geometry studied experimentally by Forrester and Young, 1970b).

Since the velocity profiles in Eq. (6.97) are composed of a parabolic term along with a term that is multiplied by the slope at the wall, they will have a parabolic form for all Reynolds numbers at $x/\lambda = -0.5$, 0, and 0.5 where the slope is zero. The axial position of $x/\lambda = 0.2175$ also has significance because this position marks the incipient separation point for the given geometry and occurs at a critical Reynolds number of 127.5, as will be shown later.

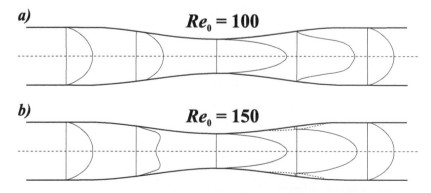

Fig. 6.15. Velocity profiles

At a Reynolds number of 150, separation exists in the diverging section, but the thickness of the separation region is relatively thin when compared with the layer moving upstream, as can be seen from the dotted line in Fig. 6.15b. As Reynolds number increases, the velocity profile in the converging section tends to flatten in the central portion of the tube. The dip in the velocity profile is due to the approximate nature of the imposed conditions on the second derivative of the velocity profile at $r = 0$. On account of the simplifying assumptions made in order to obtain a closed-form solution, its validity cannot be quantitatively stretched for Reynolds numbers in the range of values where flow separation occurs.

Forrester (1968) has shown that for slowly changing boundary forms of the flow considered herein, the wall shear stress can be written as

$$\tau_w \sim \mu (\partial u / \partial r)_R. \tag{6.98}$$

Thus, combining Eqs. (6.97) and (6.98) gives the expressions for wall shearing stress as

$$\frac{\tau_w}{\rho U_{e0}^2} = \frac{616}{1575} \frac{R_0^4}{R^4} \frac{dR}{dx} - \frac{8}{Re_0} \frac{R_0^3}{R^3}. \tag{6.99}$$

In Fig. 6.16 the dimensionless form of the wall shear stress predicted by the above Eq. (6.99) is plotted against the axial position along the stenosis for the tube geometry $\lambda = 8R_0 = 48a$. It is seen that the shearing stress reverses direction between the points $x/\lambda = 0.0925$ and 0.3675 and $x/\lambda = 0.0425$ and 0.4376 for Reynolds numbers of 200 and 400, respectively. These positions represent the separation and reattachment points for these particular Reynolds numbers. The same does not hold for a Reynolds number of 100 where separation does not occur. Such anomalies are seen because the analysis is based on an approximate solution.

Hence, the qualitative results and the trends should be given more importance than quantitative values. The separation–reattachment condition can be determined by equating the wall shear stress in the above Eq. (6.99) to zero, to give

$$(Re_0) \frac{R_0}{R} \frac{dR}{dx} = 20.4. \tag{6.100}$$

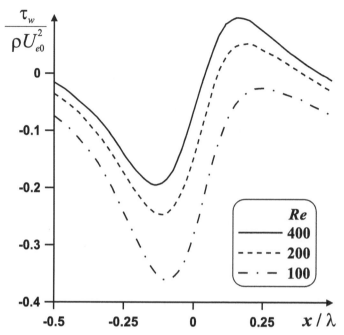

Fig. 6.16. Variation of wall shear stress with x/λ

In order to determine the point in the diverging portion of the tube where separation will begin, $(1/R)(dR/dx)$ is maximized with respect to x as follows:

$$\frac{d}{dx}\left(\frac{1}{R}\frac{dR}{dx}\right) = \frac{R(d^2R/dx^2) - (dR/dx)^2}{R^2} = 0. \qquad (6.101)$$

Using Eq. (6.79) in the above Eq. (6.101) gives the expression for locating the point of separation as

$$\frac{x}{\lambda} = \frac{1}{2\pi}\mathrm{arcsec}\left(\frac{R_0}{a} - 1\right). \qquad (6.102)$$

For the geometry studied experimentally by Forrester and Young (1970b), $R_0 = 6a$ and then $x/\lambda = 0.2175$. From this value, the critical Reynolds number required to produce the separation can be determined from Eq. (6.100), which for the geometry under consideration with $\lambda = 8R_0 = 48a$ is 127.5.

As the Reynolds number increases above the critical value, Eq. (6.100) is satisfied for two values of x — the smaller value giving the separation point while the larger value gives the reattachment point. As the Reynolds number increases, the separation point moves upstream and the reattachment point moves downstream from the initial separation point.

Forrester and Young (1970b) have carried out an experimental program to verify the theoretical findings. A plastic PCT is used for simulating the flow through the stenosis whose geometry is defined by Eq. (6.79), with the uniform diameter in the straight section being 0.75 ins, the converging-diverging section being 3 ins long, and the minimum at the throat being 0.5 ins, so that $\lambda = 8R_0 = 48a$. This selected geometry has been used in the theoretical analysis for the predictions and this facilitates a comparison between theory and experiment. The upstream and downstream portions of the PCT are connected to 0.75 in brass tubing. The brass upstream section is long enough to ensure fully developed conditions and is connected to a constant head tank. The downstream portion of the brass tubing is connected to a reservoir, from which it is pumped back to the constant head tank. All experiments are performed at room temperature ranging from 22 to 27°C over the entire period of test. Water and blood are used in the experiments, and their viscosities are determined using a Cannon–Fenske viscometer. The chosen human blood is a combination of 4 pints of type O positive and one pint of type O negative, and is about three months old with a hematocrit of 32. Since normal human blood has approximately 35–45 per cent by volume of red blood cells, it is obvious that deterioration of the red blood cells has occurred due to aging. Besides this three-month-old blood, fresh bovine blood is also used in the experiment. Also, 300 mg of sodium heparin is added in two installments of 150 mg each to six liters of this blood to prevent clotting.

The main purpose of the experimental investigation has been to determine the minimum Reynolds number for separation and to observe the separated region at larger Reynolds numbers. In order to do this, series of 0.026-in diameter holes are drilled along the PCT so that a dye can be injected at various stations to study the separation phenomena and to measure the pressure variation along the wall. The dyes used are made by mixing aniline blue or potassium permanganate with water or saline, depending on whether water or blood is flowing through the converging–diverging section. The dye is injected into the separation zone slowly, at a rate just sufficient to produce a thin filament along the tube wall. It was found that a well-defined separation region could be observed at a Reynolds number of approximately 900. This value was very close to the upper limit when such a well-defined laminar separation region could be observed. This fact was confirmed when, at the higher value of Reynolds number of approximately 1200, the dye began to become somewhat wavy near the position of x/λ of 0.65. In fact, if the Reynolds number was further increased to about 1800, turbulent separation was observed by Forrester and Young (1970b) to extend both upstream to the separation point and downstream.

The theoretical separation–reattachment curve from Eq. (6.100) for the experimental tube geometry, along with experimental curves for distilled water, fresh bovine blood, and three-month-old human blood are shown in Fig. 6.17.

It can be seen that experimental curves for the three fluids agree very well, except for the reattachment points for the fresh bovine blood, at Reynolds number greater than 450. The critical Reynolds number to produce incipient separation is seen to lie in the range of 200–250 for all three fluids as against the theoretical predicted value of 127.5.

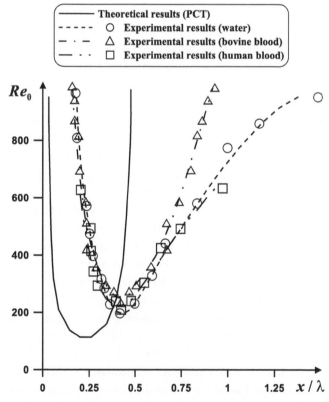

Fig. 6.17. Theoretical and experimental results

So also, the axial position at which separation occurs is approximately $x/\lambda = 0.425$, as compared to the theoretical predicted value of 0.2175. One reason for these discrepancies is that the values of a/λ and a/R_0 of the actual experimental PCT are not small enough to emulate the conditions set up in the assumptions of the theoretical derivation. Further, the theoretical development requires the velocity profile to be parabolic near the centerline of the tube, which is not necessarily true during experiments. Moreover, the experimental tube, though prepared carefully, did not match Eq. (6.79) exactly. So also, there is difficulty in tracing the dye filament to the actual points of separation or reattachment. Thus, it is very likely that true separation and reattachment points occur further upstream and downstream, respectively, than those that are actually determined experimentally.

The theoretical and experimental results for the pressure drop have also been obtained by Forrester and Young (1970b).

Based on theoretical predictions and experimental findings, Forrester and Young (1970b) concluded that the increase in average pressure in the diverging portion of a stenosis cannot be the cause of post-stenotic dilation, as suggested by some investigators. Since the time-averaged wall shearing stress is expected to be small in the converging portion of the stenosis, post-stenotic dilation due to increased wall shearing stress is unlikely. The possibility exists, however, that the reversed direction of the wall shearing stresses in a separation region may, somehow, affect the endothelial lining and cause the artery to dilate due to either autoregulatory mechanisms or to a weakening of the arterial wall. The abnormal action of pressure and shearing stress on the arterial wall may lead to atherosclerosis due to hypertension in the arterial wall. In fact, the abnormal flow conditions in a stenotic obstruction can certainly be a contributing factor in the development and progression of arterial disease. Forrester and Young (1970a,b) have shown that even a mild "collarlike" stenosis in a small artery can create abnormalities in the flow, including chances of observing the phenomenon of separation.

It should be noted that Forrester and Young (1970a,b) limited their analysis to flow in tubes with local constriction. In reality, of course, stenoses may be present in a number of places at the same time. In order to simulate these conditions, Chow and Soda (1972) considered flow through a tube with continuous constriction or obstruction as shown in Fig. 1.13, with the specific boundary form of the stenoses described by the following equation:

$$r_w = r_m + a\sin(2\pi x/\lambda). \tag{6.103}$$

They obtained an analytical solution for the case where the spread of roughness is large compared with the mean radius of the tube. Physically, their solution corresponds to the existence of patchy mild stenoses in the blood vessel. The boundary irregularities are caused by intravascular plaques or the impingement of ligaments or spur on the vessel wall. The important flow characteristics in an arterial system are the pressure, shear stress, possible separation, and reattachment, as these are all related to arterial diseases. Chow and Soda (1972) have presented results for streamlines, velocity, and vorticity distributions, pressure, energy dissipation, and separation and reattachment points for PCT with sinusoidal wall variations.

It is assumed, as done earlier by Forrester and Young (1970a), that blood behaves like a homogeneous Newtonian fluid and the flow field is steady. The governing equation of motion for the steady laminar flow of an incompressible Newtonian fluid in an axisymmetric PCT in terms of the non-dimensional stream function ψ_1 is written (Chow and Soda, 1972) as

$$Re\,r_{1m}\frac{\partial\psi_1}{\partial x_1}\left(\nabla^2\frac{\partial\psi_1}{\partial r_1} - \frac{2}{r_1}\nabla^2\psi_1 + \frac{1}{r_1^2}\frac{\partial\psi_1}{\partial r_1}\right) - \frac{\partial\psi_1}{\partial r_1}\nabla^2\psi_1$$

$$= \nabla^4\psi_1 \tag{6.104}$$

where

$$\nabla^2 = r_{1m}^2\frac{\partial^2}{\partial x_1^2} + \frac{\partial^2}{\partial r_1^2} - \frac{1}{r_1}\frac{\partial}{\partial r_1}$$

$$x_1 = \frac{x}{\lambda}, \quad r_1 = \frac{r}{\lambda}, \quad r_{1m} = \frac{r_m}{\lambda} \tag{6.105}$$

$$Re = \frac{U_e r_m}{\nu}, \quad \psi_1 = \frac{\psi}{U_e r_m^2}.$$

The boundary conditions are the no-slip condition at the wall, the axisymmetric condition, and the condition of constant volume flux along the tube.

Thus,

$$\frac{\partial \psi_1}{\partial r_1} = \frac{\partial \psi_1}{\partial x_1} = 0 \text{ at the tube wall} \tag{6.106}$$

$$\frac{\partial \psi_1}{\partial x_1} = 0 \text{ at the axis of symmetry} \tag{6.107}$$

$$\int_0^{r_{1w}} -2\frac{\partial \psi_1}{\partial r_1} dr_1 = Q_1 \text{ for all } x_1 \tag{6.108}$$

where $Q_1 = Q/\pi r_m^2 U_e$, and the tube boundary is given by

$$r_{1w} = 1 + a_1 \sigma_r(x_1) \tag{6.109}$$

where $\sigma_r(x_1)$ describes the wall variation relative to the mean radius and $a_1 = a/r_m$ with a denoting the average height of the waviness of the wall.

Solution of Eq. (6.104) under the imposed boundary conditions (6.107)–(6.109) is sought by expanding the dimensionless stream function in terms of a series expansion, as follows:

$$\psi_1(x_1, r_1; Re, a_1, r_{1m}) \sim \sum_{i=0}^{N} r_{1m}^i \psi_{1,i}(x_1, r_1; Re, a_1). \tag{6.110}$$

Substitution of Eq. (6.110) into Eqs. (6.104), (6.106), and (6.107), and subsequent collection of terms with equal power of r_{1m}, result in the following sets of perturbed equations up to the second order:

Zeroth Order:

$$L^2 \psi_{1,0} = 0 \text{ with } L = \frac{\partial^2}{\partial r_1^2} - \frac{1}{r_1}\frac{\partial}{\partial r_1} \tag{6.111}$$

with the boundary conditions satisfying

$$\psi_{1,0} = -\frac{1}{2} \text{ and } \frac{\partial \psi_{1,0}}{\partial r_1} = \frac{\partial \psi_{1,0}}{\partial x_1} = 0 \text{ at } r_1 = r_{1w} \tag{6.112}$$

$$\frac{\partial \psi_{1,0}}{\partial x_1} = 0 \text{ at } r_1 = 0. \tag{6.113}$$

First Order:

$$\frac{1}{Re} L^2 \psi_{1,1} = \frac{1}{r_1}\frac{\partial \psi_{1,0}}{\partial x_1}\left(\frac{\partial^3 \psi_{1,0}}{\partial r_1^3} - \frac{3}{r_1}\frac{\partial^2 \psi_{1,0}}{\partial r_1^2} + \frac{3}{r_1^2}\frac{\partial \psi_{1,0}}{\partial r_1}\right)$$
$$- \frac{1}{r_1}\frac{\partial \psi_{1,0}}{\partial r_1} L \frac{\partial \psi_{1,0}}{\partial x_1} \tag{6.114}$$

with the boundary conditions satisfying

$$\psi_{1,1} = 0 \text{ and } \frac{\partial \psi_{1,1}}{\partial r_1} = \frac{\partial \psi_{1,1}}{\partial x_1} = 0 \text{ at } r_1 = r_{1w} \tag{6.115}$$

$$\frac{\partial \psi_{1,1}}{\partial x_1} = 0 \text{ at } r_1 = 0. \tag{6.116}$$

Second Order:

$$\frac{1}{Re} L^2 \psi_{1,2} = -\frac{2}{Re} L \frac{\partial^2 \psi_{1,0}}{\partial x_1^2}$$

$$+ \frac{1}{r_1} \frac{\partial \psi_{1,0}}{\partial x_1} \left(\frac{\partial^3 \psi_{1,0}}{\partial r_1^3} - \frac{3}{r_1} \frac{\partial^2 \psi_{1,0}}{\partial r_1^2} + \frac{3}{r_1^2} \frac{\partial \psi_{1,0}}{\partial r_1} \right)$$

$$+ \frac{1}{r_1} \frac{\partial \psi_{1,0}}{\partial x_1} \left(\frac{\partial^3 \psi_{1,1}}{\partial r_1^3} - \frac{3}{r_1} \frac{\partial^2 \psi_{1,1}}{\partial r_1^2} + \frac{3}{r_1^2} \frac{\partial \psi_{1,1}}{\partial r_1} \right) \tag{6.117}$$

with the boundary conditions satisfying

$$\psi_{1,2} = 0 \text{ and } \frac{\partial \psi_{1,2}}{\partial r_1} = \frac{\partial \psi_{1,2}}{\partial x_1} = 0 \text{ at } r_1 = r_{1w} \tag{6.118}$$

$$\frac{\partial \psi_{1,2}}{\partial x_1} = 0 \text{ at } r_1 = 0. \tag{6.119}$$

The solutions of the stream functions $\psi_{1,0}$, $\psi_{1,1}$, and $\psi_{1,2}$ are obtained by integrating their respective governing equations, satisfying the corresponding boundary conditions to give

$$\psi_{1,0} = \left(\frac{Q_1}{2} \right) (r_2^4 - 2r_2^2) \tag{6.120}$$

$$\psi_{1,1} = \left(\frac{Q_1}{2} \right)^2 \left(\frac{d\sigma_1}{dx_1} \right) \left(\frac{Re\, a_1}{9 r_{1w}} \right) (r_2^8 - 6r_2^6 + 9r_2^4 - 4r_2^2) \tag{6.121}$$

$$\psi_{1,2} = -\left(\frac{Q_1}{2} \right) \left(\frac{a_1}{3} \right) \left[5a_1 \left(\frac{d\sigma_r}{dx_1} \right)^2 - r_{1w} \left(\frac{d^2 \sigma_r}{dx_1^2} \right) \right] (r_2^2 - 1)^2 r_2^2$$

$$- \left(\frac{Q_1}{2} \right)^3 \left(\frac{d\sigma_r}{dx_1} \right)^2 \left(\frac{Re\, a_1}{60 r_{1w}} \right)^2$$

$$\times (32 r_2^{12} - 305 r_2^{10} + 750_2^0 - 713 r_2^6 + 236 r_2^4) \tag{6.122}$$

where

$$r_2 = r_1 / r_{1w}. \tag{6.123}$$

It has been established by Chow and Soda (1972) that the expansion series provides a valid approximation up to the second order, provided

$$Re\, a_1 r_{1m}(1 - a_1^2)^{1/2} \ll 1.433 \text{ and } (1 + 0.987a_1)a_1 r_{1m}^2 \ll 1.5. \quad (6.124)$$

The second condition in Eq. (6.124) is usually satisfied since a_1 varies from zero to one, which is a singular point and corresponds to the complete blockage of the conduit. The solution for the conduits with sinusoidal wall variation as shown in Fig. 1.13 is now presented.

Representative flow patterns for $\psi_{1,0}$, $\psi_{1,1}$, and $\psi_{1,2}$ and the combined streamlines are shown in the upper half and lower half of Fig. 6.18, respectively, for $Re = 12.5$, $a_1 = 0.2$, and $r_{1m} = 0.1$.

The zeroth-order solution corresponds to flow with a vanishing wall slope, and reduces to the flow in a straight tube for $a_1 = 0$. The first-order solution induces clockwise and counterclockwise rotational motion, respectively, in the convergent and divergent sections of the tube. The second-order solution reinforces the first-order solution in the divergent part of the tube, except near the throat. The separation would, therefore, occur at a lower Reynolds number for the given a_1 and r_{1m} as compared to the solution up to the first order.

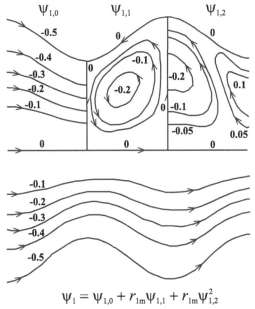

$$\Psi_1 = \Psi_{1,0} + r_{1m}\Psi_{1,1} + r_{1m}\Psi_{1,2}^2$$

Fig. 6.18. Flow patterns and the combined streamlines

The streamlines and vorticity contours when $Re = 25$, $a_1 = 0.4$, and $r_{1m} = 0.1$ are shown in Fig. 6.19, clearly depicting the occurrence and subsequent reattachment in the divergent part of the tube.

Since shear stress is proportional to the vorticity on the wall, it can be seen from Fig. 6.19 that the maximum shear stresses occur near the upstream side of the throat.

Figure 6.20 shows positions of separation and reattachment as a function of Re and a_1 for $r_{1m} = 0.1$. The separation Re is seen to decrease markedly as the tube constriction decreases.

By increasing Re or a_1 for any given r_{1m}, the separation point would move down toward the throat in the divergent portion of the tube with subsequent enlargement of the region of separation. This is truly unfavorable physiologically and it is certainly desirable to check through experiment whether the flow becomes unstable at much lower Reynolds number after the separation point reaches the throat.

For a tube with a cosine-shaped constriction with $a_1 = 0.4$ and $r_{1m} = 0.104$, which is nearly the same as that used by Forrester and Young (1970a,b) in theory and experiment, Chow and Soda (1972) predict the lowest separation Reynolds number based on the tube diameter to be 221.6.

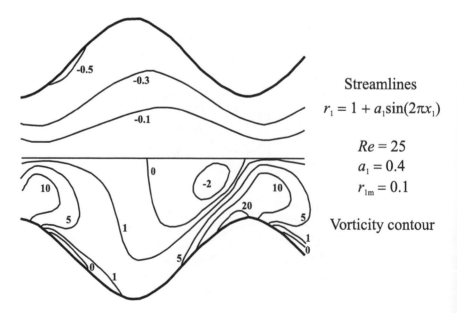

Streamlines

$$r_1 = 1 + a_1\sin(2\pi x_1)$$

$Re = 25$
$a_1 = 0.4$
$r_{1m} = 0.1$

Vorticity contour

Fig. 6.19. Streamlines and vorticity contours

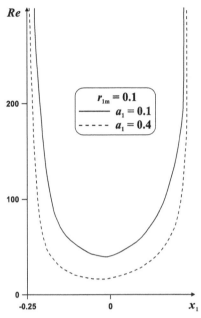

Fig. 6.20. Positions of separation and reattachment as a function
of the Reynolds number

This is a closer match to the experimental data (Forrester and
Young, 1970b), using distilled water, fresh bovine blood, and three-
months-old human blood, wherein the lowest separation Reynolds number is found to be between 200 and 250. The theory of Forrester and
Young (1970a) predicts a much lower value of 127.5 for this Reynolds
number. In that sense, the theoretical analysis of Chow and Soda (1972)
is superior, but it would still not be precise because the solution has
been obtained only up to the second order of r_{1m}.

The hydrodynamic solutions obtained by Chow and Soda (1972)
have been used by Chow and Soda (1973) to get temperature distribution and heat fluxes from the isothermal wall of conduits with constriction through which a Newtonian fluid of constant fluid properties
is flowing. It is assumed that viscous dissipation effects are negligible
and that the wall temperature is suddenly heated from a temperature
$T = T_e$ to a temperature $T = T_w$ and maintained constant thereafter.
The non-dimensional energy equation governing the temperature field
can be written (Chow and Soda, 1973) in terms of the transformed
variable r_2 as

$$Pr\,Re\,r_{1m}\left(\frac{\partial\psi_1}{\partial r_2}\frac{\partial\theta}{\partial x_1} - \frac{\partial\psi_1}{\partial x_1}\frac{\partial\theta}{\partial r_2}\right) = \frac{\partial}{\partial r_2}\left(r_2\frac{\partial\theta}{\partial r_2}\right) \tag{6.125}$$

where

$$\theta = \frac{T - T_w}{T_e - T_w} \tag{6.126}$$

and the rest of the terms in Eq. (6.125) are defined in Eqs. (6.105) and (6.123).

The initial and boundary conditions are the conditions specifying the uniform temperature across the conduit at $x_1 = x_{10}$, the equality of temperature on the wall, and the symmetric conditions on the axis of symmetry. Thus,

$$\theta = 1 \text{ at } x_1 \le x_{10} \tag{6.127}$$

$$\theta = 0 \text{ at } r_2 = 1, \ x_1 > x_{10} \tag{6.128}$$

$$\frac{\partial\theta}{\partial r_2} = 0 \text{ at } r_2 = 0, \ x_1 > x_{10}. \tag{6.129}$$

The solutions of Eq. (6.125) satisfying Eqs. (6.127)–(6.129) are first expanded in series in terms of r_{1m} and the asymptotic solutions are sought in the limit of $r_{1m} \to 0$. Thus, the following equations for the temperature field and the stream functions are used:

$$\theta(x_1, r_2; Pr, Re, a_1, r_{1m}) \sim \sum_{i=0}^{N} r_{1m}^{i}\theta_i(x_1, r_2; Pr, Re, a_1), \tag{6.130}$$

$$\psi_1(x_1, r_2; Re, a_1, r_{1m}) \sim \sum_{i=0}^{N} r_{1m}^{i}\psi_{1,i}(x_1, r_2; Re, a_1). \tag{6.131}$$

Substitution of Eqs. (6.130) and (6.131) into Eqs. (6.125), and subsequent collection of terms with equal power of r_{1m}, result in the sets of perturbed equations similar to those obtained as Eqs. (6.111) to (6.119). The equations governing the different orders of the temperature field, which are thus obtained, are solved by first satisfying the homogeneous part of the equation using separation of variables, and then followed by constructing Green's function to satisfy the inhomogeneous part.

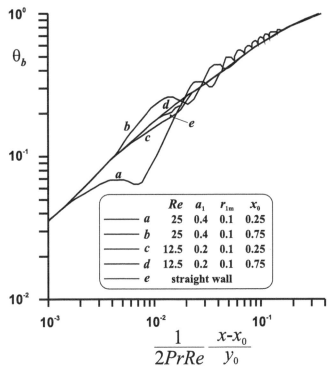

Fig. 6.21. Temperature profiles with dimensionless axial distance

The bulk fluid temperature difference, defined as follows:

$$\theta_b = \frac{\displaystyle\int_0^1 \theta \frac{\partial \psi_1}{\partial r_2} dr_2}{\displaystyle\int_0^1 \frac{\partial \psi_1}{\partial r_2} dr_2} \qquad (6.132)$$

and the Nusselt number based on mean radius are both plotted versus nondimensional axial distance, for conduits with sinusoidal wall variations, and shown in Figs. 6.21 and 6.22, respectively.

It can be readily seen that at low Reynolds number and in tubes with small conduit constriction, the bulk fluid temperature difference varies very little from the corresponding straight-wall conduit except that the curves are oscillatory in nature and their oscillations decrease in amplitude as the fluid moves away from the thermal entrance.

At higher Reynolds numbers and in tubes that have larger constriction, the bulk temperature difference behaves essentially in the same way except that the amplitude of the oscillations becomes much more pronounced.

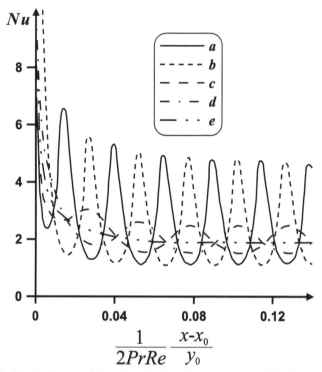

Fig. 6.22. Variations of the local Nusselt number with dimensionless
axial distance

The Nusselt number also fluctuates markedly with respect to its
mean when the Reynolds number and the tube constrictions are larger.
The sinusoidal variation of the wall, however, makes the Nusselt number stay oscillatory throughout the conduit. At low Reynolds number
and in tubes with small constriction, the mean coincides with its corresponding straight-wall conduit, but is much higher when the Reynolds
number and the tube constriction are larger. Due to the oscillatory
nature of the local Nusselt number, the thermal entrance length can
be defined as the distance from the entrance when the amplitudes of
the local Nusselt number become constant. The thermal entry length
changes very little from the corresponding straight-wall conduit. The
reason for this is that the concave part of the conduit enhances the
heat transfer rate, while the concave part decreases it, thereby leaving
the position of the thermal entry unaltered. Thus, it can be concluded
that there are large temperature and Nusselt number fluctuations in
a sinusoidal tube but very little change in the thermal entry length
compared with a straight-wall tube.

It should be noted that Neira and Payatakes (1979) presented a very nice collocation solution for creeping flow of Newtonian fluids through sinusoidal tubes. Such a collocation solution is particularly useful when dealing with geometries where the amplitude-to-wavelength ratio is large. Another desirable feature is that the collocation method gives the analytical form of the approximate solution. However, in order to save space, we will not present this solution here.

2. Iterative Finite Difference Solution

Deiber and Schowalter (1979) have developed an iterative technique to solve the equations of motion for flow of an incompressible Newtonian fluid through a circular tube whose radius varies sinusoidally in the axial direction. The problem is first solved assuming an arbitrarily large wavelength of diameter change such that locally Poiseuille flow is obtained. Using suitable decrements, the wavelength is decreased until a solution is found for the wavelength of interest. In terms of computer time, their iteration scheme has been found to be very economical. Further, their solutions are valid in the Reynolds number range wherein inertial effects are not negligible. Deiber and Schowalter (1979) have also conducted experiments and have found that their theoretical computations of friction factor versus Reynolds number function are in good agreement with their experiments. They have found that the shape of the toroidal vortex that appears as a consequence of inertial forces is also accurately predicted. The work of Deiber and Schowalter (1979) is outlined below and discussed.

The equation for the wall used by Deiber and Schowalter (1979) is basically the same as the equation that was used by Fedkiw and Newman (1977). However, this equation is further transformed as follows:

$$r_{2w} = 1 - a_2 \cos x_2 \qquad (6.133)$$

where

$$r_{2w} = r_{1w}/r_{1m}, \ a_2 = a_1/r_{1m}, \ x_2 = 2\pi x_1. \qquad (6.134)$$

The continuity and vorticity conservation equations are written in the non-dimensional form as

$$r_{2m}\frac{\partial u_1}{\partial x_2} + \frac{1}{r_2}\frac{\partial(r_2 v_1)}{\partial r_2} = 0 \qquad (6.135)$$

$$Re\left(u_1 r_{2m}\frac{\partial\omega_1}{\partial x_2} + v_1\frac{\partial\omega_1}{\partial r_2} - \frac{v_1}{r_2}\omega_1\right) = \frac{\partial}{\partial r_2}\left[\frac{1}{r_2}\frac{\partial}{\partial r_2}(r_2\omega_1)\right]$$

$$+ r_{2m}^2\frac{\partial^2\omega_1}{\partial x_2^2} \qquad (6.136)$$

where

$$u_1 = \frac{ur_m^2}{Q}, \quad v_1 = \frac{vr_m^2}{Q}, \quad r_2 = \frac{r_1}{r_{1m}}, \quad Re = \frac{Q}{\nu r_m} \qquad (6.137)$$

and the dimensionless vorticity is defined as

$$\omega_1 = \frac{\partial v_1}{\partial r_2} - r_{2m}\frac{\partial u_1}{\partial x_2}. \qquad (6.138)$$

It is assumed that the flow is axisymmetric and that there is no tangential velocity. The boundary conditions imposed on Eqs. (6.135) and (6.136) are

$$
\begin{aligned}
&u_1 = v_1 = 0 && \text{at } r_2 = r_{2w}\\
&\frac{\partial u_1}{\partial r_2} = v_1 = 0 && \text{at } r_2 = 0\\
&u_1(x_2, r_2) = u_1(x_2 + 2\pi, r_2), && 0 \le r_2 \le r_{2w}\\
&v_1(x_2, r_2) = v_1(x_2 + 2\pi, r_2), && 0 \le r_2 \le r_{2w}.
\end{aligned}
\qquad (6.139)
$$

A transformation of the variable is now introduced as

$$r_3 = \frac{r_2}{r_{2w}} \qquad (6.140)$$

so that the radial coordinate is $0 \le r_3 \le 1$ at all axial locations. The dimensionless stream function is defined such that

$$u_1 = \frac{1}{r_2}\frac{\partial\psi_1}{\partial r_2} = \frac{1}{r_3 r_{2w}^2}\frac{\partial\psi_1}{\partial r_3}$$

$$v_1 = -\frac{r_{2m}}{r_2}\frac{\partial\psi_1}{\partial x_2} = -\frac{r_{2m}}{r_3 r_{2w}}\left[\frac{\partial\psi_1}{\partial x_2} - \frac{r_3}{r_{2w}}\left(\frac{\partial r_{2w}}{\partial x_2}\right)\left(\frac{\partial\psi_1}{\partial r_3}\right)\right]. \qquad (6.141)$$

Combining Eqs. (6.136) and (6.141) gives the vorticity conservation equation as

$$r_{2m}Re\left\{r_3\left(\frac{\partial\psi_1}{\partial r_3}\frac{\partial\omega_1}{\partial x_2} - \frac{\partial\psi_1}{\partial x_2}\frac{\partial\omega_1}{\partial r_3}\right)\right.$$

$$+ \left[\frac{\partial \psi_1}{\partial x_2} - \frac{r_3}{r_{2w}} \left(\frac{\partial r_{2w}}{\partial x_2} \right) \left(\frac{\partial \psi_1}{\partial r_3} \right) \right] \omega_1 \Big\}$$

$$= r_3^2 \frac{\partial^2 \omega_1}{\partial r_3^2} \left\{ 1 + \left[r_{2m} r_3 \left(\frac{\partial r_{2w}}{\partial x_2} \right) \right]^2 \right\}$$

$$+ r_3 \frac{\partial \omega_1}{\partial r_3} \left\{ 1 - (r_{2m} r_3)^2 \left[r_{2w} \frac{\partial^2 r_{2w}}{\partial x_2^2} - 2 \left(\frac{\partial r_{2w}}{\partial x_2} \right)^2 \right] \right\}$$

$$+ (r_{2m} r_3 r_{2w})^2 \frac{\partial^2 \omega_1}{\partial x_2^2} - 2 r_{2m}^2 r_3^3 r_{2w} \frac{\partial r_{2w}}{\partial x_2} \frac{\partial^2 \omega_1}{\partial x_2 \partial r_3} - \omega_1 \qquad (6.142)$$

$$\omega_1 = \frac{1}{r_3^2 r_{2w}^3} \left[r_3 \frac{\partial^2 \psi_1}{\partial r_3^2} \left\{ 1 + \left[r_{2m} r_3 \left(\frac{\partial r_{2w}}{\partial x_2} \right) \right]^2 \right\} \right.$$

$$- \frac{\partial \psi_1}{\partial r_3} \left\{ 1 + (r_{2m} r_3)^2 \left[r_{2w} \frac{\partial^2 r_{2w}}{\partial x_2^2} - 2 \left(\frac{\partial r_{2w}}{\partial x_2} \right)^2 \right] \right\}$$

$$+ r_3 (r_{2m} r_{2w})^2 \frac{\partial^2 \psi_1}{\partial x_2^2} - 2 r_{2m}^2 r_3^3 r_{2w} \frac{\partial r_{2w}}{\partial x_2} \frac{\partial^2 \psi_1}{\partial x_2 \partial r_3} \right]. \qquad (6.143)$$

The boundary conditions on the dimensionless stream function and the dimensionless vorticity are

$$\text{at } r_3 = 0: \quad \psi_1 = \frac{\partial \psi_1}{\partial r_3} = \omega_1 = 0$$

$$\text{at } r_3 = 1: \quad \psi_1 = \frac{1}{2\pi}, \quad \frac{\partial \psi_1}{\partial r_3} = 0, \qquad (6.144)$$

$$\omega_1 = \frac{\partial^2 \psi_1}{\partial r_3^2} \frac{1}{r_{2w}^3} \left[1 + r_{2m}^2 \left(\frac{\partial r_{2w}}{\partial x_2} \right)^2 \right]$$

and the condition of periodicity requires

$$\psi_1(x_2, r_3) = \psi_1(x_2 + 2\pi, r_3), \qquad 0 \le r_3 \le 1$$

$$\frac{\partial \psi_1}{\partial x_2}(x_2, r_3) = \frac{\partial \psi_1}{\partial x_2}(x_2 + 2\pi, r_3) \quad 0 \le r_3 \le 1 \qquad (6.145)$$

$$\omega_1(x_2, r_3) = \omega_1(x_2 + 2\pi, r_3), \qquad 0 \le r_3 \le 1.$$

Equations (6.142) and (6.143) are solved by rewriting them in finite difference form and then solving (Franzen, 1977) by means of an iterative technique, wherein the iteration is in the geometry rather than the values of ψ_1 and ω_1 at various grid locations. Central differences for the derivatives in r_3 and backward differences for derivatives in x_2 are used. As a base state, a solution is first obtained in the limit of $r_{2m} \to 0$, which corresponds to a flow that is locally a Poiseuille flow. Thus,

$$\psi_1^{(0)} = (1 - r_3^2/2)r_3^2/\pi \tag{6.146}$$

$$\omega_1^{(0)} = -4r_3/(\pi r_{2w}^3). \tag{6.147}$$

Now a tube with a very small value of r_{2m} is considered, thereby keeping the increment step to a value that is sufficiently small. Using the values determined by Eqs. (6.146) and (6.147), the next set of solutions are obtained as $\psi_1^{(1)}$ and $\omega_1^{(1)}$ from the finite difference equations. This procedure is systematically followed using one small increment at a time. At each stage, the boundary conditions are also evaluated, and care is taken to see that the periodicity residuals at each iteration is minimized.

Once the values of $\psi_1(x_2, r_3)$ and $\omega_1(x_2, r_3)$ are obtained, computation of velocity profiles can be done. Further, the product of friction factor–Reynolds number can be obtained from the following equation:

$$
\begin{aligned}
f_{PCT} \cdot Re_{PCT} = \int_0^{2\pi} \Big\{ & r_{2m} r_3 u_1 \omega_1 Re \frac{\partial r_{2w}}{\partial x_2} \\
& - r_3 r_{2m}^2 \frac{\partial r_{2w}}{\partial x_2} \left[\frac{\partial \omega_1}{\partial x_2} - \frac{r_3}{r_{2w}} \frac{\partial r_{2w}}{\partial x_2} \frac{\partial \omega_1}{\partial r_3} \right] \\
& - v_1 \omega_1 \frac{\partial r_{2w}}{\partial x_2} + \frac{1}{r_{2w}} \frac{\partial \omega_1}{\partial r_3} + \frac{\omega_1}{r_{2w} r_3} \Big\} dx_2.
\end{aligned} \tag{6.148}
$$

Among the various solutions discussed so far, the approach of Deiber and Schowalter (1979) is the most attractive due to its relative simplicity, and also because it would be the easiest to extend to nonlinear equations, such as those which describe the flow of non-Newtonian fluids. However, the solution can be expected to fail for large wall amplitudes at high Reynolds numbers. In addition to the finite difference approximation errors, there is also an error associated with the iteration with respect to the parameter a_1, and this error can be expected to become dominant at high values of a_1.

3. Other Numerical Solutions

Lahbabi and Chang (1986) have used a global Galerkin/spectral method for solving the Navier–Stokes equation in PCT. Using the global method presents some advantages over the conventional methods, like finite difference, collocation, and finite elements. Firstly, the primitive variables are retained in the low-order Navier–Stokes equation, and secondly, the pressure term is eliminated without using the stream function–vorticity formulation, thus leaving the velocity vector as the only dependent variable. This is achieved by exploiting the geometric periodicity using the weak formulation of Leray, which is described by Ladyzhenskaya (1969) and Temam (1977) for the mathematical theories of the Navier–Stokes equations. Such a reduction of dependent variables allows the study of the full three-dimensional and time-dependent higher dimension problem. However, the major difficulty in the spectral method is the requirement of large storage and numerous operations for high-order non-linearities. For the Navier–Stokes equation, the highest non-linearity being the bilinear inertial term, the spectral method only requires the computation of three- and lower-dimensional tensors. Though the construction of the tensors is tedious, once constructed they can be used repeatedly for stationary solution, linear stability analysis, and non-linear bifurcation analysis at all Reynolds numbers. Another point of concern in the global spectral method is the proper construction of the basis vectors and finding methods to reduce the operations required to compute the inner products associated with the bilinear inertial term. The details of the problem formulation and the method of solution are available in Lahbabi and Chang (1985, 1986).

Table 6.3. Numerical Results of Lahbabi and Chang (1986)
for Flow through Sinusoidal Periodically Constricted Tubes
with $r_{1m} = 0.16$ and $a_2 = 0.3$ Corresponding
to the Points (O) in Fig. 6.22 [reprinted from Lahbabi and Chang
(1986) with permission from Elsevier]

Re_{PCT}	f_{PCT}	$f_{PCT} \cdot Re_{PCT}$
12.0	2.26	27.1
22.6	1.26	28.5
51.0	0.62	31.7
73.0	0.46	33.4
132.0	0.28	36.7
207.4	0.19	38.9
264.0	0.15	39.7
397.2	0.10	40.6
783.0	0.05	41.2

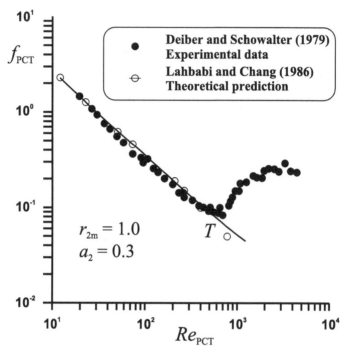

Fig. 6.23. Comparison between theoretical and experimental results

The numerical results of Lahbabi and Chang (1986) given in Table 6.3, when compared with the experimental data of Deiber and Schowalter (1979), show excellent agreement, as can be seen from Fig. 6.23. Deviations in predicted values are less than 3% for Reynolds numbers lower than 500.

Further, the range of validity of numerical results (Lahbabi and Chang, 1986) also extends beyond that of Deiber and Schowalter (1979) using the iterative finite difference scheme. A further test of the results can be seen from Table 6.4, where creeping flow results from various numerical studies are compared.

Results of Lahbabi and Chang (1986) match closely with those of Tilton and Payatakes (1984). Both these works disagree quantitatively and qualitatively with the study of Fedkiw and Newman (1977), especially for $a_2 \geq 0.3$. However, this happens because the friction factors for packed beds, as presented by Fedkiw and Newman (1977), are incorrectly used for comparison for a PCT. In a subsequent analysis, Fedkiw and Newman (1987) address this problem and show that the trends are quantitatively and qualitatively correct when properly interpreted.

Table 6.4. Comparison of Numerical Results of Various
Workers for Creeping Flow through Sinusoidal
Periodically Constricted Tubes of Different Geometries

		$f_{PCT} \cdot Re_{PCT}$				
r_{1m}	a_2	Fedkiw and Newman (1977)	Deiber and Schowalter (1979)	Tilton and Payatakes (1984)	Lahbabi and Chang (1986)	Fedkiw and Newman (1987)
0.1042	0.2	–	–	19.9	19.76	–
0.5000	0.2	19.3	–	23.4	23.40	23.3
0.2333	0.286	27.1	–	26.0	25.85	–
0.1592	0.3	27.3	25.12	26.6	26.40	–
0.3333	0.3	26.6	–	30.8	29.90	–
0.5000	0.3	25.6	–	33.2	34.50	–
0.2500	0.5	–	–	71.3	71.98	–
0.5000	0.5	64.6	–	96.0	99.66	109.3
0.3025	0.6	–	120.00	147.5	152.30	–

Some of the typical values of Fedkiw and Newman (1987) are also
shown in Table 6.4. The largest difference between the results of Fedkiw and Newman (1987) and those of Lahbabi and Chang (1986) occurs
at $r_{1m} = 0.5$ and $a_2 = 0.5$, but this 10% discrepancy is not surpris-
ing, considering the fact that this parameter set requires the great-
est numerical resolution. The deviations of the results of Deiber and
Schowalter (1979) from those of Lahbabi and Chang (1986) and Tilton
and Payatakes (1984) at $r_{1m} = 0.3$ and $a_2 = 0.6$ are probably due to
the combined effect of finite difference approximation errors and those
associated with their iteration with respect to a_2.

Table 6.4 compares the numerical results of Lahbabi and Chang
(1986) in the inertial region, with the only available experimental data
of Forrester and Young (1970b), who used a PCT to simulate blood flow
in vessels. It can be seen that, except for fresh bovine blood, which is
a non-Newtonian fluid, the results of Lahbabi and Chang (1986) are
within 10% of the experimental data of Forrester and Young (1970b).
Considering that there is an error of 5.1% in the match of the tube
radius of the stenoses (Forrester and Young, 1970b) and a true PCT,
this agreement is excellent.

The actual flow fields are depicted in Figs. 6.24 for various Reynolds
numbers.

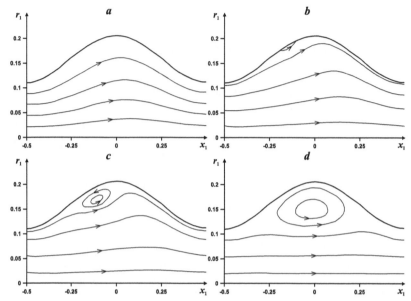

Fig. 6.24. Streamlines for: $Re = 12$ (a); $Re = 51$ (b); $Re = 73$ (c);
$Re = 397$ (d)

The creeping flow field is symmetric with respect to the x_1-plane. With increasing Reynolds numbers, the symmetry is broken, a vortex begins to form at a finite Reynolds number, and a separated flow region appears to the left of the x_1-plane. As the Reynolds number increases further, the vortex increases in size and its center shifts downstream towards the x_1-plane. Finally, at the potential flow limit, the $x_1 = 0$ symmetry is again regained and a large symmetric vortex exists inside the entire hollow with straight streamlines in the tube center. At such a high Reynolds number, the flow field will be unstable.

The onset of separation and the size of the separation region can be seen from Fig. 6.25.

In the creeping flow regime, the wall vorticity reflects the maximum symmetries of the flow field. As the Reynolds number increases, the curve lowers and shifts upstream. It crosses the x-axis at $Re = 51$, which corresponds to the flow separation, as is evident from Fig. 6.24b. It is to be noted that the separation region is represented by the portion of wall vorticity curve that is negative. As Reynolds number increases further, this region increases in size and shifts downstream, showing a trend that is consistent with the flow field of Figs. 6.24.

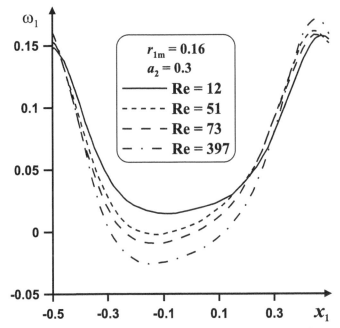

Fig. 6.25. Vorticity profiles for different values of Re

The onset of development of the vortex at the diverging portion of the tube is also confirmed experimentally by Deiber and Schowalter (1979), who observed the first vortex at $Re = 75$.

When considering the application of the result in a PCT to those in a porous medium, one has to refer to Darcy's law and the departures from it. There is a common belief that the onset of flow separation in a PCT signifies the dominance of inertial forces and the departure from Darcy's law. Fig. 6.26 shows variation of the product of friction factor and Reynolds number with increasing values of Reynolds number.

The Darcy region, where the flow rate is proportional to the pressure drop, is represented by the flat segment below a Reynolds number of 5. For a Reynolds number greater than 5, the inertial forces become dominant, and Darcy's law is no longer valid. The numerical results indicate that the flow separation does not occur until Reynolds number of 51, and hence, it can be seen that inertial effects are important at much lower Reynolds numbers. The second flat segment that occurs beyond a certain critical value of Reynolds number is inconsistent with experimental data of nonsteady flow (Deiber and Schowalter, 1979), which shows a sharp increase in pressure at a Reynolds number of about 500.

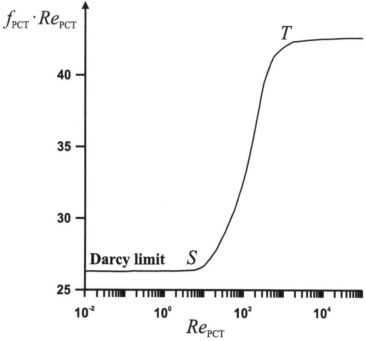

Fig. 6.26. Variation of $f_{PCT} \cdot Re_{PCT}$ with Re_{PCT}

It has been shown by Lahbabi and Chang (1986) that their stationary solution loses its linear stability before this critical value of Reynolds number, and hence, is irrelevant beyond this value. Their stationary solution is stable to axisymmetric disturbance at all Reynolds numbers, but destabilizes to azimuthal-dependent disturbance at this critical Reynolds number. Due to the rotational symmetry, the Hopf bifurcation that occurs at the critical Reynolds number is a degenerate one with two identical pairs of neutrally stable eigenvalues. Nonlinear analysis based on the center manifold theorem and normal form analysis is carried out (Lahbabi and Chang, 1986) on the reduced version of the full equation to characterize the Hopf bifurcation.

A numerical solution of the Navier–Stokes equations for steady axisymmetric flows in tubes with sinusoidal walls has also been obtained by Ralph (1987). The Reynolds number used by him is different from the one conventionally used by prior workers in this area. It is based on the tube radius and mean velocity at the constriction. Ralph (1987) asserts that the use of Reynolds number based on the conditions at the constriction is superior to the more usual definition in terms of conditions at the cross-section of mean flow area, unless of course the

Reynolds number is small. With this definition of Reynolds number, the pressure drop was found to be less sensitive to changes in geometry than if the "mean cross-section" Reynolds number was held constant, except when Re was less than about 10. Numerical calculations have been performed (Ralph, 1987) for Reynolds number up to 500 and for varying depth and wavelength of the wall perturbations. Results for the highest Reynolds number show features suggestive of flow based on the boundary layer theory. In the other Reynolds number limit, it is found that creeping flow solutions can exhibit flow reversal if the perturbation depth is large enough. Experimentally measured pressure drops for a particular tube geometry are in agreement with computed predictions up to a Reynolds number of about 300, at which transitional effects begin to disturb the experiments. The dimensionless mean pressure gradient is found to decrease with increasing Reynolds number, although the rate of decrease is less rapid than in a straight-walled tube. It has also been shown (Ralph, 1987) by numerical results that the mean pressure gradient decreases as perturbation wavelength and amplitude both increase. Higher Reynolds number is found to be more influenced by wavelength, while lower Reynolds number is more affected by the amplitude.

Chapter 7

Natural Convection Flow Saturated with Nanoparticles in Wavy-Walled Cavities

A. Magnetic Field Effect on the Unsteady Natural Convection in a Wavy-Walled Cavity Filled with a Nanofluid: Buongiorno's Mathematical Model

1. Introduction

The study of magnetohydrodynamic (MHD) flow and heat transfer of an electrically conducting fluid is of considerable interest in the modern metallurgical and metal-working processes. This study is also considered to be of great interest due to the effect of the magnetic field on the boundary layer flow control and on the performance of many systems using electrically conducting fluids. Some of the engineering applications are in MHD generators, plasma studies, nuclear reactors, geothermal energy extractions, purifications of metal from non-metal enclosures, polymer technology and metallurgy, MHD heat and mass transfer systems, design of MHD power generators, etc., and can be found in the books by Shercliff (1965), Branover and Tinober (1970), and Cramer and Pai (1973), and in the classical papers such as, for example, Goldsworthy (1961), Apelblat (1969), Ingham (1973), Liron and Wilhelm (1974), Watanabe and Pop (1993), Chandran et al. (1996), Hossain and Pop (1996), and Yih (1999).

Wavy geometries are used in many engineering systems as a means of enhancing the transport performance (Benjamin, 1959; Inger, 1971; Goldstein and Sparrow, 1977; Bordner, 1978; Thien-Phan, 1980, 1981; Caponi et al., 1982; Sparrow and Comb, 1983; Saidi et al., 1987; Wang, 1987; Blancher et al., 1990; Wang and Vanka, 1995; Hadjadj and Kyal, 1999; Rush et al., 1999; Adjlout et al., 2002; Mahmud et al., 2002; Wang and Chen, 2002; Das and Mahmud, 2003; Zhang et al., 2003, 2004; Tashtoush and Al-Odat, 2004; Mahmud and Fraser, 2004; Bahaidarah et al., 2005; Zhang, 2005; Pozrikidis, 2006; Wang, 2006a,b; Dalal and Das, 2006; Chen and Cho, 2007; Al-Amiri et al., 2007; Junqi et al., 2007; Tao et al., 2007; Sheik et al., 2009; Peterson, 2010; Cho et al., 2012a,b,c,d; Nasrin et al., 2012; Prince et al., 2013; Siddiqa and Hossain, 2013; Siddiqa et al., 2013, 2014, 2015; Khanafer, 2014; Karami et al., 2015; Nayak et al., 2015, 2016).

For example, Adjlout et al. (2002) studied numerically the effect of a hot wavy wall in an inclined differentially heated square cavity for different inclination angles, amplitudes, and Rayleigh numbers, while the Prandtl number was kept constant. Their results concluded that the wavy wall affected the flow and heat transfer rate in the enclosure. The flow and heat transfer characteristics inside an isothermal vertical wavy-walled enclosure bounded by two adiabatic straight walls has been investigated numerically by Mahmud et al. (2002). Simulation was carried out for a range of wave ratio (defined by amplitude/average width) $0.0-0.4$, aspect ratio (defined by height/average width) $1.0-2.0$, Grashof number $Gr = 10^0 - 10^7$, and a fixed value of the Prandtl number equal to 0.7.

Further, Das and Mahmud (2003) studied the hydrodynamic and thermal behaviors of fluid inside two wavy and two straight-walled enclosures. The top and the bottom walls were wavy and kept isothermal while the vertical straight walls were considered adiabatic. Results were presented for the local and average Nusselt number distributions for a range of Grashof numbers (10^3–10^7). It has been shown that the amplitude–wavelength ratio affected local Nusselt, but it had no significant influence on average Nusselt number.

A numerical analysis of the natural convection inside a two-dimensional cavity with a wavy right vertical wall has been performed by Dalal and Das (2006). The bottom wall was heated by a spatially varying temperature and the other three walls were kept at constant lower temperature.

Results were presented in the form of local and average Nusselt numbers for a selected range of Rayleigh numbers (10^0–10^6). The results showed that the presence of undulation in the right wall affected the flow and heat transfer characteristics.

On the other hand, it should be pointed out that several authors, such as, Al-Najem et al. (1998), have conducted a numerical study of laminar natural convection in a tilted enclosure with a transverse magnetic field, Hossain and Alim (2014) have studied the MHD free convection within a trapezoidal cavity with a non-uniformly heated bottom wall, while Revnic et al. (2011) have performed an analysis of the magnetic field effect on the unsteady free convection flow in a square cavity filled with a porous medium with a constant heat generation, while Groşan et al. (2009) studied the magnetic field and internal heat generation effects on the free convection in a rectangular cavity filled with a porous medium.

As is well-known, the cooling of electronic devices is the major industrial requirement due to the fast technology nowadays (Kuznetsov and Sheremet, 2008, 2010), but the low thermal conductivity rate of ordinary base fluids including water, ethylene glycol, and oil is the basic limitation. To overcome such limitations, the nanoscale solid particles are submerged into host fluids, which change the thermophysical characteristics of these fluids and enhance the heat transfer rate dramatically (see Choi, 1995; Kang et al., 2006; Ding et al., 2007; Wang and Mujumdar, 2007; Yu et al., 2008; Kuznetsov and Nield, 2010a,b,c,d; Muthtamilselvan et al., 2010; Selvan, 2010; Khanafer and Vafai, 2011; Salari et al., 2012; Lomascolo et al., 2015; Noghrehabadi et al., 2015). The recent developments in nanofluids and their mathematical modelling play a vital role in industrial and nanotechnology. The nanofluids are used in applications such as cooling of electronics, heat exchangers, nuclear reactor safety, hyperthermia, biomedicine, engine cooling, vehicle thermal management, and many others (see Das et al., 2007; Edel and deMello, 2008; Schaefer, 2010; Rao, 2010; Groşan and Pop, 2011a,b, 2012; Groşan et al., 2015; Kleinstreuer, 2013; Kasaeian et al., 2015; Sarkar et al., 2015). Further, the magneto nanofluids are useful in the manufacturing processes of industries and biomedicine applications (Brust et al., 1994; Chengara et al., 2004; Papazoglou and Parthasarathy, 2007; Chen et al., 2008; Choi, 2009; Rees et al., 2015; Bahiraei and Hangi, 2015). Examples include gastric medications, biomaterials for wound treatment, sterilized devices, etc. The magneto nanoparticles can be employed in the elimination of tumors with hyperthermia, targeted drug release, and magnetic resonance imaging.

Different mathematical models have been employed by several authors to describe heat transfer in nanofluids. Among all these models, the most used are those where the concentration of the nanoparticles is constant and the addition of the nanoparticles into the base fluid improved their physical properties (Khanafer et al., 2003; Tiwari and Das, 2007; Bondareva et al., 2015). Moreover, there are other models that are based on the variation of the physical properties, including thermal dispersion (Groşan, 2011) or Brownian motion (Kleinstreuer et al., 2008). A more complex mathematical nanofluid model has been developed by Buongiorno (2006) to explore the thermal properties of base fluids. Here the Brownian motion and thermophoresis are two dominant particle transfer mechanisms in nanofluids. The Brownian motion force tends to uniform nanoparticles in the fluid. The thermophoresis force originates from the temperature gradient in the base fluid. When the size of a particle is very fine (on the order of nanometers), the particle receives more momentum impacts from the fluid molecules on the hot side than that on the cold side; hence, the particle tends to move in a direction opposite to the temperature gradient (the particle moves from hot to cold). Therefore, there are mass transfer mechanisms in nanofluids because of the Brownian motion and thermophoresis forces. The migrated nanoparticles carry energy and transfer it to the surrounding medium. After that, the researchers investigated the flow of nanofluid under different conditions and different types of nanoparticles. Recently, Celli (2013) used Buongiorno's model (2006) in combination with a non-homogeneous model to study the free convection flow in a side-heated square cavity filled with a nanofluid. The thermophysical properties of the nanofluid are assumed to be functions of the average volume fraction of nanoparticles dispersed inside the cavity. Sheikholeslami et al. (2014) have analyzed MHD natural convection of Cu–water nanofluid in a square cavity heated from below and cooled from the side walls. It has been found that the heat transfer rate is an increasing function of heat source length, nanoparticle volume fraction, and Rayleigh number, and a decreasing function of Hartmann number. Mehrez et al. (2015) have studied MHD mixed convection flow of Cu–water nanofluid in an open cavity with an isothermal horizontal wall. The authors have shown that an increase in the Hartmann number leads to an attenuation of the recirculation inside the cavity for horizontal and inclined magnetic fields. Also, an inclusion of the nanoparticles leads to an increase in the heat transfer rate and the enhancement rate of heat transfer is higher than the increase rate of entropy generation. We mention, to this end, the very recent numerical study presented by Elshehabey and Ahmed (2015) to study the MHD mixed convection of a lid-driven cavity permeated by an inclined uniform magnetic field and filled with nanofluid using Buongiorno's (2006) nanofluid model. A sinusoidal temperature and nanoparticle volume fraction distributions on both vertical sides is considered, while the horizontal walls are kept adiabatic.

As a result, the heat transfer characteristics of nanofluids confined in cavities with wavy surfaces have also attracted increasing attention in recent times. Abu-Nada and Öztop (2011) performed a numerical investigation into the natural convection heat transfer characteristics of Al_2O_3-water nanofluid in a cavity with wavy walls. The natural convection performance of Al_2O_3-water nanofluid in a cavity with wavy side walls has also been studied by Nikfar and Mahmoodi (2012). Cho et al. (2012b) investigated the natural convection heat transfer performance of Al_2O_3-water nanofluid in a complex-wavy wall cavity.

Further, a very good paper that should be mentioned is that by Esmaeilpour and Abdollahzadeh (2012), who studied the natural convection heat transfer behavior and entropy generation of Cu–water nanofluid confined in a cavity with vertical wavy walls.

We also mention to this end the paper by Dhanai et al. (2015), who have studied the multiple solutions for the problem of slip flow and heat transfer of non-Newtonian nanofluid utilizing a heat source/sink and variable magnetic field. Selimefendigil and Öztop (2015) have numerically analyzed free convection in a nanofluid cavity with obstacles of different shapes, such as circular, square, and diamond, in conditions of uniform magnetic field and uniform heat generation.

Overall, these studies showed that the addition of nanoparticles to traditional working fluids results in a significant improvement in the flow and heat transfer characteristics. In addition, the results showed that the wavy-wall geometry parameters (e.g., the amplitude, wavelength, and waveform) have a critical effect on the flow and temperature fields and therefore directly affect the heat transfer performance.

This section presents a numerical investigation on the unsteady natural convection of water-based nanofluid within a wavy-walled cavity under the influence of a uniform inclined magnetic field, using the mathematical nanofluid model proposed by Buongiorno (2006).

2. Basic Equations

Let us study the unsteady natural convection in a two-dimensional wavy-walled cavity filled with an electrically, conducting nanofluid based on water and nanoparticles. The domain of interest is presented in Fig. 7.1, where the $\bar{x}$ axis is measured in the horizontal direction along the lower wall of the cavity and the $\bar{y}$ axis is measured along the left vertical wavy wall of the cavity. It is assumed that the left vertical wall is maintained at temperature T_h, while the right vertical wall is kept at a temperature T_c, where we assume that $T_h > T_c$. It is also assumed that the top and bottom walls of the cavity are adiabatic. We consider that the wavy-walled cavity is permeated by a uniform inclined magnetic field. The two vertical walls of the cavity are assumed to be impermeable. It is assumed the left wavy wall and right flat wall of the cavity are described by the relations

$$\bar{x}_1 = L - L[a + b\cos(2\pi\kappa\bar{y}/H)] \text{ and } \bar{x}_2 = L, \text{ respectively,}$$

where $\bar{\Delta} = \bar{x}_2 - \bar{x}_1 = L[a + b\cos(2\pi\kappa\bar{y}/H)]$ is the distance between vertical walls.

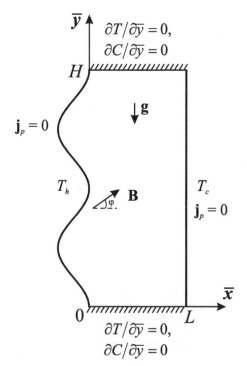

Fig. 7.1. Physical model and coordinate system

Except for the density, the properties of the fluid are taken to be constant. It is further assumed that the effect of buoyancy is included through the Boussinesq approximation. The viscous, radiation, and Joule heating effects are neglected. The magnetic Reynolds number is assumed to be small so that the induced magnetic field can be neglected compared to the applied magnetic field.

Under the above assumptions, the conservation equations for mass, momentum, thermal energy, electric transfer, and nanoparticles can be written as follows (see Buongiorno, 2006; Hossain and Alim, 2014):

$$\nabla \cdot \mathbf{V} = 0 \tag{7.1}$$

$$\rho_f \left[\frac{\partial \mathbf{V}}{\partial t} + (\mathbf{V} \cdot \nabla)\mathbf{V} \right] = -\nabla p + \mu \nabla^2 \mathbf{V}$$

$$+[C\rho_p + (1 - C)\rho_{f0}(1 - \beta(T - T_c))]\mathbf{g} + \mathbf{I} \times \mathbf{B} \tag{7.2}$$

$$\nabla \cdot \mathbf{I} = 0 \tag{7.3}$$

$$\mathbf{I} = \sigma(-\nabla\zeta + \mathbf{V} \times \mathbf{B}) \tag{7.4}$$

$$\frac{\partial T}{\partial t} + (\mathbf{V} \cdot \nabla)T = \alpha\nabla^2 T + \delta[D_B \nabla C \cdot \nabla T + (D_T/T_c)\nabla T \cdot \nabla T] \tag{7.5}$$

$$\rho_p \left(\frac{\partial C}{\partial t} + (\mathbf{V} \cdot \nabla)C \right) = -\nabla \cdot \mathbf{j}_p. \tag{7.6}$$

Here $-\nabla\zeta$ is the associated electric field. As discussed by Revnic et al. (2011), Eqs. (7.3) and (7.4) reduce to $\nabla^2\zeta = 0$. The unique solution is $\nabla\zeta = 0$ since there is always an electrically insulating boundary around the cavity. It follows that the electric field vanishes everywhere (see Groşan et al., 2009; Revnic et al., 2011).

Equations (7.1)–(7.6) can be written taking into account the dilute nanoparticle concentration without the pressure in Cartesian coordinates $\bar{x}$ and $\bar{y}$. For this purpose we formulate a mathematical model in dimensionless variables such as stream function $\left(u = \dfrac{\partial\bar{\psi}}{\partial\bar{y}}, \; v = -\dfrac{\partial\bar{\psi}}{\partial\bar{x}} \right)$, vorticity $\left(\bar{\omega} = \dfrac{\partial\bar{v}}{\partial\bar{x}} - \dfrac{\partial\bar{u}}{\partial\bar{y}} \right)$, and temperature, using the following dimensionless variables:

$$\tau = \alpha t/L^2, \; x = \bar{x}/L, \; y = \bar{y}/L, \; \psi = \bar{\psi}/\alpha,$$
$$\omega = \bar{\omega}L^2/\alpha, \; \theta = (T - T_c)/(T_h - T_c), \; \phi = C/C_0. \tag{7.7}$$

The governing equations of MHD natural convective heat transfer in dimensionless variables stream function–vorticity become (see Groşan et al., 2009; Revnic et al., 2011; Sheremet and Pop, 2014a–c, 2015a–e):

$$\frac{\partial^2 \psi}{\partial x^2} + \frac{\partial^2 \psi}{\partial y^2} = -\omega \tag{7.8}$$

$$\frac{\partial \omega}{\partial \tau} + \frac{\partial \psi}{\partial y}\frac{\partial \omega}{\partial x} - \frac{\partial \psi}{\partial x}\frac{\partial \omega}{\partial y} = Pr\left(\frac{\partial^2 \omega}{\partial x^2} + \frac{\partial^2 \omega}{\partial y^2}\right) + Ra \cdot Pr\left(\frac{\partial \theta}{\partial x} - Nr\frac{\partial \phi}{\partial x}\right)$$

$$+Ha^2 \cdot Pr\left[\frac{\partial^2 \psi}{\partial x^2}\cos^2\varphi + 2\frac{\partial^2 \psi}{\partial x \partial y}\sin\varphi \cdot \cos\varphi + \frac{\partial^2 \psi}{\partial y^2}\sin^2\varphi\right] \tag{7.9}$$

$$\frac{\partial \theta}{\partial \tau} + \frac{\partial \psi}{\partial y}\frac{\partial \theta}{\partial x} - \frac{\partial \psi}{\partial x}\frac{\partial \theta}{\partial y} = \frac{\partial^2 \theta}{\partial x^2} + \frac{\partial^2 \theta}{\partial y^2}$$

$$+Nb\left(\frac{\partial \phi}{\partial x}\frac{\partial \theta}{\partial x} + \frac{\partial \phi}{\partial y}\frac{\partial \theta}{\partial y}\right) + Nt\left[\left(\frac{\partial \theta}{\partial x}\right)^2 + \left(\frac{\partial \theta}{\partial y}\right)^2\right] \tag{7.10}$$

$$\frac{\partial \phi}{\partial \tau} + \frac{\partial \psi}{\partial y}\frac{\partial \phi}{\partial x} - \frac{\partial \psi}{\partial x}\frac{\partial \phi}{\partial y} = \frac{1}{Le}\left(\frac{\partial^2 \phi}{\partial x^2} + \frac{\partial^2 \phi}{\partial y^2}\right)$$

$$+\frac{Nt}{Le \cdot Nb}\left(\frac{\partial^2 \theta}{\partial x^2} + \frac{\partial^2 \theta}{\partial y^2}\right). \tag{7.11}$$

Following Kuznetsov and Nield (2013), the corresponding initial and boundary conditions can be written as follows:

$$\psi = 0, \quad \theta = 1, \quad \tilde{\mathbf{j}}_p = 0 \left(\text{or } Nb\frac{\partial \phi}{\partial \mathbf{n}} + Nt\frac{\partial \theta}{\partial \mathbf{n}} = 0\right)$$
$$\text{on } x = x_1$$

$$\psi = 0, \quad \theta = 0, \quad \tilde{\mathbf{j}}_p = 0 \left(\text{or } Nb\frac{\partial \phi}{\partial x} + Nt\frac{\partial \theta}{\partial x} = 0\right) \tag{7.12}$$
$$\text{on } x = x_2$$

$$\psi = 0, \quad \frac{\partial \theta}{\partial y} = 0, \quad \frac{\partial \phi}{\partial y} = 0 \text{ on } y = 0 \text{ and } y = A.$$

Taking into account the considered dimensionless variables (7.7), the vertical walls of the cavity are described by the following relations:

$$x_1 = 1 - a - b\cos(2\pi\kappa y/A) \text{ and } x_2 = 1, \text{ respectively,}$$

where $\Delta = x_2 - x_1 = a + b\cos(2\pi\kappa y/A)$ is the distance between vertical walls.

3. Numerical Solution

For numerical simulation of the formulated equations (7.8)–(7.11), we introduce new independent variables ξ and η:

$$\xi = \frac{x - x_1}{\Delta} = \frac{x - 1 + a + b\cos(2\pi\kappa y/A)}{a + b\cos(2\pi\kappa y/A)}, \quad \eta = y. \tag{7.13}$$

Taking into account these variables (7.13), the governing equations (7.8)–(7.13) are rewritten in the following form:

$$\left[\left(\frac{\partial\xi}{\partial x}\right)^2 + \left(\frac{\partial\xi}{\partial y}\right)^2\right]\frac{\partial^2\psi}{\partial\xi^2} + 2\frac{\partial\xi}{\partial y}\frac{\partial^2\psi}{\partial\xi\partial\eta} + \frac{\partial^2\psi}{\partial\eta^2} + \frac{\partial^2\xi}{\partial y^2}\frac{\partial\psi}{\partial\xi} = -\omega \tag{7.14}$$

$$\frac{\partial\omega}{\partial\tau} + \frac{\partial\xi}{\partial x}\frac{\partial\psi}{\partial\eta}\frac{\partial\omega}{\partial\xi} - \frac{\partial\xi}{\partial x}\frac{\partial\psi}{\partial\xi}\frac{\partial\omega}{\partial\eta}$$

$$= Pr\left\{\left[\left(\frac{\partial\xi}{\partial x}\right)^2 + \left(\frac{\partial\xi}{\partial y}\right)^2\right]\frac{\partial^2\omega}{\partial\xi^2} + 2\frac{\partial\xi}{\partial y}\frac{\partial^2\omega}{\partial\xi\partial\eta} + \frac{\partial^2\omega}{\partial\eta^2} + \frac{\partial^2\xi}{\partial y^2}\frac{\partial\omega}{\partial\xi}\right\}$$

$$+ Ra\cdot Pr\left(\frac{\partial\xi}{\partial x}\frac{\partial\theta}{\partial\xi} - Nr\frac{\partial\xi}{\partial x}\frac{\partial\phi}{\partial\xi}\right) + Ha^2\cdot Pr\left\{\left(\frac{\partial\xi}{\partial x}\right)^2\frac{\partial^2\psi}{\partial\xi^2}\cos^2(\varphi)\right.$$

$$+ 2\left(\frac{\partial\xi}{\partial x}\frac{\partial\xi}{\partial y}\frac{\partial^2\psi}{\partial\xi^2} + \frac{\partial\xi}{\partial x}\frac{\partial^2\psi}{\partial\xi\partial\eta} + \frac{\partial^2\xi}{\partial x\partial y}\frac{\partial\psi}{\partial\xi}\right)\sin(\varphi)\cos(\varphi)$$

$$+ \left.\left[\left(\frac{\partial\xi}{\partial y}\right)^2\frac{\partial^2\psi}{\partial\xi^2} + 2\frac{\partial\xi}{\partial y}\frac{\partial^2\psi}{\partial\xi\partial\eta} + \frac{\partial^2\psi}{\partial\eta^2} + \frac{\partial^2\xi}{\partial y^2}\frac{\partial\psi}{\partial\xi}\right]\sin^2(\varphi)\right\} \tag{7.15}$$

$$\frac{\partial\theta}{\partial\tau} + \frac{\partial\xi}{\partial x}\frac{\partial\psi}{\partial\eta}\frac{\partial\theta}{\partial\xi} - \frac{\partial\xi}{\partial x}\frac{\partial\psi}{\partial\xi}\frac{\partial\theta}{\partial\eta} = \left[\left(\frac{\partial\xi}{\partial x}\right)^2 + \left(\frac{\partial\xi}{\partial y}\right)^2\right]\frac{\partial^2\theta}{\partial\xi^2}$$

$$+ 2\frac{\partial\xi}{\partial y}\frac{\partial^2\theta}{\partial\xi\partial\eta} + \frac{\partial^2\theta}{\partial\eta^2} + \frac{\partial^2\xi}{\partial y^2}\frac{\partial\theta}{\partial\xi} + Nb\left\{\left[\left(\frac{\partial\xi}{\partial x}\right)^2 + \left(\frac{\partial\xi}{\partial y}\right)^2\right]\frac{\partial\phi}{\partial\xi}\frac{\partial\theta}{\partial\xi}\right.$$

$$+ \frac{\partial\xi}{\partial y}\left[\frac{\partial\phi}{\partial\xi}\frac{\partial\theta}{\partial\eta} + \frac{\partial\theta}{\partial\xi}\frac{\partial\phi}{\partial\eta}\right] + \left.\frac{\partial\theta}{\partial\eta}\frac{\partial\phi}{\partial\eta}\right\}$$

$$+ Nt\left[\left(\frac{\partial\xi}{\partial x}\right)^2\left(\frac{\partial\theta}{\partial\xi}\right)^2 + \left(\frac{\partial\xi}{\partial y}\right)^2\left(\frac{\partial\theta}{\partial\xi}\right)^2\right.$$

$$+ \left.2\frac{\partial\xi}{\partial y}\frac{\partial\theta}{\partial\xi}\frac{\partial\theta}{\partial\eta} + \left(\frac{\partial\theta}{\partial\eta}\right)^2\right] \tag{7.16}$$

$$\frac{\partial \phi}{\partial \tau} + \frac{\partial \xi}{\partial x}\frac{\partial \psi}{\partial \eta}\frac{\partial \phi}{\partial \xi} - \frac{\partial \xi}{\partial x}\frac{\partial \psi}{\partial \xi}\frac{\partial \phi}{\partial \eta} = \frac{1}{Le}\left\{\left[\left(\frac{\partial \xi}{\partial x}\right)^2 + \left(\frac{\partial \xi}{\partial y}\right)^2\right]\frac{\partial^2 \phi}{\partial \xi^2}\right.$$

$$\left. + 2\frac{\partial \xi}{\partial y}\frac{\partial^2 \phi}{\partial \xi \partial \eta} + \frac{\partial^2 \phi}{\partial \eta^2} + \frac{\partial^2 \xi}{\partial y^2}\frac{\partial \phi}{\partial \xi}\right\} + \frac{Nt}{Le \cdot Nb}\left\{\left[\left(\frac{\partial \xi}{\partial x}\right)^2 + \left(\frac{\partial \xi}{\partial y}\right)^2\right]\frac{\partial^2 \theta}{\partial \xi^2}\right.$$

$$\left. + 2\frac{\partial \xi}{\partial y}\frac{\partial^2 \theta}{\partial \xi \partial \eta} + \frac{\partial^2 \theta}{\partial \eta^2} + \frac{\partial^2 \xi}{\partial y^2}\frac{\partial \theta}{\partial \xi}\right\}. \tag{7.17}$$

The initial and boundary conditions for these equations can be written as follows:

$$\psi = 0, \quad \theta = 0.5, \quad \phi = 1 \text{ for } \tau = 0$$

$$\psi = 0, \quad \theta = 1.0, \quad Nb\frac{\partial \phi}{\partial \xi} + Nt\frac{\partial \theta}{\partial \xi} = 0 \text{ on } \xi = 0$$

$$\psi = 0, \quad \theta = 0.0, \quad Nb\frac{\partial \phi}{\partial \xi} + Nt\frac{\partial \theta}{\partial \xi} = 0 \text{ on } \xi = 1 \tag{7.18}$$

$$\psi = 0, \quad \frac{\partial \theta}{\partial \eta} = 0, \quad \frac{\partial \phi}{\partial \eta} = 0 \text{ on } \eta = 0$$

$$\psi = 0, \quad \frac{\partial \theta}{\partial \eta} = 0, \quad \frac{\partial \phi}{\partial \eta} = 0 \text{ on } \eta = A.$$

The local and average heat and mass transfer rates are defined as

$$Nu = -\left(\frac{\partial \theta}{\partial \xi}\right)_{\xi=0}, \quad Sh = -\left(\frac{\partial \phi}{\partial \xi}\right)_{\xi=0},$$

$$\overline{Nu} = \frac{1}{A}\int_0^A Nu\,d\eta, \quad \overline{Sh} = \frac{1}{A}\int_0^A Sh\,d\eta. \tag{7.19}$$

It is worth noting that for an analysis of the Sherwood number along the wavy wall, it is possible to study only Nusselt number along this wall, because at the wavy wall we have

$$\frac{\partial \phi}{\partial \xi} = -\frac{Nt}{Nb}\frac{\partial \theta}{\partial \xi}$$

taking into account boundary conditions for ϕ (Eqs. (7.18)). Therefore the further analysis concerning integral parameters will be conducted only for the average Nusselt number, because

$$Sh = -\frac{Nt}{Nb}Nu \text{ and } \overline{Sh} = -\frac{Nt}{Nb}\overline{Nu}.$$

The partial differential equations (7.14)–(7.17) with corresponding boundary conditions (7.18) were solved using the finite difference method of the second order accuracy. A detailed description of the numerical technique used is presented by Sheremet and Trifonova (2013), Sheremet and Pop (2014a–c, 2015a–e), Sheremet et al. (2014, 2015a–g, 2016a,b).

The present models, in the form of an in-house computational fluid dynamics (CFD) code, have been validated successfully against the work by Al-Najem et al. (1998), Sarris et al. (2006), and Pirmohammadi and Ghassemi (2009) for steady-state MHD natural convection in a square cavity for Rayleigh numbers $7 \cdot 10^3$ and $7 \cdot 10^5$, and for Hartmann numbers in the range 0 to 100. Figures 7.2 and 7.3 show a good agreement between the obtained streamlines and isotherms for different Rayleigh and Hartmann numbers and the numerical results of Sarris et al. (2006), and Pirmohammadi and Ghassemi (2009).

The comparison between the results of the present model and those by Al-Najem et al. (1998) and Sarris et al. (2006) are shown in Table 7.1.

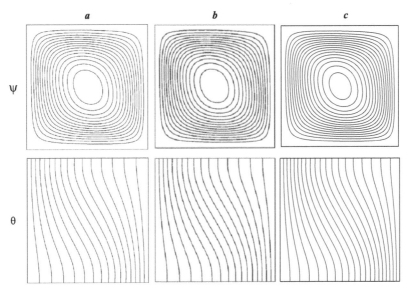

Fig. 7.2. Comparison of streamlines Ψ and isotherms Θ at $Ra = 7 \cdot 10^3$, $Ha = 25$: numerical data of Sarris et al. (2006) (a), numerical data of Pirmohammadi and Ghassemi (2009) (b), present results (c) [reprinted from Pirmohammadi and Ghassemi (2009) with permission from Elsevier]

Table 7.1. Comparison of present calculations with those
of Al-Najem et al. (1998) and Sarris et al. (2006)

Authors	$Ha = 10$		$Ha = 50$	
	$U_{c,\max}$	Nu	$U_{c,\max}$	Nu
Al-Najem et al. (1998)	0.135	1.75	0.026	1.05
Sarris et al (2006)	0.138	1.738	0.025	1.022
Present results	0.138	1.683	0.026	1.019

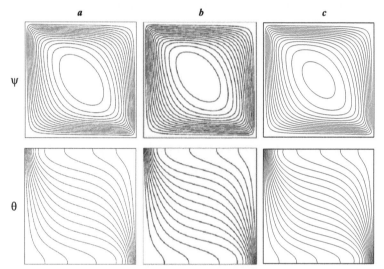

Fig. 7.3. Comparison of streamlines Ψ and isotherms Θ at
$Ra = 7 \cdot 10^5$, $Ha = 100$: numerical data of Sarris et al. (2006) (a),
numerical data of Pirmohammadi and Ghassemi (2009) (b),
present results (c) [reprinted from Pirmohammadi and Ghassemi
(2009) with permission from Elsevier]

The maximum horizontal velocity in the midsection ($U_{c,\max}$) and
the average Nusselt number are practically the same for every value of
Ha for the case of $Ra = 7 \cdot 10^3$.

For the purpose of obtaining the grid-independent solution, a grid
sensitivity analysis is performed. The grid-independent solution was
performed by preparing the solution for unsteady natural convection
in a wavy-walled cavity filled with a nanofluid at $Ra = 10^5$, $Pr = 6.26$,
$Le = 10$, $Nr = Nb = Nt = 0.1$, $Ha = 20$, $A = 1$, $\kappa = 3$, $a = 0.8$,
$\varphi = 0$. Four cases of the uniform grid are tested: a grid of 100×100
points, a grid of 150×150 points, a grid of 200×200 points, and a
much finer grid of 250×250 points. Figure 7.4 shows an effect of the
mesh on the average Nusselt number of the wavy wall.

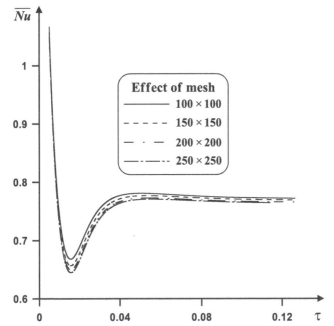

Fig. 7.4. Variation of the average Nusselt number versus
the dimensionless time and the mesh parameters

On the basis of the conducted verifications, the uniform grid of 150×150 points has been selected for the following analysis.

4. Results and Discussion

Numerical study has been conducted at the following values of the governing parameters: Rayleigh number ($Ra = 10^5$), Lewis number ($Le = 10$), Prandtl number ($Pr = 6.26$), buoyancy ratio parameter ($Nr = 0.1$), Brownian motion parameter ($Nb = 0.1$), thermophoresis parameter ($Nt = 0.1$), aspect ratio parameter ($A = 1$), Hartmann number ($Ha = 0 - 100$), undulation number ($\kappa = 1 - 3$), wavy contraction ratio ($b = 0.1 - 0.3$), inclination angle of the external magnetic field vector ($\varphi = 0 - \pi$), dimensionless time ($\tau = 0 - 0.13$). Particular efforts have been focused on the effects of the Hartmann number, inclination angle of the external magnetic field vector, undulation number, wavy contraction ratio, and dimensionless time on the fluid flow and heat transfer. Streamlines, isotherms, isoconcentrations, and average Nusselt number at the left wavy wall for different values of key parameters mentioned above are illustrated in Figs. 7.5–7.13.

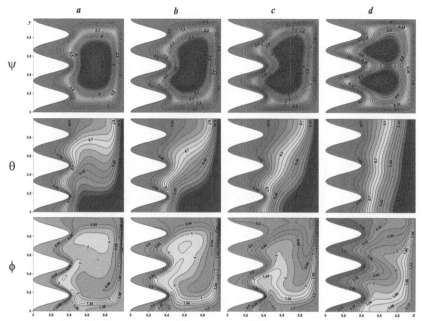

Fig. 7.5. Streamlines ψ, isotherms θ, isoconcentrations ϕ for $\kappa = 3.0$, $b = 0.2$, $\varphi = 0$, $\tau = 0.13$: $Ha = 0$ (a), $Ha = 30$ (b), $Ha = 50$ (c), $Ha = 100$ (d)

Figure 7.5 shows contours of the stream function ψ, temperature θ, and nanoparticle volume fraction ϕ for $\kappa = 3.0$, $b = 0.2$, $\varphi = 0$, $\tau = 0.13$, and different values of the Hartmann number. In the case of $Ha = 0$, when the external magnetic field is absent, one can find inside the cavity a convective cell of a single core.

This motion characterizes a formation of ascending flows along the hot wavy wall and descending flows near the right vertical wall. It should be noted that the dominant heat transfer mechanism is a heat convection due to the significant circulation of the fluid $|\psi|_{\max}^{Ha=0} = 10.97$. Such a convective heat transfer regime is also reflected in the isotherms distributions. Taking into account the contours of nanoparticle volume fraction, it is possible to consider the distribution of nanoparticles as weakly homogeneous. It is worth noting that the area between isoconcentrations $\phi = 0.98$ and $\phi = 1.02$ in the central part of the cavity refers to the area where the nanoparticle volume fraction values belong to the neighborhood of the value $\phi = 1$. When this area occupies an essential part of the cavity, the distribution of nanoparticles has to be considered as homogeneous. On the other hand, when this area occupies a negligible part of the enclosure, the distribution of the nanoparticles can be considered as non-homogeneous. A presence of a wavy wall leads to a distortion of the streamlines and more essential heating of the wavy troughs. The latter is due to a weak fluid flow in these parts. A presence of the external magnetic field of low intensity $Ha = 30$ and $\varphi = 0$ (see Fig. 7.5b) first of all leads to modification of streamlines and isotherms.

One can find in this regime a displacement of the convective cell core to the bottom part of the cavity with an essential attenuation of the convective flow $|\psi|_{\mathrm{max}}^{Ha=30} = 5.01$. At the same time, isotherms begin to flatten due to an intensification of conduction heat transfer. Isoconcentrations reflect insignificant changes in the distributions of nanoparticle volume fraction. Further increase in the Hartmann number ($Ha = 50$) leads to a formation of the convective cell with two cores of low intensity, such as $|\psi|_{\mathrm{max}}^{Ha=30} = 2.93$ for the bottom core and $|\psi|_{\mathrm{max}}^{Ha=30} = 2.73$ for the upper core. It is interesting to note that the locations of these cores correspond to the locations of the central wavy troughs. Isotherms become less curved and more parallel to the isothermal vertical walls. The conduction regime enhances the effect of the thermophoresis phenomenon (Celli, 2013) and the nanoparticles distribution here becomes less homogeneous. For high value of the Hartmann number ($Ha = 100$) a double-core convective cell is formed in the cavity where these cores are located opposite the wavy troughs. Also, a presence of a horizontal magnetic field leads to an elongation of these cores along the horizontal axis. Distributions of nanoparticle volume fraction can be considered as non-homogeneous for the developed heat conduction regime. An effect of the dimensionless time and Hartmann number on the average Nusselt number at a vertical hot wavy wall (see Eq. (7.19)) is depicted in Figure 7.6.

An increase in the Hartmann number leads to a decrease in the average Nusselt number. Taking into account time dependence for $\overline{Nu}$, it is possible to indicate the presence of three time levels. The first time level (initial level) is a heat conduction regime that characterizes a decrease in the average Nusselt number owing to a heating of the surrounding fluid through the heat conduction mechanism. As $\overline{Nu}$ approaches local minimum value, the first time level is completed and the second time level is begun. The second time level can be considered as a convection level due to an intensive fluid flow that leads to an intensive heat removal from the isothermal wavy wall, and as a result, to an increase in the average Nusselt number. As $\overline{Nu}$ approaches the constant value the second time level is completed.

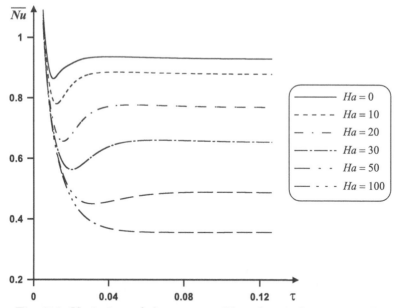

Fig. 7.6. Variation of the average Nusselt number versus the
dimensionless time and Hartmann number for $\kappa = 3.0$, $b = 0.2$, $\varphi = 0$

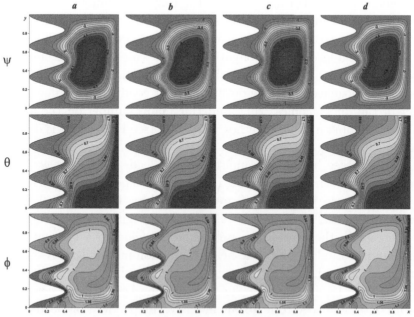

Fig. 7.7. Streamlines ψ, isotherms θ, isoconcentrations ϕ for $Ha = 20$,
$\kappa = 3.0$, $b = 0.2$, $\tau = 0.13$: $\varphi = 0$ (a), $\varphi = \pi/4$ (b), $\varphi = \pi/2$ (c),
$\varphi = \pi$ (d)

The third time level can be considered a steady-state regime, with constant value of the average Nusselt number. As has been mentioned above, an increase in the Hartmann number leads to an intensification of conduction heat transfer, and as a result one can find a formation of long-lasting initial time level with small convective level. In the case of $Ha = 100$ the second time level has vanished due to domination of the heat conduction regime.

Figure 7.7 illustrates streamlines, isotherms, and isoconcentrations at different values of the inclination angle of the external magnetic field vector for $Ha = 20$, $\kappa = 3.0$, $b = 0.2$, $\tau = 0.13$. An increase in the inclination angle from 0 up to $\pi/2$ leads to a formation of an intensive convective cell ($|\psi|_{\max}^{\varphi=0} = 6.8 < |\psi|_{\max}^{\varphi=\pi/4} = 8.26 < |\psi|_{\max}^{\varphi=\pi/2} = 8.39$) with the stable core. At the same time, one can find less intensive heating of the cavity, taking into account the location of the isotherm $\theta = 0.95$ and more homogeneous distribution of the nanoparticle volume fraction, taking into account the location of the isoline $\phi = 0.94$. An increase in the inclination angle from $\pi/2$ to π leads to an attenuation of convective flow and domination of the conduction heat transfer. An effect of the dimensionless time and the inclination angle of the external magnetic field vector on the average Nusselt number at the vertical hot wavy wall is depicted in Figure 7.8. It is necessary to note that an increase in φ from 0 up to $\pi/2$ leads to an increase in the average Nusselt number, while an increase in φ from $\pi/2$ up to π leads to a decrease in $\overline{Nu}$. Also, it is possible to conclude that an increase in the average Nusselt number is attended with a decrease in the duration of the first heat conduction time level and an increase in the duration of the second heat convection time level. More intensive growth of the average Nusselt number is in a range from $\varphi = \pi/6$ to $\varphi = \pi/3$.

Figure 7.9 demonstrates streamlines, isotherms, and isoconcentrations at different values of the wavy contraction ratio for $Ha = 20$, $\varphi = 0$, $\kappa = 3.0$, $\tau = 0.13$. An increase in the wavy contraction ratio leads to an increase in the wave amplitude and a decrease in the fluid flow cavity. Such reduction of the cavity is reflected in an attenuation of the convective flow ($|\psi|_{\max}^{b=0.1} = 7.24 > |\psi|_{\max}^{b=0.2} = 6.8 > |\psi|_{\max}^{b=0.3} = 5.67$) and more intensive heating of the wavy troughs. Moreover, an increase in the wavy amplitude leads to formation of a double-core convective cell due to the obstacle influence. At the same time, isotherms become less curved and reflect a domination of conduction heat transfer in a narrow cavity.

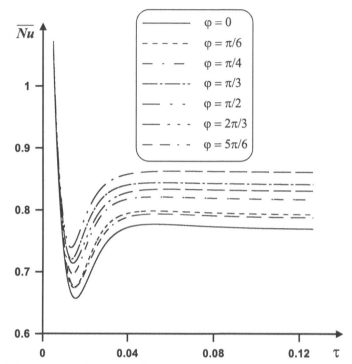

Fig. 7.8. Variation of the average Nusselt number versus
the dimensionless time and inclination angle of the external
magnetic field vector for $Ha = 20$, $\kappa = 3.0$, $b = 0.2$

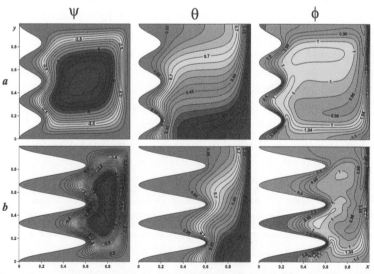

Fig. 7.9. Streamlines ψ, isotherms θ, isoconcentrations ϕ for $Ha = 20$,
$\varphi = 0$, $\kappa = 3.0$, $\tau = 0.13$: $b = 0.1$ (a), $b = 0.3$ (b)

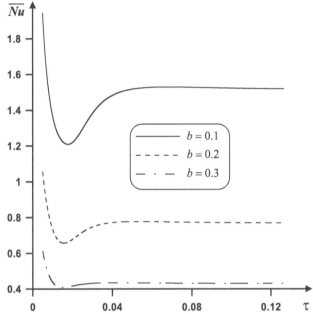

Fig. 7.10. Variation of the average Nusselt number versus the
dimensionless time and wavy contraction ratio for $Ha = 20$, $\varphi = 0$,
$\kappa = 3.0$

Changes in the nanoparticle volume fraction can be described by a
variation of the domain of interest where a weakly homogeneous mode
is presented.

An effect of the dimensionless time and wavy contraction ratio on
the average Nusselt number at the vertical hot wavy wall is shown in
Figure 7.10. An increase in b leads to a decrease in $\overline{Nu}$ due to more
intensive heating of the wavy troughs, and as a result, less intensive
heat removal from the hot wall. Moreover, an increase in the wavy
amplitude leads to a decrease in the duration of the second convective
time level.

Figure 7.11 shows streamlines, isotherms, and isoconcentrations at
different values of the undulation number for $Ha = 20$, $\varphi = 0$, $b = 0.2$,
$\tau = 0.13$. An increase in the undulation number from 1 to 2 leads to an
intensification of convective flow ($|\psi|_{\max}^{\kappa=1} = 6.47 < |\psi|_{\max}^{\kappa=2} = 7.25$). Such
a result can be explained by the presence of the wavy trough in the
central part of the cavity that reflects a large space for the convective
core. Distributions of isotherms characterize an essential heating of the
central and upper wavy troughs.

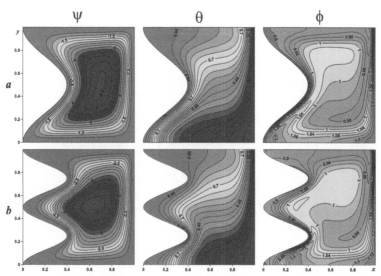

Fig. 7.11. Streamlines ψ, isotherms θ, isoconcentrations ϕ for
$Ha = 20$, $\varphi = 0$, $b = 0.2$, $\tau = 0.13$: $\kappa = 1.0$ (a), $\kappa = 2.0$ (b)

An effect of the Hartmann number and undulation number on the
average Nusselt number at the vertical hot wavy wall is shown in Figure
7.12. An increase in κ leads to a decrease in $\overline{Nu}$ due to more intensive
heating of the wavy troughs. It should be noted that an essential decrease in the average Nusselt number with κ is for $Ha = 0$. An increase
in Ha leads to a reduction of the decreasing rate of the average Nusselt number. An effect of the inclination angle of the external magnetic
field and undulation number on the average Nusselt number at the vertical hot wavy wall is shown in Figure 7.13. An increase in φ leads to
non-monotonic variation of $\overline{Nu}$. It is interesting to note that for $\kappa = 2$
and $\kappa = 3$, the average Nusselt number is an increasing function of the
inclination angle in a range from 0 to $\pi/2$ and a decreasing function in
a range from $\pi/2$ to π.

While for $\kappa = 1$ the average Nusselt number is an increasing function of the inclination angle in a range from 0 to $\pi/3$, it is a decreasing
function in a range from $\pi/2$ to $5\pi/6$.

Based on the findings in this study, we conclude the following:

(1) An increase in the Hartmann number leads to an attenuation
of convective flow and heat transfer and a formation of a double-core
convective cell for high values of Ha. Also, an increase in the Hartmann
number leads to a formation of long-lasting initial time level with small
convective level.

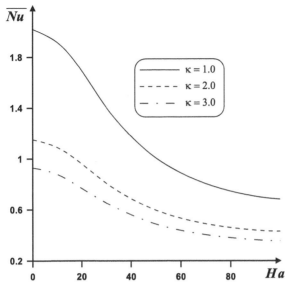

Fig. 7.12. Variation of the average Nusselt number versus the
Hartmann number and undulation number for $\varphi = 0$, $b = 0.2$,
$\tau = 0.13$

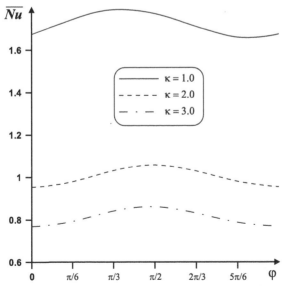

Fig. 7.13. Variation of the average Nusselt number versus
the inclination angle of the external magnetic field vector
and undulation number for $Ha = 20$, $b = 0.2$, $\tau = 0.13$

(2) An increase in the inclination angle from 0 up to $\pi/2$ leads to a formation of an intensive convective cell with a stable core. The average Nusselt number at a vertical hot wavy wall is an increasing function of φ $(0 < \varphi < \pi/2)$ and a decreasing function of φ $(\pi/2 < \varphi < \pi)$ for $\kappa = 2$ and $\kappa = 3$.

(3) An increase in the wavy contraction ratio leads to an increase in the wave amplitude and an attenuation of the convective flow with more intensive heating of the wavy troughs where heat conduction is a dominating heat transfer regime.

(4) An increase in the undulation number leads to a decrease in the average Nusselt number due to more intensive heating of the wavy troughs, while an increase in the Hartmann number leads to a reduction of the decreasing rate of the average Nusselt number.

B. Analysis of Entropy Generation in Natural Convection of Nanofluid inside a Wavy-Walled Cavity with a Non-Uniform Heating: Tiwari and Das' Model

1. Introduction

The second law of thermodynamics analysis is used to predict the performance and also the conditions of engineering processes. This law is more reliable than the first law of thermodynamics because of the limitation of the efficiency of the first law in heat transfer engineering systems (Öztop and Al-Salem, 2012). The second law is applied to achieve the optimal design of thermal systems by minimizing the irreversibility and entropy generation, which can improve the efficiency of the systems. The contemporary trend in the field of heat transfer and thermal design is the second law (of thermodynamics) analysis and its design-related concept of entropy generation and its minimization. This trend is important and, at the same time, necessary, if the heat transfer community is to contribute to a viable engineering solution to the energy problems. Entropy generation is associated with thermodynamic irreversibility, which is present in all types of heat transfer processes. Different sources of irreversibility are responsible for heat transfer's generation of entropy, like heat transfer across finite temperature gradients, characteristics of convective heat transfer, viscous effects, etc. Entropy generation is a criterion of the destruction of the available system work. The evaluation of the entropy generation is carried out to improve system performance. Heat transfer, mass transfer, viscous dissipation, etc., can be used as sources of entropy generation. In many engineering and industrial processes, entropy production destroys the available energy in the system. It is, therefore, very important to determine the rate of entropy generation in a system, in order to optimize the energy and efficient operation in the system. According to the second law of thermodynamics, all the flow and heat transfer processes undergo changes that are irreversible. This process causes the entropy of the system to increase. For a detailed analysis of the entropy generation process, see Woods (1975), Sayegh and Vera (1980), Bejan (1980, 1982, 1994, 1996), Baytas (2000), Krishna et al. (2002), Mahmud and Fraser (2003), Mahmud and Sadrul Islam (2003), Cho (2014), and Wenming et al. (2015).

In recent years, many papers have been published on the applications and entropy generation rates of the second law of thermodynamics. Odat et al. (2004) have explored the entropy generation effects in the laminar flow past a flat plate under the influence of the magnetic field. It was found that the entropy generation increases by increasing Hartmann number, Eckert number, and Joule heating parameter. Arikoglu et al. (2008) studied the effect of slip on the entropy generation in MHD flow over a rotating disk. Saouli and Aiboud-Saouli (2004) investigated second law analysis of laminar falling liquid film along an inclined heated plate. Aiboud and Saouli (2010) dealt with the analysis of entropy for viscoelastic magneto hydrodynamic flow over a stretching surface. An analysis of the boundary layer entropy generation past a flat plate has been done by Esfahani and Jafarian (2005).

In a series of papers, Makinde (2006, 2010, 2011, 2012) studied the gravity-driven non-Newtonian liquid film along an inclined isothermal plate, the entropy generation on magnetohydrodynamic flow, and heat transfer over a flat plate with a convective boundary condition. He has also studied a variable viscosity boundary layer flow over a flat plate under the effects of thermal radiation and Newtonian heating, as well the entropy generation effects in this flow. Butt and Ali (2012, 2014a,b) have illustrated the effects of magnetic field on entropy generation in the flow and heat transfer due to a radially stretching surface, and flow and heat transfer caused by a moving surface. The effect of viscous dissipation and thermal radiation on entropy generation in Blasius flow has been displayed numerically by Butt et al. (2012a,b,c). Finally, there is the paper by Rashidi el al. (2013) where the entropy generation in steady MHD flow due to a rotating porous disk in a nanofluid has been analyzed. The results showed that as the thermal radiation parameter increases, the produced entropy decreases. Papers by Cîmpean et al. (2008) and Chen et al. (2011) on the entropy generation in a vertical channel are also worth noting.

Most of the studies on convective flow of viscous fluids and flows in porous media have used the base fluid with a low thermal conductivity, which, in turn, limits the heat transfer enhancement. An innovative technique to enhance heat transfer by using nano-scale particles in the base fluid has been proposed by Choi (1995), and shows good promise in significantly increasing heat transfer rate. In an illuminating paper by Khanafer et al. (2003), the heat transfer enhancement in a two-dimensional enclosure utilizing nanofluids for various pertinent parameters, and taking into account the solid particle dispersion, has been investigated. In this paper, the effect of suspended ultrafine metallic nanoparticles on the fluid flow and heat transfer processes is analyzed and effective thermal conductivity enhancement maps are developed for various controlling parameters. It is shown that the variances within different models have substantial effects on the results. A heat transfer correlation of the average Nusselt number for various Grashof numbers and volume fractions is also presented.

Jou and Tzeng (2006) numerically investigated the heat transfer performance of nanofluids inside the two-dimensional rectangular enclosures and found that increasing the volume fraction causes a significant increase in average heat transfer rate. Heat transfer augmentation in a two-sided lid-driven differentially heated square cavity utilizing nanofluids has been studied by Tiwari and Das (2007) using a simpler nanofluid model than that proposed by Khanafer et al. (2003).

Very good literature reviews on convective flow and applications of nanofluids have been done by Buongiorno (2006), Das et al. (2007), CEA (2007), Daungthongsuk and Wongwises (2007), Chopkar et al. (2008), Kleinstreuer et al. (2008), Das and Choi (2009), Kakaç and Pramuanjaroenkij (2009), Habibzadeh et al. (2010), Wong and Leon (2010), Abu-Nada and Chamkha (2010, 2014), Mansour et al. (2010), Nemati et al. (2010), Alinia et al. (2011), Wen et al. (2011), Arefmanesh and Mahmoodi (2011), Aminossadati et al. (2012), Chamkha and Abu-Nada (2012), Jaluria et al. (2012), Mahian et al. (2013), Moghaddam et al. (2013), Nield and Bejan (2013), and Sreeremya et al. (2014). Flow and heat transfer in wavy geometries are encountered in many engineering systems as a means of enhancing the transport performance (see Rahman, 2001; Cho et al., 2012a,b,c,d). On the other hand, the heat transfer characteristics of nanofluids confined in cavities with wavy surfaces have attracted increasing attention in recent times. Several authors, such as Abu-Nada and Öztop (2009), Nikfar and Mahmoodi (2012), and Cho et al. (2013a,b; 2014) have performed numerical studies on natural convection heat transfer characteristics of Al_2O_3-water nanofluid in a cavity with wavy walls, while Esmaeilpour and Abdollahzadeh (2012) examined the natural convection heat transfer behavior and entropy generation of Cu–water nanofluid confined in a cavity with vertical wavy walls. Motivated by these studies, the objective of the present work is to numerically study the entropy generation in natural convection of nanofluid inside a wavy-walled cavity with a non-uniform heating using the mathematical nanofluid model proposed by Tiwari and Das (2007). A systematic study of the effects of the various pertinent parameters on flow and heat transfer characteristics is carried out with the help of graphs and tables.

2. Basic Equations

Consider the natural convection in a nanofluid based on water and solid nanoparticles located in a cavity with left wavy, right, bottom and top flat solid walls. A schematic geometry of the problem under investigation is shown in Fig. 7.14, where the $\bar{x}$ axis is measured in the horizontal direction along the lower wall of the cavity and the $\bar{y}$ axis is measured along the left vertical wavy wall of the cavity, while L is the bottom wall length, and H is the height of the vertical wall. It is assumed that the left vertical wall is heated sinusoidally

$$T_w(\bar{y}) = T_c + \frac{\Delta T}{2}\left(1 - \cos\left(\frac{2\pi\bar{y}}{H}\right)\right), \qquad (7.20)$$

where ΔT is the temperature difference between the maximum and minimum temperatures, while the right vertical wall is kept at a temperature T_c. It is also assumed that the top and bottom walls of the cavity are adiabatic. The walls of the cavity are assumed to be impermeable. It is considered that the left wavy wall and right flat wall of the cavity are described by the relations

$$\bar{x}_1 = L - L[a + b\cos(2\pi\kappa\bar{y}/H)] \text{ and } \bar{x}_2 = L, \text{ respectively}, \qquad (7.21)$$

where $\overline{\Delta} = \bar{x}_2 - \bar{x}_1 = L[a + b\cos(2\pi\kappa\bar{y}/H)]$ is the distance between vertical walls.

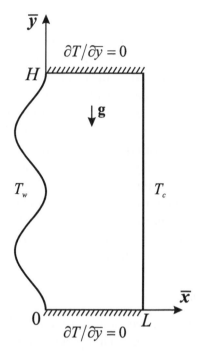

Fig. 7.14. Physical model and coordinate system

Except for the density, the properties of the fluid are taken to be constant. It is further assumed that the effect of buoyancy is included through the Boussinesq approximation. The viscous, radiation, and Joule heating effects are neglected. Under the above assumptions, the conservation equations for mass, momentum, and thermal energy can be written as follows:

$$\nabla \cdot \mathbf{V} = 0 \tag{7.22}$$

$$\rho_{nf}\left[\frac{\partial \mathbf{V}}{\partial t} + (\mathbf{V}, \nabla)\mathbf{V}\right] = -\nabla p + \mu_{nf}\nabla^2\mathbf{V} - (\rho\beta)_{nf}(T - T_c)\mathbf{g} \tag{7.23}$$

$$(\rho C_p)_{nf}\left[\frac{\partial T}{\partial t} + (\mathbf{V}, \nabla)T\right] = k_{nf}\left(\frac{\partial^2 T}{\partial \bar{x}^2} + \frac{\partial^2 T}{\partial \bar{y}^2}\right). \tag{7.24}$$

The used physical properties of the nanofluid were described in detail previously (see Sheremet et al., 2015a–g).

Equations (7.22)–(7.24) can be written in Cartesian coordinates as

$$\frac{\partial \bar{u}}{\partial \bar{x}} + \frac{\partial \bar{v}}{\partial \bar{y}} = 0 \tag{7.25}$$

$$\rho_{nf}\left(\frac{\partial \bar{u}}{\partial t} + \bar{u}\frac{\partial \bar{u}}{\partial \bar{x}} + \bar{v}\frac{\partial \bar{u}}{\partial \bar{y}}\right) = -\frac{\partial p}{\partial \bar{x}} + \mu_{nf}\left(\frac{\partial^2 \bar{u}}{\partial \bar{x}^2} + \frac{\partial^2 \bar{u}}{\partial \bar{y}^2}\right) \tag{7.26}$$

$$\rho_{nf}\left(\frac{\partial \overline{v}}{\partial t} + \overline{u}\frac{\partial \overline{v}}{\partial \overline{x}} + \overline{v}\frac{\partial \overline{v}}{\partial \overline{y}}\right)$$

$$= -\frac{\partial p}{\partial \overline{y}} + \mu_{nf}\left(\frac{\partial^2 \overline{v}}{\partial \overline{x}^2} + \frac{\partial^2 \overline{v}}{\partial \overline{y}^2}\right) + (\rho\beta)_{nf}g(T - T_c) \qquad (7.27)$$

$$\frac{\partial T}{\partial t} + \overline{u}\frac{\partial T}{\partial \overline{x}} + \overline{v}\frac{\partial T}{\partial \overline{y}} = \alpha_{nf}\left(\frac{\partial^2 T}{\partial \overline{x}^2} + \frac{\partial^2 T}{\partial \overline{y}^2}\right). \qquad (7.28)$$

The existence of irreversibility sources in the flow field, such as the viscous dissipation effect and heat transfer, causes the entropy generation. In Cartesian coordinates, the dimensional local entropy generation $\overline{S}_{gen}$ determined by Woods can be expressed as follows in the present study:

$$\overline{S}_{gen} = \frac{k_{nf}}{T_0^2}\left[\left(\frac{\partial T}{\partial \overline{x}}\right)^2 + \left(\frac{\partial T}{\partial \overline{y}}\right)^2\right]$$

$$+\frac{\mu_{nf}}{T_0}\left[2\left(\frac{\partial u}{\partial \overline{x}}\right)^2 + 2\left(\frac{\partial v}{\partial \overline{y}}\right)^2 + \left(\frac{\partial u}{\partial \overline{y}} + \frac{\partial v}{\partial \overline{x}}\right)^2\right] \qquad (7.29)$$

where $T_0 = T_c + \frac{\Delta T}{2}$.

Equation (7.29) constitutes two terms: The first is the local entropy generation due to the heat transfer $(\overline{S}_{gen,ht})$; and the second is the dimensional local entropy generation due to fluid friction $(\overline{S}_{gen,ff})$.

Further, we introduce the following dimensionless variables:

$$x = \overline{x}/L, \; y = \overline{y}/L, \; \tau = t\alpha_{nf}/L^2, \; u = \overline{u}L/\alpha_{nf},$$
$$v = \overline{v}L/\alpha_{nf}, \; \theta = (T - T_c)/\Delta T. \qquad (7.30)$$

One can introduce a dimensionless stream function ψ and vorticity ω defined by

$$u = \frac{\partial \psi}{\partial y}, \; v = -\frac{\partial \psi}{\partial x}, \; \omega = \frac{\partial v}{\partial x} - \frac{\partial u}{\partial y} \qquad (7.31)$$

so that Eq. (7.25) is satisfied identically. We are then left with the following equations:

$$\frac{\partial^2 \psi}{\partial x^2} + \frac{\partial^2 \psi}{\partial y^2} = -\omega \qquad (7.32)$$

$$\frac{\partial \omega}{\partial \tau} + \frac{\partial \psi}{\partial y}\frac{\partial \omega}{\partial x} - \frac{\partial \psi}{\partial x}\frac{\partial \omega}{\partial y} = Pr \cdot H_1(\varphi)\left(\frac{\partial^2 \omega}{\partial x^2} + \frac{\partial^2 \omega}{\partial y^2}\right)$$

$$+Ra \cdot Pr \cdot H_2(\varphi)\frac{\partial \theta}{\partial x} \qquad (7.33)$$

$$\frac{\partial \theta}{\partial \tau} + \frac{\partial \psi}{\partial y}\frac{\partial \theta}{\partial x} - \frac{\partial \psi}{\partial x}\frac{\partial \theta}{\partial y} = \frac{\partial^2 \theta}{\partial x^2} + \frac{\partial^2 \theta}{\partial y^2}. \tag{7.34}$$

Taking into account the considered dimensionless variables (7.30), the left wavy and the right flat walls of the cavity are described by the following relations: $x_1 = 1 - a - b\cos(2\pi\kappa y/A)$ is the left wavy wall; $x_2 = 1$ is the right flat wall; $\Delta = x_2 - x_1 = a + b\cos(2\pi\kappa y/A)$ is the dimensionless distance between vertical walls.

The corresponding boundary conditions for these equations are given by

$$\begin{aligned}
\psi &= 0, \quad \theta = 0.5[1 - \cos(2\pi y/A)] \text{ on } x = x_1 \\
\psi &= 0, \quad \theta = 0 \text{ on } x = x_2 \\
\psi &= 0, \quad \partial\theta/\partial y = 0 \text{ on } y = 0 \\
\psi &= 0, \quad \partial\theta/\partial y = 0 \text{ on } y = A.
\end{aligned} \tag{7.35}$$

Here,

$Pr = \mu_f (\rho C_p)_f / (\rho_f k_f)$ is the Prandtl number,

$Ra = g(\rho\beta)_f \Delta T L^3 / (\alpha_f \mu_f)$ is the Rayleigh number,

$A = L/D$ is the aspect ratio of the cavity,

and the functions $H_1(\varphi)$ and $H_2(\varphi)$ are given by

$$H_1(\varphi) = \frac{1 - \varphi + \varphi(\rho C_p)_p / (\rho C_p)_f}{(1-\varphi)^{2.5}[1 - \varphi + \varphi\rho_p/\rho_f]\left[\frac{k_p + 2k_f - 2\varphi(k_f - k_p)}{k_p + 2k_f + \varphi(k_f - k_p)}\right]} \tag{7.36}$$

$$H_2(\varphi) = \frac{[1-\varphi+\varphi(\rho\beta)_p/(\rho\beta)_f][1-\varphi+\varphi(\rho C_p)_p/(\rho C_p)_f]^2}{[1 - \varphi + \varphi\rho_p/\rho_f]\left[\frac{k_p + 2k_f - 2\varphi(k_f - k_p)}{k_p + 2k_f + \varphi(k_f - k_p)}\right]^2}. \tag{7.37}$$

These functions depend on the nanoparticles concentration φ and physical properties of the fluid and solid nanoparticles.

The dimensionless local entropy generation S_{gen} is obtained by using the dimensionless parameters presented in Eq. (7.30), given as:

$$S_{gen} = \overline{S}_{gen}\frac{T_0^2 L^2}{k_f (\Delta T)^2} = \frac{k_{nf}}{k_f}\left[\left(\frac{\partial\theta}{\partial x}\right)^2 + \left(\frac{\partial\theta}{\partial y}\right)^2\right]$$

$$+\chi\left[4\left(\frac{\partial^2\psi}{\partial x\partial y}\right)^2 + \left(\frac{\partial^2\psi}{\partial y^2} - \frac{\partial^2\psi}{\partial x^2}\right)^2\right] = S_{gen,ht} + S_{gen,ff}. \tag{7.38}$$

In Eq. (7.38), χ is the irreversibility factor. It is expressed as:

$$\chi = \frac{\mu_{nf} T_0}{k_f} \left[\frac{\alpha_{nf}}{L \Delta T} \right]^2. \tag{7.39}$$

The integration of Eq. (7.38) in the entire computational domain gives the dimensionless average entropy generation, $S_{gen,avg}$, expressed as follows:

$$S_{gen,avg} = \frac{1}{\vartheta} \int S_{gen} d\vartheta = S_{gen,ht,avg} + S_{gen,ff,avg}. \tag{7.40}$$

Further, the Bejan number Be is a parameter that shows the importance of heat transfer irreversibility in the domain and is defined as

$$Be = \frac{S_{gen,ht}}{S_{gen,ht} + S_{gen,ff}}. \tag{7.41}$$

The relative global dominance of heat transfer irreversibility is predicted by Be_{avg} (average Bejan number), which can be defined as

$$Be_{avg} = \frac{S_{gen,ht,avg}}{S_{gen,ht,avg} + S_{gen,ff,avg}}. \tag{7.42}$$

It may be noted that $Be_{avg} > 0.5$ shows that irreversibility due to heat transfer dominates in the flow. Fluid friction irreversibility dominates when $Be_{avg} < 0.5$, and when $Be_{avg} = 0.5$, the heat transfer and fluid friction entropy generation are equal.

3. Numerical Method and Validation

The cavity in the x and y plane, i.e., physical domain, is transformed into a rectangular geometry in the computational domain using an algebraic coordinate transformation by introducing new independent variables ξ and η. The left and right walls of the cavity become coordinate lines having constant values of ξ. The independent variables in the physical domain are transformed to independent variables in the computational domain by the following equations:

$$x = x(\xi, \eta), \quad y = y(\xi, \eta). \tag{7.43}$$

The channel geometry is mapped into a rectangle on the basis of the following transformation:

$$\begin{cases} \xi = \dfrac{x - x_1}{\Delta} = \dfrac{x - 1 + a + b\cos(2\pi\kappa y/A)}{a + b\cos(2\pi\kappa y/A)}, \\ \eta = y. \end{cases} \tag{7.44}$$

Taking into account transformation (7.44), the governing equations (7.32)–(7.34) will be rewritten in the following form:

$$\left[\left(\frac{\partial\xi}{\partial x}\right)^2+\left(\frac{\partial\xi}{\partial y}\right)^2\right]\frac{\partial^2\psi}{\partial\xi^2}+2\frac{\partial\xi}{\partial y}\frac{\partial^2\psi}{\partial\xi\partial\eta}+\frac{\partial^2\psi}{\partial\eta^2}+\frac{\partial^2\xi}{\partial y^2}\frac{\partial\psi}{\partial\xi}=-\omega \quad (7.45)$$

$$\frac{\partial\omega}{\partial\tau}+\frac{\partial\xi}{\partial x}\frac{\partial\psi}{\partial\eta}\frac{\partial\omega}{\partial\xi}-\frac{\partial\xi}{\partial x}\frac{\partial\psi}{\partial\xi}\frac{\partial\omega}{\partial\eta}$$

$$=Pr\cdot H_1(\varphi)\left\{\left[\left(\frac{\partial\xi}{\partial x}\right)^2+\left(\frac{\partial\xi}{\partial y}\right)^2\right]\frac{\partial^2\omega}{\partial\xi^2}+2\frac{\partial\xi}{\partial y}\frac{\partial^2\omega}{\partial\xi\partial\eta}+\frac{\partial^2\omega}{\partial\eta^2}+\frac{\partial^2\xi}{\partial y^2}\frac{\partial\omega}{\partial\xi}\right\}$$

$$+Ra\cdot Pr\cdot H_2(\varphi)\frac{\partial\xi}{\partial x}\frac{\partial\theta}{\partial\xi} \quad (7.46)$$

$$\frac{\partial\theta}{\partial\tau}+\frac{\partial\xi}{\partial x}\frac{\partial\psi}{\partial\eta}\frac{\partial\theta}{\partial\xi}-\frac{\partial\xi}{\partial x}\frac{\partial\psi}{\partial\xi}\frac{\partial\theta}{\partial\eta}$$

$$=\left[\left(\frac{\partial\xi}{\partial x}\right)^2+\left(\frac{\partial\xi}{\partial y}\right)^2\right]\frac{\partial^2\theta}{\partial\xi^2}+2\frac{\partial\xi}{\partial y}\frac{\partial^2\theta}{\partial\xi\partial\eta}+\frac{\partial^2\theta}{\partial\eta^2}+\frac{\partial^2\xi}{\partial y^2}\frac{\partial\theta}{\partial\xi}. \quad (7.47)$$

The corresponding boundary conditions of these equations are given by

$$\psi=0, \quad \theta=0.5[1-\cos(2\pi\eta/A)] \text{ on } \xi=0$$
$$\psi=0, \quad \theta=0 \text{ on } \xi=1$$
$$\psi=0, \quad \partial\theta/\partial\eta=0 \text{ on } \eta=0$$
$$\psi=0, \quad \partial\theta/\partial\eta=0 \text{ on } \eta=A.$$

$$(7.48)$$

It should be noted here that

$$\frac{\partial\xi}{\partial x}=\frac{1}{a+b\cos(2\pi\kappa y/A)},$$

$$\frac{\partial\xi}{\partial y}=\frac{2\pi\kappa b(x-1)\sin(2\pi\kappa y/A)}{A[a+b\cos(2\pi\kappa y/A)]^2}, \quad \frac{\partial^2\xi}{\partial x^2}=0,$$

$$\frac{\partial^2\xi}{\partial y^2}=\frac{4\pi^2\kappa^2 b(x-1)[a\cos(2\pi\kappa y/A)+b+b\sin^2(2\pi\kappa y/A)]}{A^2[a+b\cos(2\pi\kappa y/A)]^3}.$$

The physical quantities of interest are the local Nusselt numbers Nu_l and Nu_r, which are defined as

$$Nu_l=-\frac{k_{nf}}{k_f}\sqrt{\left(\frac{\partial\xi}{\partial x}\right)^2\left(\frac{\partial\theta\theta}{\partial\xi}\right)^2+\left(\frac{\partial\xi}{\partial y}\frac{\partial\theta}{\partial\xi}+\frac{\partial\theta}{\partial\eta}\right)^2},$$

$$Nu_r=-\frac{k_{nf}}{k_f}\left(\frac{\partial\xi}{\partial x}\right)\left(\frac{\partial\theta}{\partial\xi}\right)_{\xi=1}.$$

$$(7.49)$$

and the average Nusselt numbers $\overline{Nu_l}$ and $\overline{Nu_r}$, which are given by

$$\overline{Nu_l} = \frac{1}{A} \int_0^A Nu_l d\eta, \quad \overline{Nu_r} = \frac{1}{A} \int_0^A Nu_r d\eta. \qquad (7.50)$$

At the same time, the dimensionless local entropy generation S_{gen} using the new independent variables will be given as:

$$S_{gen} = \frac{k_{nf}}{k_f} \left[\left(\frac{\partial \xi}{\partial x} \frac{\partial \theta}{\partial \xi} \right)^2 + \left(\frac{\partial \xi}{\partial y} \frac{\partial \theta}{\partial \xi} + \frac{\partial \theta}{\partial \eta} \right)^2 \right]$$

$$+ \chi \left\{ 4 \left(\frac{\partial^2 \xi}{\partial x \partial y} \frac{\partial \psi}{\partial \xi} + \frac{\partial \xi}{\partial x} \frac{\partial \xi}{\partial y} \frac{\partial^2 \psi}{\partial \xi^2} + \frac{\partial \xi}{\partial x} \frac{\partial^2 \psi}{\partial \xi \partial \eta} \right)^2 \right.$$

$$\left. + \left[\left(\left(\frac{\partial \xi}{\partial y} \right)^2 - \left(\frac{\partial \xi}{\partial x} \right)^2 \right) \frac{\partial^2 \psi}{\partial \xi^2} + 2 \frac{\partial \xi}{\partial y} \frac{\partial^2 \psi}{\partial \xi \partial \eta} + \frac{\partial^2 \psi}{\partial \eta^2} + \frac{\partial^2 \xi}{\partial y^2} \frac{\partial \psi}{\partial \xi} \right]^2 \right\}$$

$$= S_{gen,ht} + S_{gen,ff}. \qquad (7.51)$$

The partial differential equations (7.45)–(7.47) with corresponding boundary conditions (7.48) were solved using the finite difference method with the second-order differencing schemes. Detailed description of the numerical technique used is presented by Sheremet et al. (2014, 2015a–g), Aleshkova and Sheremet (2010), Sheremet and Trifonova (2013), and Sheremet and Pop (2014a–c).

The part of the model that deals with performance of entropy generation was tested against the results of Ilis et al. (2008) and Bhardwaj et al. (2015) for steady-state natural convection in a differentially heated square cavity filled with the regular fluid for Prandtl number 0.7. Figures 7.15 and 7.16 show a good agreement between the obtained fields of local entropy generation due to heat transfer and fluid friction with $\chi = 10^{-4}$ for different Rayleigh numbers, and the numerical data of Ilis et al. (2008) and Bhardwaj et al. (2015).

For the purpose of obtaining a grid-independent solution, a grid sensitivity analysis may be performed. The grid independent solution was performed by preparing the solution for unsteady free convection in a square wavy cavity filled with a Cu–water nanofluid at $Ra = 10^5$, $Pr = 6.82$, $\varphi = 0.02$, $\kappa = 2$, $a = 0.9$, $A = 1$.

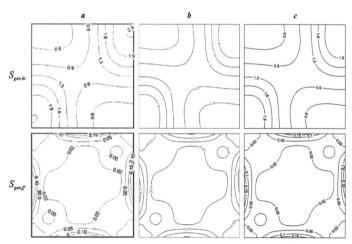

Fig. 7.15. Comparison of local entropy generation due to heat transfer $S_{gen,ht}$ and fluid friction $S_{gen,ff}$ for $Ra = 10^3$: numerical data of Ilis et al. (2008) (a) [reprinted from Ilis et al. (2008) with permission from Elsevier], numerical data of Bhardwaj et al. (2015) (b) [reprinted from Bhardwaj et al. (2015) with permission from Elsevier], present results (c)

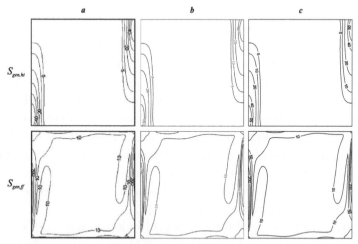

Fig. 7.16. Comparison of local entropy generation due to heat transfer $S_{gen,ht}$ and fluid friction $S_{gen,ff}$ for $Ra = 10^5$: numerical data of Ilis et al. (2008) (a) [reprinted from Ilis et al. (2008) with permission from Elsevier], numerical data of Bhardwaj et al. (2015) (b) [reprinted from Bhardwaj et al. (2015) with permission from Elsevier], present results (c)

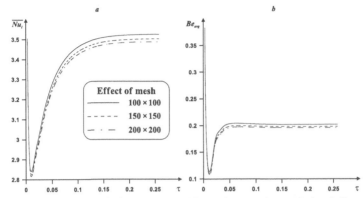

Fig. 7.17. Variation of the average Nusselt number of the left vertical wall (a) and the average Bejan number (b) versus the dimensionless time and the mesh parameters

Three cases of the uniform grid are tested: a grid of 100×100 points, a grid of 150×150 points, and a much finer grid of 200×200 points. Figure 7.17 shows an effect of the mesh parameters on the average Nusselt number of the left vertical wall and the average Bejan number. On the basis of the conducted verifications, the uniform grid of 150×150 points has been selected for the following analysis.

4. Results and Discussion

Numerical studies have been conducted at the following values of the governing parameters: Rayleigh number ($Ra = 10^3 - 10^5$), Prandtl number ($Pr = 6.82$), solid volume fraction parameter of nanoparticles ($\varphi = 0.0 - 0.05$), aspect ratio parameter ($A = 1$), undulation number ($\kappa = 1 - 3$), wavy contraction ratio ($b = 0.1 - 0.3$), dimensionless time ($\tau = 0 - 0.27$). Particular efforts have been focused on the effects of the Rayleigh number, nanoparticle volume fraction, undulation number, wavy contraction ratio, and dimensionless time on the fluid flow and heat transfer. Streamlines, isotherms, local entropy generations due to heat transfer and fluid friction, average Nusselt and Bejan numbers, and average entropy generation for different values of key parameters mentioned above are illustrated in Figs. 7.18–7.25.

Figure 7.18 presents streamlines ψ, isotherms θ, local entropy generation due to heat transfer $S_{gen,ht}$, and local entropy generation due to fluid friction $S_{gen,ff}$ for $\kappa = 2.0$, $b = 0.1$, $\varphi = 0.03$, $\tau = 0.27$, and different values of the Rayleigh number.

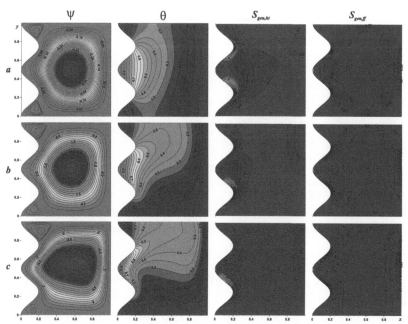

Fig. 7.18. Streamlines ψ, isotherms θ, local entropy generation due to heat transfer $S_{gen,ht}$, local entropy generation due to fluid friction $S_{gen,ff}$ for $\kappa = 2.0$, $b = 0.1$, $\varphi = 0.03$, $\tau = 0.27$: $Ra = 10^3$ (a), $Ra = 10^4$ (b), $Ra = 10^5$ (c)

Regardless of the Rayleigh number values, there are three convective cells inside the wavy-walled cavity. The major convective cell is located in the central part of the cavity, and small recirculations are situated in the bottom and top wavy troughs. The main reason for an appearance of these vortices is the non-uniform heating of the wavy wall. An increase in the Rayleigh number leads to an intensification of central convective flow $|\psi|_{\max}^{Ra=10^3} = 0.52 < |\psi|_{\max}^{Ra=10^4} = 3.49 < |\psi|_{\max}^{Ra=10^5} = 9.37$, an expansion of the upper recirculation flow, and a narrowing of the bottom minor convective cell. At the same time one can find that an increase in Ra also leads to change of the central convective cell core, which elongates along the horizontal axis with more intensive motion inside the central trough.

The temperature fields are changed with Ra. For small value of the Rayleigh number ($Ra = 10^3$ in Fig. 7.18a) the main heat transfer regime is heat conduction. This regime is characterized by isotherms that are parallel to the vertical isothermal walls. Moreover, it is possible to define uniform heating of the central wavy trough from the isothermal wavy wall. An increase in Ra leads to a formation of convective plume over the top wavy crest. Taking into account the obtained isotherms, it is possible to explain a decrease in the size of the bottom recirculation and an increase in the size of the upper vortex with Ra in the following way. An increase in the Rayleigh number leads to an intensification of major convective flow inside the cavity, with essential ascending flows along the hot central part of the wavy wall and descending flows along the cold right vertical wall. These intensive cold convective flows begin to deform the bottom recirculation, while the separation point of the hot ascending convective flows drifts to the upper wavy crest with the thin thermal boundary layers. The latter characterizes a formation of convective plume under the upper wavy crest that leads to an expansion of the upper vortex. It is worth noting that for high values of the Rayleigh number, the thermal stratification is formed in the central part of the cavity.

For the analyzed problem it is interesting to define the entropy generation areas inside the wavy-walled cavity. Figure 7.18 also shows the local entropy generation due to heat transfer and fluid friction. For the small values of the Rayleigh number, distributions of $S_{gen,ht}$ and $S_{gen,ff}$ are uniform and reflect a formation of diffusive mechanism. For this Rayleigh number value, the intensive entropy generation places due to heat transfer are the wavy crests. An increase in the Rayleigh number leads to attenuation of such places located on the upper crest.

Moreover, for high values of Ra close to the place of intensive entropy generation, one can find a formation of entropy generation trace that replicates the formation of thermal plume. The main reason for the formation of entropy generation places on the wavy crests is the high temperature gradients in these places due to thin thermal boundary layers. This also leads to a formation of the entropy generation place along the upper part of the right cold wall, owing to an interaction between the hot ascending flow from the left hot wall and cold right wall. As a result of such interaction, one can find an increase in the temperature gradient in this zone for high values of the Rayleigh number. Distributions of the local entropy generation due to fluid friction are defined by the Rayleigh dissipative function. In this case an increase in Ra leads to an intensification of entropy generation in the same places as in the case of the entropy generation due to heat transfer, but the more intensive zones are the internal walls of the central wavy trough and the opposite part of the right vertical wall. Such distributions can be explained by thin velocity boundary layers.

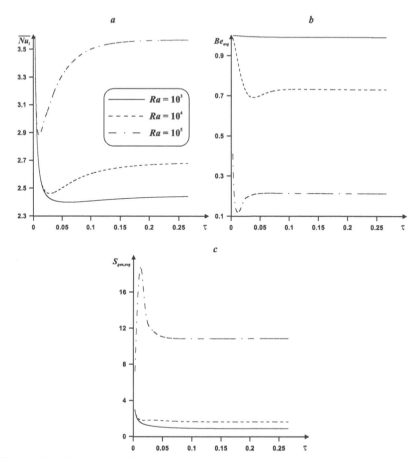

Fig. 7.19. Variation of the average Nusselt number of the left vertical wall (a), the average Bejan number (b), and the average entropy generation (c) versus the dimensionless time and the Rayleigh number for $\kappa = 2.0$, $b = 0.1$, $\varphi = 0.03$

One can find in these places more dense distribution of the stream-lines. An effect of the dimensionless time and Rayleigh number on the average Nusselt number at the vertical hot wavy wall (see Eqs. (7.49), (7.50)), average Bejan number (see Eq. (7.42)), and the average entropy generation (see Eqs. (7.40), (7.51)) is shown in Fig. 7.19. An increase in the Rayleigh number leads to an increase in the heat transfer rate, decrease in the average Bejan number, and an increase in the average entropy generation. It should be noted that more intensive increase in $\overline{Nu}_l$ and $S_{gen,avg}$ and decrease in Be_{avg} are for an increase in Ra from 10^4 to 10^5.

Also, it is possible to conclude that for $Ra = 10^4$ and 10^5 $Be_{avg} >$ 0.5, irreversibility due to heat transfer dominates in the flow, while for high values of Ra ($\geq 10^5$) the fluid friction irreversibility dominates. Therefore, for $Ra = 10^5$, the average entropy generation is defined essentially by the fluid friction. As for an effect of the dimensionless time, one can find non-linear dependencies of the average parameters for high values of Ra. In the case of $Ra = 10^3$ we have monotonic variation of the considered parameters. An increase in the Rayleigh number leads to non-monotonic changes of the average Nusselt and Bejan numbers. A formation of local minimum in these dependencies with dimensionless time reflects a presence of the conductive–convective threshold in the heat transfer process. For $Ra = 10^4$ this threshold is not so essential; therefore, the average entropy generation does not reflect this threshold. In the case of high values of Ra, one can find small duration of the conduction regime and essential effect of the convective heat transfer mode. Therefore, we have local extremums in the obtained dependencies.

Figure 7.20 shows streamlines ψ, isotherms θ, local entropy generation due to heat transfer $S_{gen,ht}$, and local entropy generation due to fluid friction $S_{gen,ff}$ for $Ra = 10^5$, $b = 0.1$, $\varphi = 0.03$, $\tau = 0.27$, and different values of the undulation number. An increase in the undulation number leads to an intensification of major convective flow inside the cavity: $|\psi|_{max}^{\kappa=1} = 9.25 < |\psi|_{max}^{\kappa=2} = 9.37 < |\psi|_{max}^{\kappa=3} = 9.53$. The main reason for such intensification of convective flow is a more essential narrowing of the cavity due to an increase in quantity of wavy crests. At the same time, one can find more essential changes in the temperature field. More intensive cooling of the domain of interest occurs in the case of high value of the undulation number due to significant interaction between the hot thermal plume from the central wavy crest and cold temperature waves from the upper and bottom wavy crests and the right cold vertical wall. In this case a presence of upper wavy crest is a natural obstacle for an evolution of thermal plume. Local entropy generation due to heat transfer characterizes a decrease in the sizes of the places, with high values of $S_{gen,ht}$. Such distributions also can be explained by small thickness of the thermal boundary layer along the peak of the wave crest.

In the case of single undulation one can find a peak of the wave crest of high sizes. Distributions of the local entropy generation due to fluid friction also replicate the distributions of $S_{gen,ht}$.

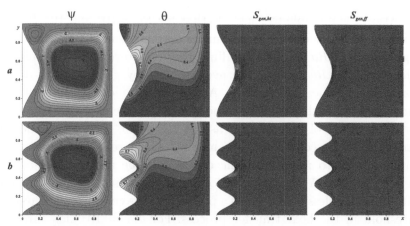

Fig. 7.20. Streamlines ψ, isotherms θ, local entropy generation due to
heat transfer $S_{gen,ht}$, local entropy generation due to fluid friction
$S_{gen,ff}$ for $Ra = 10^5$, $b = 0.1$, $\varphi = 0.03$, $\tau = 0.27$:
$\kappa = 1.0$ (a), $\kappa = 3.0$ (b)

An effect of the dimensionless time and undulation number on the
average Nusselt number, average Bejan number, and average entropy
generation is presented in Fig. 7.21. An increase in κ leads to a decrease
in all considered parameters. It should be noted that an increase in
the undulation number leads to more essential domination of the fluid
friction irreversibility and at the same time one can find a reduction
of the heat transfer rate at the wavy wall. Taking into account these
considered dependencies, it is possible to conclude that an increase in
the undulation number leads to a decrease in the heat transfer rate and
average entropy generation due to heat transfer.

Figure 7.22 demonstrates profiles of stream function ψ, temperature θ, local entropy generation due to heat transfer $S_{gen,ht}$, and local
entropy generation due to fluid friction $S_{gen,ff}$ for $Ra = 10^5$, $\kappa = 2.0$,
$\varphi = 0.03$, $\tau = 0.27$, and different values of wavy contraction ratio.
An increase in b leads to an increase in the wavy wall amplitude that
characterizes an essential narrowing of the internal fluid space. Such
variation of the wavy-walled cavity form leads to an attenuation of the
convective flow $|\psi|^{b=0.1}_{\max} = 9.37 > |\psi|^{b=0.2}_{\max} = 9.09 > |\psi|^{b=0.3}_{\max} = 7.64$,
also due to an intensification of the secondary vortices located in the
bottom and upper wavy troughs. An increase in b also characterizes
more intensive cooling of the cavity with secondary vortice areas of the
dominating heat conduction mechanism.

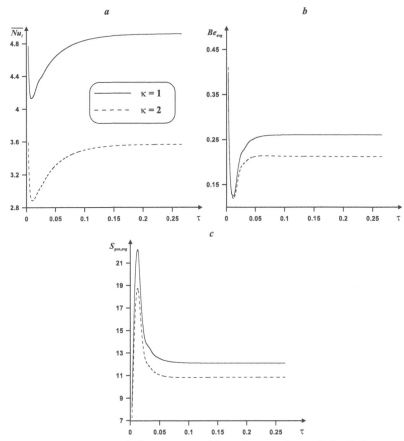

Fig. 7.21. Variation of the average Nusselt number of the left vertical wall (a), the average Bejan number (b) and the average entropy generation (c) versus the dimensionless time and the undulation number for $Ra = 10^5$, $b = 0.1$, $\varphi = 0.03$

In the case of entropy generation, an increase in the wavy contraction ratio leads to an intensification of entropy generation due to fluid friction in the central part of the cavity.

Taking into account the results presented in Fig. 7.23, it is possible to conclude that an increase in the wavy contraction ratio leads to a reduction of the heat transfer rate and average Bejan number, while the average entropy generation increases with b. The observed changes in the case of the entropy generation are due to an intensification of fluid friction with small thickness of the velocity boundary layers.

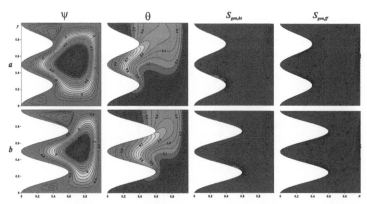

Fig. 7.22. Streamlines ψ, isotherms θ, local entropy generation due to heat transfer $S_{gen,ht}$, local entropy generation due to fluid friction $S_{gen,ff}$ for $Ra=10^5$, $\kappa=2.0$, $\varphi=0.03$, $\tau=0.27$: $b=0.2$ (a), $b=0.3$ (b)

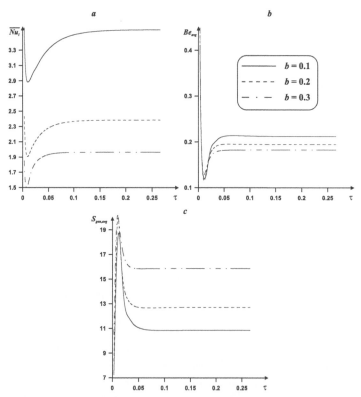

Fig. 7.23. Variation of the average Nusselt number of the left vertical wall (a), the average Bejan number (b), and the average entropy generation (c) versus the dimensionless time and wavy contraction ratio for $Ra = 10^5$, $\kappa = 2.0$, $\varphi = 0.03$

In the case of the average Nusselt number it is worth noting that we see more intensive heating of the central wavy trough and more intensive cooling of the bottom and top wavy troughs. Such changes lead to a reduction of the temperature differences.

An effect of the nanoparticle volume fraction is presented in Figs. 7.24 and 7.25. An insertion of nanoparticles inside the base fluid (Fig. 7.24) leads to an attenuation of the convective flow inside the cavity, and more intensive cooling of the cavity, with unessential variation of the local entropy generation due to heat transfer, and also with an attenuation of entropy generation due to fluid friction. Such changes are due to an increase in the dynamic viscosity and thermal conductivity of nanofluid. All observed variations in the local parameters lead to both an increase in the average Nusselt number and the average Bejan number, and a decrease in the average entropy generation. Changes of the average Bejan number and entropy generation are due to a decrease in the entropy generation due to fluid friction.

The main findings can be listed as:

(1) An increase in the Raleigh number leads to an intensification of convective flow and heat transfer, and a reduction of the average Bejan number due to an enhancement of fluid friction. Also, an increase in Ra leads to both a formation of convective plume over the top wavy crest and non-monotonic changes of the average Nusselt and Bejan numbers.

(2) An increase in the undulation number leads to an intensification of major convective flow inside the cavity and more intensive cooling of the domain of interest. An increment in κ leads to a reduction in the heat transfer rate and average entropy generation due to heat transfer that is reflected in a decrease in all considered integral parameters.

(3) An increase in the wavy contraction ratio leads to an attenuation of the convective flow due to an intensification of the secondary vortices located in the bottom and upper wavy troughs. An increase in b also characterizes more intensive cooling of the cavity with secondary vortice areas of the dominating heat conduction mechanism.

At the same time the average Nusselt number and the average Bejan number are decreasing functions of the wavy contraction ratio while the average entropy generation is an increasing function of b.

(4) The solid volume fraction suppresses the fluid motion. At the same time an increase in the nanoparticle volume fraction leads to an enhancement of the heat transfer rate and the average Bejan number, while the average entropy generation decreases.

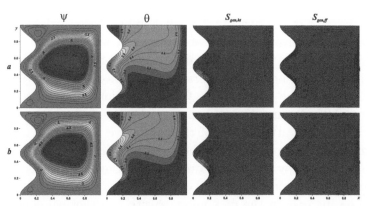

Fig. 7.24. Streamlines ψ, isotherms θ, local entropy generation due to heat transfer $S_{gen,ht}$, local entropy generation due to fluid friction $S_{gen,ff}$ for $Ra = 10^5$, $\kappa = 2.0$, $b = 0.1$, $\tau = 0.27$: $\varphi = 0.0$ (a), $\varphi = 0.05$ (b)

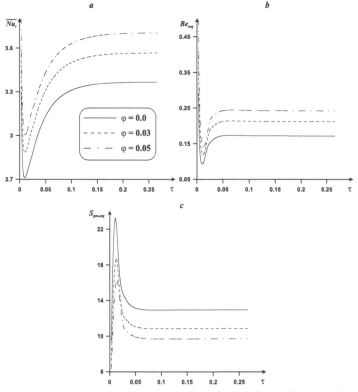

Fig. 7.25. Variation of the average Nusselt number of the left vertical wall (a), the average Bejan number (b), and the average entropy generation (c) versus the dimensionless time and the nanoparticle volume fraction for $Ra = 10^5$, $\kappa = 2.0$, $b = 0.1$

Chapter 8

Natural Convection Flow Saturated with Nanoparticles in Wavy-Walled Porous Cavities

A. Natural Convection in a Wavy Porous Cavity with Sinusoidal Temperature Distributions on Both Side Walls Filled with a Nanofluid: Buongiorno's Mathematical Model

1. Introduction

Buoyancy-driven phenomena in enclosures filled with porous media have been actively under investigation for several years. They have received a great deal of attention due to a large number of technical applications such as thermal insulating systems, solar power collectors, geothermal applications, nuclear reactors, matrix heat exchangers, separation processes in chemical industries, dispersion of chemical contaminants through water-saturated soil, solidification of casting, oil extraction, storage of nuclear waste materials, grain storage systems, groundwater hydrology, and crude oil production, etc. These applications of porous media in many practical problems can be found in the well-known books by Pop and Ingham (2001), Bejan et al. (2004),

Ingham and Pop (2005), Vafai (2005), Vadasz (2008), de Lemos (2012), and Nield and Bejan (2013). Aydin et al. (1999) and Sarris et al. (2002) have indicated that there exist only a limited amount of works published on convective flow in fluid-filled cavities (non-porous medium) with a more complex case of cooling from the top wall, mainly with periodic temperature conditions imposed upon the bottom or side walls. For example, Poulikakos (1985) studied an enclosure with its left side wall differentially heated: one half of the wall is heated and the other half is cooled, and the remaining walls are insulated. He showed that a penetrating thermal layer is formed, the size of which is a function of Rayleigh number and aspect ratio of the enclosure. Bilgen et al. (1995) used a system of discrete temperature sources placed periodically on the bottom wall of a shallow cavity. Lage and Bejan (1993) studied enclosures with one side wall heated using a pulsating heat flux and the other side wall cooled at constant temperature. They showed that at high Rayleigh numbers, the buoyancy-driven flow has the tendency to resonate to the periodic heating that has been supplied from the side. Studies of natural convection in molten glass cells with periodic heating from above and specified temperature on the side walls for Rayleigh numbers up to 10^7 were made by Wright and Rawson (1973) and Burley et al. (1978). Periodic heating from above has strong implications for the glass industry, where the main objective is to increase the mixing of the glass melt.

In all these studies, the case of natural convection flow in a porous cavity filled with a nanofluid has not been investigated. It is well known that conventional heat transfer liquids have low thermal conductivity. Significant features of nanofluids over base fluids include enhanced thermal conductivity, greater viscosity, and enhanced value of critical heat flux. Of these the most important is the enhanced thermal conductivity, a phenomenon which was first reported by Masuda et al. (1993).

Many modern industries deal with heat transfer in some or an other way, and thus have a strong need for improved heat transfer mediums. This could possibly be nanofluids (see Öztop and Abu-Nada, 2008; Sun and Pop, 2011, 2014; Haddad et al., 2012a,b; Mahdy and Ahmed, 2012; Ghalambaz et al., 2015; Zargartalebi et al., 2015), because of some potential benefits over normal fluids — large surface area provided by nanoparticles for heat exchange, reduced pumping power due to enhanced heat transfer, minimal clogging, and innovation of miniaturized systems leading to savings of energy and cost. The prediction of heat transfer from irregular surfaces is a topic of fundamental importance for some heat transfer devices (Misirlioglu et al., 2005), such as flat plate solar collectors, flat plate condensers in refrigerators, double-wall thermal insulation, underground cable systems, electric machinery, cooling systems of micro-electronic devices, natural circulation in the atmosphere, the molten core of the Earth, etc. In addition, roughened surfaces could be used in the cooling of electrical and nuclear components where the wall heat flux is known. Surfaces are sometimes intentionally roughened to enhance heat transfer, and hence an understanding of natural convection of nanofluids in a porous cavity with wavy walls attains importance. Therefore, the principal aim of the present section is to present a numerical analysis of the natural convection flow in a porous cavity with a wavy bottom and top walls having sinusoidal temperature distributions on vertical walls filled with a nanofluid. The mathematical model has been formulated in dimensionless stream function and temperature, taking into account the Darcy–Boussinesq approximation and using the nanofluid mathematical model proposed by Buongiorno (2006). It should be noted that nanoparticles participate in complex motion where base fluid motion is a transportation motion and a slip motion of nanoparticles inside the base fluid is a relative motion. Therefore the Buongiorno's nanofluid model used considers the nanoparticle absolute velocity as the sum of the base fluid velocity and a relative (slip) velocity. Taking into account the nanofluid as a two-component mixture (base fluid plus nanoparticles) with the following assumptions in convective flow: incompressible flow, no chemical reactions, negligible external forces, dilute mixture, negligible viscous dissipation, negligible radiative heat transfer and nanoparticles, and base fluid locally in thermal equilibrium, Buongiorno has shown that Brownian diffusion and thermophoresis are the two most important slip mechanisms. Therefore, the Buongiorno's nanofluid model is a two-component non-homogeneous equilibrium model for transport phenomena in nanofluids that can be used for description of convective flow in different industrial applications, taking into account the assumptions used.

2. Basic Equations

Consider the free convection in a wavy porous cavity filled with a nanofluid based on water and nanoparticles. It is assumed that nanoparticles are suspended in the nanofluid using either surfactant or surface charge technology.

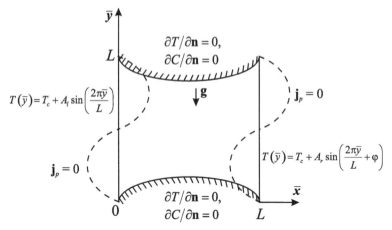

Fig. 8.1. Physical model and coordinate system [reprinted from
Sheremet and Pop (2015a) with permission from ASME]

This prevents nanoparticles from agglomeration and deposition on
the porous matrix (Kuznetsov and Nield, 2013; Nield and Kuznetsov,
2014). A schematic geometry of the problem under investigation is
shown in Fig. 8.1. The cavity is assumed to be impermeable and the
wavy walls are assumed to be thermally insulated. At the same time
the vertical walls have imposed two sinusiodally varying temperature
distributions according to the space coordinate as follows (Sarris et al.,
2002; Deng and Chang, 2008):

$$T(\overline{y}) = T_c + A_l \sin\left(\frac{2\pi\overline{y}}{L}\right) \quad \text{at } \overline{x} = 0 \qquad (8.1)$$

$$T(\overline{y}) = T_c + A_r \sin\left(\frac{2\pi\overline{y}}{L} + \varphi\right) \quad \text{at } \overline{x} = L \qquad (8.2)$$

where the reference temperatures of the sinusoidal temperature profiles
on the left and right side walls are the same, T_c, but the amplitude and
phase of the sinusoidal profiles are, respectively, A_l and 0, and A_r and
φ.

It should be noted that the upper and the lower wavy walls of the
cavity are described by the following relations (Öztop et al., 2011):

$$\overline{y}_1 = L - L\left[a + b\cos\left(\frac{2\pi\kappa\overline{x}}{L}\right)\right]$$

is the lower wavy wall;

$$\overline{y}_2 = L\left[a + b\cos\left(\frac{2\pi\kappa\overline{x}}{L}\right)\right]$$

is the upper wavy wall;

$$\overline{\Delta} = \overline{y}_2 - \overline{y}_1 = 2L\left[a + b\cos\left(\frac{2\pi\kappa\overline{x}}{L}\right)\right] - L$$

is the distance between the upper and lower wavy walls. Here a and b are constants that determine the shape and the contraction ratio of the wavy cavity ($a + b = 1$), κ is a number of undulation.

The Darcy–Boussinesq approximation is employed. Homogeneity and local thermal equilibrium in the porous medium are assumed. We consider a medium whose porosity is denoted by ε and permeability by K. The following are four field equations for the conservation of total mass, momentum, thermal energy, and nanoparticles, respectively (see Buongiorno, 2006; Kuznetsov and Nield, 2013; Nield and Kuznetsov, 2014):

$$\nabla \cdot \mathbf{V} = 0 \tag{8.3}$$

$$0 = -\nabla p - \frac{\mu}{K}\mathbf{V} + [C\rho_p + (1 - C)\rho_{f0}(1 - \beta(T - T_c))]\mathbf{g} \tag{8.4}$$

$$\sigma\frac{\partial T}{\partial t} + (\mathbf{V} \cdot \nabla)T = \alpha_m\nabla^2 T + \delta[D_B\nabla C \cdot \nabla T + (D_T/T_c)\nabla T \cdot \nabla T] \tag{8.5}$$

$$\rho_p\left(\frac{\partial C}{\partial t} + \frac{1}{\varepsilon}(\mathbf{V} \cdot \nabla)C\right) = -\nabla \cdot \mathbf{j}_p. \tag{8.6}$$

The flow is assumed to be slow so that an advective term and a Forchheimer quadratic term do not appear in the momentum equation. In keeping with the Boussinesq approximation and an assumption that the nanoparticle concentration is dilute, and with a suitable choice for the reference pressure, we can linearize the momentum equation and write Eq. (8.4) as

$$0 = -\nabla p - \frac{\mu}{K}\mathbf{V} + [C(\rho_p - \rho_{f0}) + \rho_{f0}(1 - \beta(T - T_c)(1 - C_0))]\mathbf{g}. \tag{8.7}$$

One can eliminate pressure p from Eq. (8.7) by cross-differentiation. Further, we introduce the stream function defined by

$$u = \partial\overline{\psi}/\partial\overline{y}, \quad v = -\partial\overline{\psi}/\partial\overline{x}$$

and the dimensionless variables

$$x = \overline{x}/L, \ y = \overline{y}/L, \ \psi = \overline{\psi}/\alpha_m,$$

$$\theta = (T - T_c)/\Delta T, \ \phi = C/C_0 \tag{8.8}$$

where $\Delta T = A_l$ (amplitude of the sinusoidal profile).

Thus, Eqs. (8.3) and (8.5) to (8.7) can be written as

$$\frac{\partial^2 \psi}{\partial x^2} + \frac{\partial^2 \psi}{\partial y^2} = -Ra\frac{\partial \theta}{\partial x} + Ra \cdot Nr\frac{\partial \phi}{\partial x} \tag{8.9}$$

$$\frac{\partial \psi}{\partial y}\frac{\partial \theta}{\partial x} - \frac{\partial \psi}{\partial x}\frac{\partial \theta}{\partial y} = \frac{\partial^2 \theta}{\partial x^2} + \frac{\partial^2 \theta}{\partial y^2}$$

$$+Nb\left(\frac{\partial \phi}{\partial x}\frac{\partial \theta}{\partial x} + \frac{\partial \phi}{\partial y}\frac{\partial \theta}{\partial y}\right) + Nt\left[\left(\frac{\partial \theta}{\partial x}\right)^2 + \left(\frac{\partial \theta}{\partial y}\right)^2\right] \tag{8.10}$$

$$\frac{\partial \psi}{\partial y}\frac{\partial \phi}{\partial x} - \frac{\partial \psi}{\partial x}\frac{\partial \phi}{\partial y} = \frac{1}{Le}\left(\frac{\partial^2 \phi}{\partial x^2} + \frac{\partial^2 \phi}{\partial y^2}\right)$$

$$+\frac{1}{Le}\frac{Nt}{Nb}\left(\frac{\partial^2 \theta}{\partial x^2} + \frac{\partial^2 \theta}{\partial y^2}\right) \tag{8.11}$$

where $Ra = (1 - C_0)gK\rho_{f0}\beta\Delta TL/(\alpha_m\mu)$ is the Rayleigh number.

Taking into account the considered dimensionless variables (8.8), the upper and the lower wavy walls of the cavity are described by the following relations:

$$y_1 = 1 - a - b\cos(2\pi\kappa x)$$

is the lower wavy wall;

$$y_2 = a + b\cos(2\pi\kappa x)$$

is the upper wavy wall;

$$\Delta = y_2 - y_2 = 2a + 2b\cos(2\pi\kappa x) - 1$$

is the distance between the upper and lower wavy walls.

The corresponding boundary conditions for these equations are given by

$$\psi = 0, \quad \theta = \sin(2\pi y), \quad \tilde{\mathbf{j}}_p = 0$$

$$\left(\text{or } Nb\frac{\partial \phi}{\partial x} + Nt\frac{\partial \theta}{\partial x} = 0\right) \text{ on } x = 0$$

$$\psi = 0, \quad \theta = \gamma\sin(2\pi y + \varphi), \quad \tilde{\mathbf{j}}_p = 0 \tag{8.12}$$

$$\left(\text{or } Nb\frac{\partial \phi}{\partial x} + Nt\frac{\partial \theta}{\partial x} = 0\right) \text{ on } x = 1$$

$$\psi = 0, \quad \frac{\partial \theta}{\partial \mathbf{n}} = 0, \quad \frac{\partial \phi}{\partial \mathbf{n}} = 0 \text{ on } y = y_1 \text{ and } y = y_2.$$

Here the five parameters Nr, Nb, Nt, Le, and γ denote a buoyancy ratio parameter, a Brownian motion parameter, a thermophoresis parameter, a Lewis number, and an amplitude ratio of the sinusoidal temperature on the right side wall to that on the left side wall, respectively, which are defined as

$$Nr = \frac{(\rho_p - \rho_{f0})C_0}{\rho_{f0}\beta\Delta T(1 - C_0)}, \quad Nb = \frac{\delta D_B C_0}{\alpha_m},$$

$$Nt = \frac{\delta D_T \Delta T}{\alpha_m T_c}, \quad Le = \frac{\alpha_m}{\varepsilon D_B}, \quad \gamma = \frac{A_r}{A_l}. \tag{8.13}$$

It should be noted that for $Nr = Nb = Nt = 0$ (regular fluid), Eqs. (8.9) and (8.10) reduce to those of Walker and Homsy (1978), and Baytas and Pop (1999).

The physical quantities of interest are the local Nusselt number Nu, the local Sherwood number Sh, and the average Nusselt $\overline{Nu}$ and Sherwood $\overline{Sh}$ numbers.

The local Nusselt and Sherwood numbers are defined as

$$Nu_l = -\left(\frac{\partial\theta}{\partial x}\right)_{x=0}, \quad Sh_l = -\left(\frac{\partial\phi}{\partial x}\right)_{x=0},$$

$$Nu_r = -\left(\frac{\partial\theta}{\partial x}\right)_{x=1}, \quad Sh_r = -\left(\frac{\partial\phi}{\partial x}\right)_{x=1}. \tag{8.14}$$

The average Nusselt and Sherwood numbers are defined as

$$\overline{Nu_l} = \int_0^1 Nu_l\,dy, \quad \overline{Sh_l} = \int_0^1 Sh_l\,dy,$$

$$\overline{Nu_r} = \int_0^1 Nu_r\,dy, \quad \overline{Sh_r} = \int_0^1 Sh_r\,dy. \tag{8.15}$$

It should be noted here that for an analysis of Sherwood numbers, it is possible to study only Nusselt numbers, because at the left and right vertical walls, we have

$$\frac{\partial\phi}{\partial x} = -\frac{Nt}{Nb}\frac{\partial\theta}{\partial x},$$

taking into account boundary conditions for ϕ (Eqs. (8.12)). Therefore, the further analysis concerning integral parameters will only be about average Nusselt number, because

$$Sh = -\frac{Nt}{Nb}Nu \quad \text{and} \quad \overline{Sh} = -\frac{Nt}{Nb}\overline{Nu}.$$

3. Numerical Method

The cavity in the x and y plane, i.e., physical domain, is transformed into a rectangular geometry in the computational domain, using an algebraic coordinate transformation by introducing new independent variables ξ and η. The top and bottom walls of the cavity become coordinate lines having constant values of η. The independent variables in the physical domain are transformed to independent variables in the computational domain by the following equations (Öztop et al., 2011):

$$x = x(\xi, \eta), \quad y = y(\xi, \eta). \tag{8.16}$$

The cavity geometry is mapped into a rectangle on the basis of the following transformation:

$$\begin{cases} \xi = x, \\ \eta = \dfrac{y - y_1}{\Delta} = \dfrac{y - 1 + a + b\cos(2\pi\kappa x)}{2a + 2b\cos(2\pi\kappa x) - 1}. \end{cases} \tag{8.17}$$

Taking into account transformation (8.17), the governing equations (8.9)–(8.11) are rewritten in the following form:

$$\frac{\partial^2 \psi}{\partial \xi^2} + 2\frac{\partial \eta}{\partial x}\frac{\partial^2 \psi}{\partial \xi \partial \eta} + \left[\left(\frac{\partial \eta}{\partial x}\right)^2 + \left(\frac{\partial \eta}{\partial y}\right)^2\right]\frac{\partial^2 \psi}{\partial \eta^2} + \frac{\partial^2 \eta}{\partial x^2}\frac{\partial \psi}{\partial \eta}$$

$$= -Ra\left(\frac{\partial \theta}{\partial \xi} + \frac{\partial \eta}{\partial x}\frac{\partial \theta}{\partial \eta}\right) + Ra \cdot Nr\left(\frac{\partial \phi}{\partial \xi} + \frac{\partial \eta}{\partial x}\frac{\partial \phi}{\partial \eta}\right) \tag{8.18}$$

$$\frac{\partial \eta}{\partial y}\frac{\partial \psi}{\partial \eta}\frac{\partial \theta}{\partial \xi} - \frac{\partial \eta}{\partial y}\frac{\partial \psi}{\partial \xi}\frac{\partial \theta}{\partial \eta}$$

$$= \frac{\partial^2 \theta}{\partial \xi^2} + 2\frac{\partial \eta}{\partial x}\frac{\partial^2 \theta}{\partial \xi \partial \eta} + \left[\left(\frac{\partial \eta}{\partial x}\right)^2 + \left(\frac{\partial \eta}{\partial y}\right)^2\right]\frac{\partial^2 \theta}{\partial \eta^2} + \frac{\partial^2 \eta}{\partial x^2}\frac{\partial \theta}{\partial \eta}$$

$$+Nb\left[\left(\frac{\partial \phi}{\partial \xi} + \frac{\partial \eta}{\partial x}\frac{\partial \phi}{\partial \eta}\right)\left(\frac{\partial \theta}{\partial \xi} + \frac{\partial \eta}{\partial x}\frac{\partial \theta}{\partial \eta}\right) + \left(\frac{\partial \eta}{\partial y}\right)^2\frac{\partial \phi}{\partial \eta}\frac{\partial \theta}{\partial \eta}\right]$$

$$+Nt\left[\left(\frac{\partial \theta}{\partial \xi} + \frac{\partial \eta}{\partial x}\frac{\partial \theta}{\partial \eta}\right)^2 + \left(\frac{\partial \eta}{\partial y}\frac{\partial \theta}{\partial \eta}\right)^2\right] \tag{8.19}$$

$$\frac{\partial \eta}{\partial y}\frac{\partial \psi}{\partial \eta}\frac{\partial \phi}{\partial \xi} - \frac{\partial \eta}{\partial y}\frac{\partial \psi}{\partial \xi}\frac{\partial \phi}{\partial \eta}$$

$$= \frac{1}{Le} \left\{ \frac{\partial^2 \phi}{\partial \xi^2} + 2 \frac{\partial \eta}{\partial x} \frac{\partial^2 \phi}{\partial \xi \partial \eta} + \left[\left(\frac{\partial \eta}{\partial x} \right)^2 + \left(\frac{\partial \eta}{\partial y} \right)^2 \right] \frac{\partial^2 \phi}{\partial \eta^2} + \frac{\partial^2 \eta}{\partial x^2} \frac{\partial \phi}{\partial \eta} \right\}$$

$$+ \frac{1}{Le} \frac{Nt}{Nb} \left\{ \frac{\partial^2 \theta}{\partial \xi^2} + 2 \frac{\partial \eta}{\partial x} \frac{\partial^2 \theta}{\partial \xi \partial \eta} + \left[\left(\frac{\partial \eta}{\partial x} \right)^2 + \left(\frac{\partial \eta}{\partial y} \right)^2 \right] \frac{\partial^2 \theta}{\partial \eta^2} + \frac{\partial^2 \eta}{\partial x^2} \frac{\partial \theta}{\partial \eta} \right\}.$$

$$(8.20)$$

The corresponding boundary conditions of these equations are given by

$$\psi = 0, \quad \theta = \sin(2\pi\eta), \quad \tilde{\mathbf{j}}_p = 0$$

$$\left(\text{or } Nb \frac{\partial \phi}{\partial \xi} + Nt \frac{\partial \theta}{\partial \xi} = 0 \right) \text{ on } \xi = 0$$

$$\psi = 0, \quad \theta = \gamma \sin(2\pi\eta + \varphi), \quad \tilde{\mathbf{j}}_p = 0 \qquad (8.21)$$

$$\left(\text{or } Nb \frac{\partial \phi}{\partial \xi} + Nt \frac{\partial \theta}{\partial \xi} = 0 \right) \text{ on } \xi = 1$$

$$\psi = 0, \quad \frac{\partial \theta}{\partial \eta} = 0, \quad \frac{\partial \phi}{\partial \eta} = 0 \text{ on } \eta = 0 \text{ and } \eta = 1.$$

The partial differential equations (8.18)–(8.20) with corresponding boundary conditions (8.21) were solved using the finite difference method (see Aleshkova and Sheremet, 2010; Sheremet and Trifonova, 2013; Sheremet and Pop, 2014a–c; Sheremet et al., 2014). The steady-state solution was obtained like the time limit for solution of the transient problem, where the approximation of the convective terms was conducted by Samarskii monotonic scheme (Samarski, 1983) and the approximation of the diffusion terms was conducted by the central differences. It is well known that the Samarskii monotonic scheme is the second-order difference scheme in the case of slow flow like the one considered here. The transient equations were solved on the basis of a Samarskii locally one-dimensional scheme (Samarski, 1983). The linear discretized equations were solved by Thomas algorithm. The Poisson equation for the stream function was discretized by means of the five-point difference scheme on the basis of central differences for the second derivatives. The obtained linear discretized equation was solved by the successive over-relaxation method. Optimum value of the relaxation parameter was chosen on the basis of computing experiments. The computation is terminated when the residuals for the stream function get bellow 10^{-7}. The numerical process is coded by C++ programming language.

Table 8.1. Variations of the Average Nusselt Numbers of the Left
Vertical Wall with the Uniform Grid [reprinted from Sheremet and
Pop (2015a) with permission from ASME]

Uniform grids	Nu	$\Delta = \dfrac{\left\|Nu_{i \times j} - Nu_{300 \times 300}\right\|}{Nu_{i \times j}} \times 100\%$
100×100	4.488	0.22
200×200	4.491	0.16
300×300	4.498	—
400×400	4.504	0.13

The present models, in the form of an in-house computational fluid
dynamics (CFD) code, have been validated successfully against the
works of Walker and Homsy (1978), and Baytas and Pop (1999) for
steady-state natural convection in a square porous cavity with isothermal vertical and adiabatic horizontal walls. Detailed comparison is presented in Sheremet et al. (2014).

For the purpose of obtaining a grid independent solution, a grid
sensitivity analysis is performed. The grid independent solution was
performed by preparing the solution for steady-state natural convection
in a wavy wall porous cavity with sinusoidal temperature distributions
on vertical walls filled with a nanofluid at $Ra = 100$, $Le = 10$, $Nr =
Nb = Nt = 0.1$, $\gamma = 1$, $\varphi = 0$, $\kappa = 1$, $a = 0.9$. Four cases of a uniform
grid are tested: 100×100, 200×200, 300×300, 400×400. Table 8.1
shows an effect of the mesh on the average Nusselt number of the left
vertical wall.

On the basis of the conducted verifications, the uniform grid of
300×300 points has been selected for the following analysis.

4. Results and Discussion

Numerical investigations of the boundary value problem (8.18)–
(8.21) has been carried out at the following values of dimensionless com-
plexes: Rayleigh number ($Ra = 10$–300), Lewis number ($Le = 1$–1000),
the buoyancy ratio parameter ($Nr = 0.1$–0.4), the Brownian motion
parameter ($Nb = 0.1$–0.4), the thermophoresis parameter ($Nt = 0.1$–
0.4), $\gamma = 1$, $\varphi = 0$, $\kappa = 1$, $a = 0.9$. Particular efforts have been focused
on the effects of these parameters on the fluid flow, heat, and mass
transfer characteristics. Figure 8.2 illustrates streamlines, isotherms,
and isoconcentrations at different values of the Rayleigh number at
$Le = 100$, $Nr = Nb = Nt = 0.1$. Regardless of the Rayleigh number

value, four convective cells are formed inside the cavity. Two convective cells are clockwise vortices that are located in the left bottom part and right top part of the cavity, and two convective cells are counterclockwise vortices that are located in the right bottom part and left top part of the cavity. The main reason for an appearance of these circulations is an effect of vertical walls with sinusoidally varying temperature distributions. It should be noted that an intensity of two convective cells in the top part of the cavity is greater than an intensity of two convective cells in the bottom part of the cavity that can be explained by an effect of the buoyancy force. These four vortices are separated by virtual vertical and horizontal walls that are both impervious and adiabatic. Convective cells cores are close to the vertical walls due to, on the one hand, large temperature differences in these zones, and on the other hand, a presence of solid structure of the porous medium inside the cavity. The latter leads to a formation of an extra resistance of motion and only large temperature differences can generate an intensive motion. An increase in the Rayleigh number leads to both an intensification of convective flows inside the cavity and a modification of the vortices' configuration.

The former can be confirmed by maximum absolute values of the stream function as following $|\psi|_{\mathrm{max}}^{Ra=10} = 0.23 < |\psi|_{\mathrm{max}}^{Ra=100} = 2.5 < |\psi|_{\mathrm{max}}^{Ra=300} = 8.2$. At high values of the Rayleigh number ($Ra = 300$), one can find a formation of upward flows in the central part of the cavity due to an effect of the buoyancy force.

All changes in hydrodynamic structures are related to the modification of the temperature field. This is the main difference of the natural convective mode from other convective modes. At $Ra = 10$ the dominated heat transfer mechanism is a heat conduction that defines intensive heat transfer along the horizontal coordinate owing to an interaction of the temperature sources and sinks in the vertical direction. An increase in Ra leads to a formation of ascending thermal plumes close to the heat sources and descending thermal plumes close to the heat sinks. Heat sources are the parts of the vertical walls where the dimensionless temperature is greater than zero and heat sinks are the parts of the vertical walls where the dimensionless temperature is less than zero due to the periodical thermal boundary conditions (8.21). One can find a clear confirmation of the above-mentioned results at $Ra = 300$ (Fig. 8.2c). Isoconcentrations presented in Fig. 8.2 characterize the distributions of the nanoparticles inside the cavity.

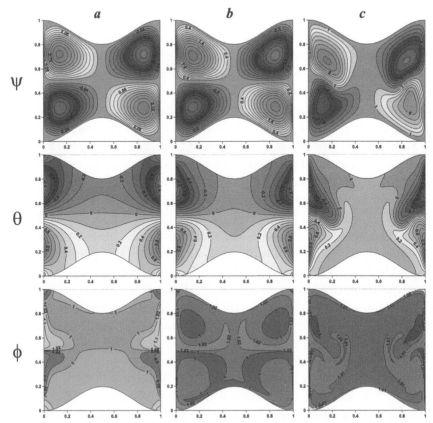

Fig. 8.2. Streamlines ψ, isotherms θ, and isoconcentrations ϕ for
$Le = 100$, $Nr = Nb = Nt = 0.1$: $Ra = 10$ (a), $Ra = 100$ (b),
$Ra = 300$ (c) [reprinted from Sheremet and Pop (2015a) with
permission from ASME]

Regardless of the Rayleigh number value, the considered regime is
defined by an increase in ϕ in the upper part of the cavity and a decrease
in ϕ in the bottom part of the cavity. At small values of $Ra = 10$
the distribution of nanoparticles is weakly homogeneous, taking into
account Fig. 8.2.

Such a description of this regime is due to a deviation of the
nanoparticles' volume fraction from the average value $\phi = 1$. Moreover,
the heat conduction regime enhances the effect of the thermophoresis
phenomenon and therefore nanoparticles distribution is weakly homogeneous with small values of the Rayleigh number $Ra = 10$ (Fig. 8.2a).
The thermophoresis effect is clearly presented in Fig. 8.2a in comparison with isoconcentrations for high values of the Rayleigh number.

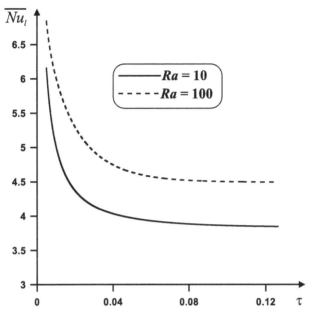

Fig. 8.3. Variation of the average Nusselt number at left vertical wall
with the Rayleigh number and dimensionless time for $Le = 100$,
$Nr = Nb = Nt = 0.1$ [reprinted from Sheremet and Pop (2015a) with
permission from ASME]

In the case of the heat conduction dominating regime, one can find
more essential changes in the nanoparticles' volume fraction close to the
heat sources and heat sinks. Nanoparticles' volume fraction deviations
from the average value $\phi = 1$ are more than 10%. Therefore it should
be noted here that the non-homogeneous areas are confined close to
the heat sources and sinks. An increase in Ra leads to intensification of
convective flow and as a result one can find homogeneous distribution of
the nanoparticles' volume fraction inside the cavity. It is worth noting
that non-homogeneous areas reflect the direction of convective heat
transfer as heatlines defined by Kimura and Bejan (1983). The main
reason for such behavior is a thermophoresis effect inside the thermal
plume. An effect of the dimensionless time and Rayleigh number on
the average Nusselt number at the left vertical wall is presented in Fig.
8.3. An increase in Ra from 10 to 100 leads to an increase in $\overline{Nu_l}$.
Taking into account boundary conditions for the nanoparticles' volume
fraction (8.21), distributions of $\overline{Sh_l}$ are similar to distributions of $\overline{Nu_l}$.

Figures 8.2b and 8.4 illustrate streamlines, isotherms, and isoconcentrations at different values of the Lewis number.

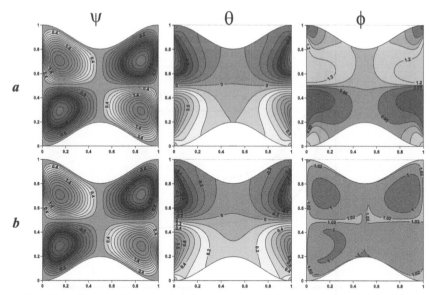

Fig. 8.4. Streamlines ψ, isotherms θ, and isoconcentrations ϕ for
$Ra = 100$, $Nr = Nb = Nt = 0.1$: $Le = 1$ (a), $Le = 1000$ (b) [reprinted
from Sheremet and Pop (2015a) with permission from ASME]

An increase in Le leads to more intensive convective heat transfer inside the cavity owing to high differences between heat and mass
transfer. It should be noted that, with respect to Fig. 8.4a, low values
of the Lewis number characterize non-homogeneous distributions of the
nanoparticles inside the cavity. Also, an increase in Lewis number leads
to an essential decrease in the thickness of concentration boundary layers at the vertical walls. It physically means that flow with a large Lewis
number prevents a spreading the nanoparticles in the nanofluid and as
a result one can find essential homogeneous distributions of nanoparticles' volume fraction inside the cavity in comparison with the case for
low values of the Lewis number (Fig. 8.4a).

An effect of the dimensionless time and Lewis number on the average Nusselt number at the left vertical wall is presented in Fig. 8.5. An
increase in Le from 1 to 1000 leads to an insignificant decrease in $\overline{Nu}_l$
and $\overline{Sh}_l$. An increase in the buoyancy ratio parameter from $Nr = 0.1$
to $Nr = 0.4$ leads to insignificant changes in all considered parameters
(streamlines, isotherms, and isoconcentrations). The average Nusselt
and Sherwood numbers non-essentially increase with Nr. At the same
time, an increase in the Brownian motion parameter from $Nb = 0.1$ to
$Nb = 0.4$ leads to more homogeneous distribution of nanoparticles.

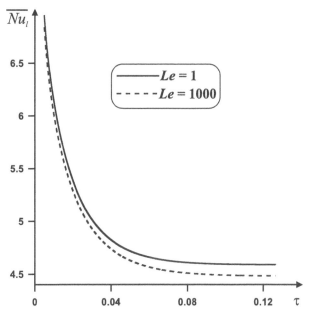

Fig. 8.5. Variation of the average Nusselt number at left vertical wall with the Lewis number and dimensionless time for $Ra = 100$, $Nr = Nb = Nt = 0.1$ [reprinted from Sheremet and Pop (2015a) with permission from ASME]

It is worth noting here that visible effect of the Brownian motion parameter occurs for the low values of the Rayleigh number where heat conduction is a dominant heat transfer mechanism (Celli, 2013). Essential changes in all considered parameters occur at changes in the thermophoresis parameter from $Nt = 0.1$ to $Nt = 0.4$.

An increase in this parameter leads to an essential intensification of convective flow inside the cavity, a more intensive heating of the domain of interest, and a less homogeneous distribution of nanoparticles. As a result, the average Nusselt and Sherwood numbers increase with the thermophoresis parameter.

Based on the findings in this study, we conclude the following:

(1) Low Rayleigh and Lewis numbers, low values of the Brownian motion parameter, and high values of thermophoresis parameter reflect non-homogeneous distribution of the nanoparticles inside the cavity. Therefore, for such values of Ra, Le, Nb, and Nt, a non-homogeneous model is more appropriate for the description of the system.

(2) Non-homogeneous areas inside the cavity reflect direction of convective heat transfer as heatlines defined by Kimura and Bejan (1983).

B. Natural Convection in a Wavy Open Porous Cavity Filled with a Nanofluid: Tiwari and Das' Nanofluid Model

1. Introduction

Transport processes through porous media play important roles in numerous practical applications in modern industry, such as the design of building components for energy consideration, control of pollutant spread in groundwater, geothermal energy technology, compact heat exchangers, solar power collectors, and the food industries, and have wide potential applications in many engineering areas including chemical, petroleum, polymer, food processing, pharmaceutical, and biochemical engineering.

Recent decades have seen a spike in the number of studies (Shenoy, 1994) devoted to the study of convective flow through porous media.

Natural convection in a porous enclosure has lots of applications in both nature and engineering, such as insulation for buildings, industrial cold-storage installations, and heat transfer improvements to heat exchanger apparatuses and petroleum reservoirs.

Excellent and comprehensive reviews have been presented by Merkin (1980), Kimura et al. (1987), Khanafer and Chamkha (1999), Ingham and Pop (1998, 2005), Vafai (2005), Nield and Bejan (2013), Jansen (2013), and Bagchi and Kulacki (2014).

Conventional heat transfer fluids such as water, ethylene glycol mixture, and engine oil have limited heat transfer capabilities due to their low thermal conductivity in enhancing the performance and compactness of many engineering electronic devices.

In contrast, metals have thermal conductivities up to three times higher than these fluids. Thus, it is naturally desirable to combine the two substances to produce a medium for heat transfer that would behave like a fluid, but has the thermal conductivity of a metal, and this combination has resulted in the so-called nanofluids. Small particles (nanoparticles) stay suspended much longer than larger particles, and help the fluid to behave more like a stable colloidal suspension rather than a settling suspension. Presence of particles alters the flow properties of the suspending medium, and a better understanding of that can be got from the treatise by Shenoy (1999). The presence of the nanoparticles in the fluids appreciably increases the effective thermal conductivity and viscosity of the base fluid and consequently enhances the heat transfer characteristics. Choi (1995) is probably the first to be credited for introducing the term "nanofluid" as a new class of fluids. These are fluids containing nanoparticles (nanometer-sized particles of metals, oxides, carbides, nitrides, or nanotubes). Choi et al. (2001) showed that the addition of a small amount (less than 1% by volume) of nanoparticles to conventional heat transfer liquids enhanced the thermal conductivity of the fluid up to approximately two times. There are engineered colloidal suspensions of nanoparticles in a base fluid.

Nanotechnology has been widely used in industry since materials with sizes of nanometers possess unique physical and chemical properties. Nanofluids may be used in various applications, which include electronic cooling, vehicle cooling transformer, and coolant for nuclear reactors. Since heat transfer occurs on the surface of a solid, this feature greatly enhances the fluid's heat conduction contribution.

Eastman et al. (2001) reported an increase of 40% in the effective thermal conductivity of ethylene–glycol with 0.3% volume of copper nanoparticles of 10nm diameter. Further, a 10–30% increase of the effective thermal conductivity in alumina-water nanofluids with 1–4% of alumina was reported by Das et al. (2003). These reports led Buongiorno and Hu (2005) to suggest the possibility of using nanofluids in advanced nuclear systems. Another recent application of the nanofluid flow is in the delivery of nano-drugs, as suggested by Kleinstreuer et al. (2008).

Many researchers (Khanafer et al., 2003; Maliga et al., 2005; Tiwari and Das, 2007; Öztop and Abu-Nada, 2008; Aminossadati and Ghasemi, 2009; Ghasemi and Aminossadati, 2009; Kuznetsov and Sheremet, 2011; Mahmoudi et al., 2011; Celli, 2013) have investigated the heat transfer characteristics of nanofluids in different cavities. Their results have shown that the presence of nanoparticles in fluids increases thermal conductivity of the fluid and consequently enhances heat transfer characteristics. Good literature reviews on convective flow and applications of nanofluids have been done by Buongiorno (2006), Das et al. (2007), Kakaç and Pramuanjaroenkij (2009), Wong and Leon (2010), Saidur et al. (2011), Wen et al. (2011), Jaluria et al. (2012), Ahmed and Abd El-Aziz (2013), Mahian et al. (2013), Sakai et al. (2014), and many others.

The above literature reviews reveal that no study investigates natural convection in a porous open cavity filled with a nanofluid having a vertical wavy wall. The present section deals with this topic.

2. Basic Equations

We analyze the natural convective heat transfer in a porous medium saturated with a Cu–water nanofluid located in a partially open cavity with left wavy, and right and bottom flat solid walls. The considered domain of interest is presented in Fig. 8.6.

It is worth noting that the left wavy wall and right flat wall of the open cavity are defined by the relations such as:

$\bar{x}_1 = D - D[a + b\cos(2\pi\kappa\bar{y}/H)]$ is the left wavy wall;

$\bar{x}_2 = D$ is the right flat wall;

$\bar{\Delta} = \bar{x}_2 - \bar{x}_1 = D[a + b\cos(2\pi\kappa\bar{y}/H)]$ is the distance between the right flat and left wavy walls.

Here a and b are constants that determine the shape and the wavy contraction ratio of the wavy wall $(a + b = 1)$.

It is assumed that the nanofluid temperature is equal to the solid matrix temperature everywhere in the homogeneous and isotropic porous medium, and the local thermal equilibrium model is used.

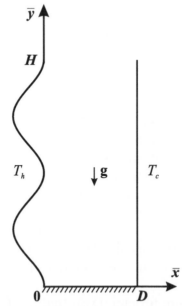

Fig. 8.6. Physical model and coordinate system

In the present study, the Darcy–Boussinesq model has been adopted in the governing equations of the problem. Taking into account these assumptions, the governing equations can be written in dimensional Cartesian coordinates (see Nield and Bejan, 2013):

$$\frac{\partial \overline{u}}{\partial \overline{x}} + \frac{\partial \overline{v}}{\partial \overline{y}} = 0 \tag{8.22}$$

$$\frac{\mu_{nf}}{K}\left(\frac{\partial \overline{u}}{\partial \overline{y}} - \frac{\partial \overline{v}}{\partial \overline{x}}\right) = -g(\rho\beta)_{nf}\frac{\partial T}{\partial \overline{x}} \tag{8.23}$$

$$\overline{u}\frac{\partial T}{\partial \overline{x}} + \overline{v}\frac{\partial T}{\partial \overline{y}} = \alpha_{nmf}\left(\frac{\partial^2 T}{\partial \overline{x}^2} + \frac{\partial^2 T}{\partial \overline{y}^2}\right). \tag{8.24}$$

The physical properties of the base fluid, nanoparticles, and solid matrix of porous medium used in the present study have been described in detail elsewhere (see Sheremet et al., 2015a–g). It should be noted that here we continue to develop the created mathematical model presented by Sheremet et al. (2015a–g) using the new empirical correlations for heat capacitance, thermal conductivity, and thermal diffusivity of the nanofluid-saturated porous medium. The correlations used include properties of the base fluid, solid particles, and solid matrix of the porous medium that allow us to analyze the fluid flow and heat transfer in the porous cavity more accurately.

Introducing the stream function ($\overline{u} = \partial\overline{\psi}/\partial\overline{y}$, $\overline{v} = -\partial\overline{\psi}/\partial\overline{x}$), and the following dimensionless variables:

$$x = \overline{x}/D, \; y = \overline{y}/D, \; \psi = \overline{\psi}/\alpha_{mnf},$$

$$\theta = (T - T_0)/(T_h - T_c), \; T_0 = 0.5(T_h + T_c), \tag{8.25}$$

the governing equations in dimensionless form can be written as follows:

$$\frac{\partial^2 \psi}{\partial x^2} + \frac{\partial^2 \psi}{\partial y^2} = -Ra \cdot H(\varphi)\frac{\partial \theta}{\partial x} \tag{8.26}$$

$$\frac{\partial \psi}{\partial y}\frac{\partial \theta}{\partial x} - \frac{\partial \psi}{\partial x}\frac{\partial \theta}{\partial y} = \frac{\partial^2 \theta}{\partial x^2} + \frac{\partial^2 \theta}{\partial y^2}. \tag{8.27}$$

Taking into account the considered dimensionless variables (8.25), the left wavy and the right flat walls of the cavity are described by the following relations: $x_1 = 1 - a - b(\cos \pi \kappa y/A)$ is the left wavy wall; $x_2 = 1$ is the right flat wall; $\Delta = x_2 - x_1 = a + b\cos(2\pi\kappa y/A)$ is the distance between the vertical walls.

The corresponding boundary conditions for these equations are

$$
\begin{aligned}
\psi = 0, \qquad & \theta = 0.5 \qquad && \text{at} \quad x = x_1 \\
\psi = 0, \qquad & \theta = -0.5 \qquad && \text{at} \quad x = x_2 \\
\psi = 0, \qquad & \partial\theta/\partial y = 0 \quad && \text{at} \quad y = 0 \\
\partial\psi/\partial y = 0, \quad & \partial\theta/\partial y = 0 \quad && \text{at} \quad y = A.
\end{aligned}
\tag{8.28}
$$

Here $Ra = gK(\rho\beta)_f(T_h - T_c)L/(\alpha_m\mu_f)$ is the Rayleigh number for the porous medium, $\alpha_m = k_m/(\rho C_p)_f$ is the thermal diffusivity of the viscous fluid-saturated porous medium, $A = H/D$ is the aspect ratio of the channel, and the function $H(\varphi)$ is given by

$$
H(\varphi) = \frac{[1 - \varphi + \varphi(\rho\beta)_p/(\rho\beta)_f][1 - \varphi + \varphi(\rho C_p)_p/(\rho C_p)_f]}{1 - \dfrac{3\varepsilon\varphi k_f(k_f - k_p)}{k_m[k_p + 2k_f + \varphi(k_f - k_p)]}}(1 - \varphi)^{2.5}.
\tag{8.29}
$$

3. Numerical Method and Validation

The domain of interest in the x and y plane has been converted into a rectangular geometry on the basis of the special algebraic coordinate transformation, with new independent variables ξ and η:

$$
\begin{cases}
\xi = \dfrac{x - x_1}{\Delta} = \dfrac{x - 1 + a + b\cos(2\pi\kappa y/A)}{a + b\cos(2\pi\kappa y/A)} \\[2mm]
\eta = y
\end{cases}
\tag{8.30}
$$

The left wavy and right flat walls of the cavity have been reflected into the lines with constant values of ξ.

Using the algebraic coordinate transformation (8.30), the governing equations (8.26) and (8.27) can be written as follows:

$$
\left[\left(\frac{\partial\xi}{\partial x}\right)^2 + \left(\frac{\partial\xi}{\partial y}\right)^2\right]\frac{\partial^2\psi}{\partial\xi^2} + 2\frac{\partial\xi}{\partial y}\frac{\partial^2\psi}{\partial\xi\partial\eta} + \frac{\partial^2\psi}{\partial\eta^2} + \frac{\partial^2\xi}{\partial y^2}\frac{\partial\psi}{\partial\xi}
$$

$$
= -Ra \cdot H(\varphi)\frac{\partial\xi}{\partial x}\frac{\partial\theta}{\partial\xi}
\tag{8.31}
$$

$$
\frac{\partial\xi}{\partial x}\frac{\partial\psi}{\partial\eta}\frac{\partial\theta}{\partial\xi} - \frac{\partial\xi}{\partial x}\frac{\partial\psi}{\partial\xi}\frac{\partial\theta}{\partial\eta} = \left[\left(\frac{\partial\xi}{\partial x}\right)^2 + \left(\frac{\partial\xi}{\partial y}\right)^2\right]\frac{\partial^2\theta}{\partial\xi^2}
$$

$$+2\frac{\partial \xi}{\partial y}\frac{\partial^2 \theta}{\partial \xi \partial \eta} + \frac{\partial^2 \theta}{\partial \eta^2} + \frac{\partial^2 \xi}{\partial y^2}\frac{\partial \theta}{\partial \xi} \tag{8.32}$$

with corresponding boundary conditions

$$\psi = 0, \ \theta = 0.5 \text{ on } \xi = 0$$
$$\psi = 0, \ \theta = -0.5 \text{ on } \xi = 1$$
$$\psi = 0, \ \partial \theta / \partial \eta = 0 \text{ on } \eta = 0 \tag{8.33}$$
$$\partial \psi / \partial \eta = 0, \ \partial \theta / \partial \eta = 0 \text{ on } \eta = A.$$

The following utilized metric coefficients worth noting are:

$$\frac{\partial \xi}{\partial x} = \frac{1}{a + b \cos(2\pi \kappa y/A)}, \quad \frac{\partial \xi}{\partial y} = \frac{2\pi \kappa b(x-1)\sin(2\pi \kappa y/A)}{A[a + b\cos(2\pi \kappa y/A)]^2}, \quad \frac{\partial^2 \xi}{\partial x^2} = 0,$$

$$\frac{\partial^2 \xi}{\partial y^2} = \frac{4\pi^2 \kappa^2 b(x-1)[a\cos(2\pi \kappa y/A) + b + b\sin^2(2\pi \kappa y/A)]}{A^2[a + b\cos(2\pi \kappa y/A)]^3}.$$

The partial differential equations (8.31) and (8.32) with corresponding boundary conditions (8.33) have been solved using an in-house computational fluid dynamics code (see Aleshkova and Sheremet, 2010; Sheremet and Trifonova, 2013; Sheremet and Pop, 2014a–c; Sheremet et al., 2014). This computational code has been described in detail and validated successfully (see Sheremet et al., 2015a–g). We have also conducted the grid independent test, analyzing the steady-state free convection in an open porous cavity filled with a Cu–water nanofluid at $Ra = 300$, $\varphi = 0.05$, $\varepsilon = 0.8$, $\kappa = 3$, $a = 0.9$, and $A = 4$ where the solid matrix of the porous medium is the aluminum foam. Three cases of the uniform grid are tested: a grid of 50×200 points, a grid of 100×400 points, and a much finer grid of 200×800 points.

Table 8.2 shows the effect of the mesh on the average Nusselt number of the left vertical wall:

$$\overline{Nu_l} = -\frac{k_{mnf}}{A \cdot k_m} \int_0^A \left(\frac{\partial \theta}{\partial \xi}\right)_{\xi=0} d\eta.$$

Table 8.2. Variations of the Average Nusselt Number
of the Left Vertical Wall with the Uniform Grid

Uniform grids	$\overline{Nu_l}$	$\Delta = \dfrac{\left|\overline{Nu_{l_{i \times j}}} - \overline{Nu_{l_{100 \times 400}}}\right|}{\overline{Nu_{l_{i \times j}}}} \times 100\%$
50×200	4.25876	3.1
100×400	4.12664	–
200×800	4.07133	1.4

Taking into account the conducted verifications, the uniform grid of 100×400 points has been selected for the further investigation.

4. Results and Discussion

A numerical study has been conducted at the following values of key parameters: Rayleigh number ($Ra = 10-500$), the solid volume fraction parameter of nanoparticles ($\varphi = 0.0-0.05$), the porosity of the porous medium ($\varepsilon = 0.1-0.8$), the aspect ratio ($A = 1-4$), the solid matrix of the porous medium (aluminum foam), undulation number ($\kappa = 1-4$), shape coefficient ($a = 0.6-1.4$). Particular efforts have been focused on the effects of these parameters on the fluid flow and heat transfer. Streamlines, isotherms, and local Nusselt number along the left vertical wavy wall $\{Nu_l = -k_{mnf}/k_m(\partial\theta/\partial\xi)_{\xi=0}\}$ for different values of key parameters mentioned above are illustrated in Figs. 8.7–8.16. Regardless of the governing parameters' values in the considered open porous cavity, one can find the main descending flow along the right vertical flat wall having low temperature and the main ascending flow along the left vertical wavy wall having high temperature. The top boundary of the porous cavity is open, and therefore, free convection is due to the presence of temperature difference between vertical walls and permanent entering and leaving of the nanofluid from the inlet/outlet boundaries. These inlet/outlet boundaries are identical and present the top open wall of the cavity. It is worth noting that the cooled nanofluid goes down to the adiabatic bottom wall, and along the left heated wall the nanofluid comes up to the upper outlet boundary.

Figure 8.7 demonstrates the effect of the Rayleigh number for the porous medium on streamlines and isotherms for Cu–water nanofluid at $\varphi = 0.04$, $\varepsilon = 0.5$, $A = 4.0$, $\kappa = 3.0$, $a = 0.9$. It should be noted that the analysis for the effect of Ra has been conducted for the wavy wall with three undulations. At low Rayleigh numbers ($Ra < 100$), the main heat transfer mechanism is heat conduction. In the case of $Ra = 10$ (Fig. 8.7a), the right-side isotherms parallel to the right vertical wall and the left-side isotherms parallel to the left vertical wall characterize the heat conduction dominance. At the same time, two vortices are formed in the center of the cavity opposite the concave part of the wavy wall.

An increase in the Rayleigh number ($Ra = 50$ and $Ra = 100$) leads to both an intensification of convective flow taking into account the values of the stream function and disappearance of one vortex in the center of the cavity.

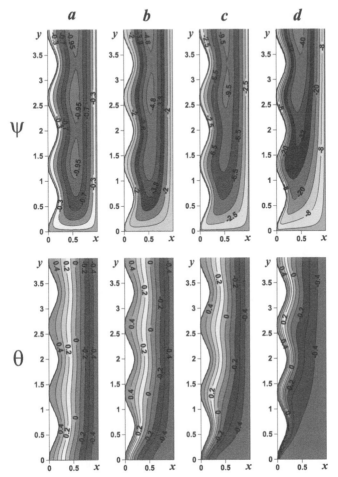

Fig. 8.7. Streamlines ψ and isotherms θ for $\varphi = 0.04$, $\varepsilon = 0.5$, $A = 4.0$, $\kappa = 3.0$, $a = 0.9$: $Ra = 10$ (a), $Ra = 50$ (b), $Ra = 100$ (c), $Ra = 500$ (d)

The latter can be explained by an essential effect of the buoyancy force and as a result the convective cores ascend in vertical direction in the very tall cavity. The utilized boundary conditions for the stream function and temperature at the top open surface can be considered like the symmetry conditions; therefore, it is possible to assume that we have the tall open cavity with left wavy wall.

Formation of the thermal boundary layer along the wavy wall is observed for $Ra \geq 50$ in the bottom part of the cavity, where one can find an increase in density of isotherms close to the wavy wall.

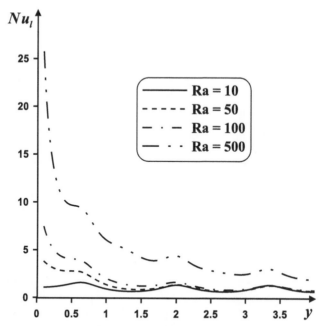

Fig. 8.8. Variation of the local Nusselt number with the Rayleigh
number at $\varphi = 0.04$, $\varepsilon = 0.5$, $A = 4.0$, $\kappa = 3.0$, $a = 0.9$

Further increase in the Rayleigh number (Fig. 8.7d) leads to both a
significant intensification of convective heat transfer with a thin thermal
boundary layer, and vanishing of the convective cores that are displaced
in the vertical direction. It should be noted that the considered range
of the Rayleigh number does not lead to formation of circulating flows
in concave parts of the vertical wall due to low motion velocities of the
nanofluid inside the porous medium.

An increase in Ra leads to an increase in the local Nusselt number
along the vertical wavy wall (Fig. 8.8). Corrugated behavior of the
local Nusselt number is due to the form of the wall where an increase
in Nu_l occurs along the undulation crests. An increase in Ra results in
an increase in the local Nusselt number close to the bottom wall due
to a formation of the thermal boundary layer that has been described
above. Therefore, these obtained distributions reflect an evolution of
the thermal boundary layer with Ra.

Figures 8.9 and 8.10 illustrate the effect of the solid volume fraction parameter of nanoparticles on streamlines, isotherms, and the local Nusselt number. A weak increase in φ does not lead to significant
changes in isotherms and streamlines.

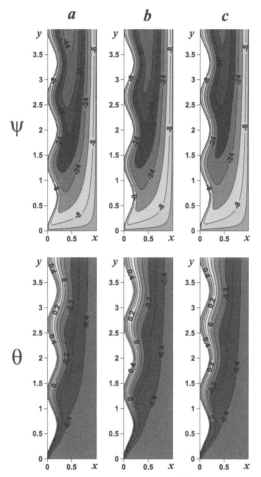

Fig. 8.9. Streamlines ψ and isotherms θ for $Ra = 500$, $\varepsilon = 0.5$, $A = 4.0$, $\kappa = 3.0$, $a = 0.9$: $\varphi = 0.0$ (a), $\varphi = 0.02$ (b), $\varphi = 0.05$ (c)

One can find only small attenuation of convective flow taking into account the absolute values of the stream function inside the porous cavity ($|\psi|_{max}^{\varphi=0.0} = 46 > |\psi|_{max}^{\varphi=0.2} = 44 > |\psi|_{max}^{\varphi=0.4} = 41$).

Figure 8.10 presents variations of the local Nusselt number along the wavy wall with the nanoparticles' volume fraction. An increase in φ leads to a small decrease in Nu_l, and as a result, in the average Nusselt number ($\overline{Nu}_{l_{\varphi=0.0}} = 6.575 > \overline{Nu}_{l_{\varphi=0.2}} = 6.309 > \overline{Nu}_{l_{\varphi=0.5}} = 5.919$). It should be noted that an increase in the porosity of the porous medium does not lead to changes in streamlines, isotherms, and Nusselt number.

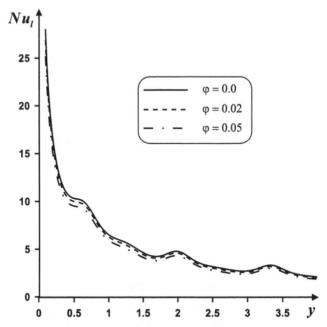

Fig. 8.10. Variation of the local Nusselt number with the solid volume fraction parameter of nanoparticles at $Ra = 500$, $\varepsilon = 0.5$, $A = 4.0$, $\kappa = 3.0$, $a = 0.9$

The effect of the aspect ratio on streamlines and isotherms are presented in Fig. 8.11. An increase in A leads to an intensification of convective flow inside the cavity, taking into account the absolute values of the stream function inside the porous cavity ($|\psi|_{\max}^{A=1.0} = 24.4 < |\psi|_{\max}^{A=2.0} = 33.2 < |\psi|_{\max}^{A=3.0} = 38.4$). Such behavior is due to an elongated cavity where an evolution of the thermal boundary layer occurs with the y-coordinate.

Changes in the local Nusselt number with A are presented in Fig. 8.12. It is interesting to note that in the case of small height of the cavity ($A = 1.0$) an effect of undulations is essential, with maximum values of Nu_l at the undulations' crests and minimum values at the center of the concave parts of the wall. An increase in A leads to a decrease in the local maximum and minimum values of Nu_l, while at the beginning of the wavy wall ($0 < y < 0.1$), an increase in A leads to an essential increase in Nu_l. Therefore, we have the following relation for the average Nusselt number: $\overline{Nu}_{l_{A=1.0}} = 5.909 < \overline{Nu}_{l_{A=2.0}} = 7.03 > \overline{Nu}_{l_{A=3.0}} = 6.676$. Such a relation characterizes a presence of the optimum value of the aspect ratio with essential heat transfer rate.

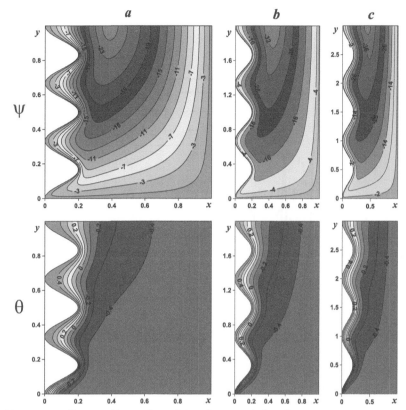

Fig. 8.11. Streamlines ψ and isotherms θ for $Ra = 500$, $\varphi = 0.04$, $\varepsilon = 0.5$, $\kappa = 3.0$, $a = 0.9$: $A = 1.0$ (a), $A = 2.0$ (b), $A = 3.0$ (c)

Figure 8.13 demonstrates the effect of the undulation number on streamlines and isotherms.

It should be noted that an increase in κ does not lead to a formation of additional recirculation zones inside the porous cavity, but this parameter certainly reflects an appearance of distortion in streamlines.

Also, one can find weak attenuation of convective flow with the undulation number ($|\psi|_{\max}^{\kappa=1.0} = 43.2 > |\psi|_{\max}^{\kappa=2.0} = 42.6 > |\psi|_{\max}^{\kappa=4.0} = 41.5 > |\psi|_{\max}^{\kappa=5.0} = 41.1$).

The effect of the undulation number on the local Nusselt number along the wavy wall is presented in Fig. 8.14. A presence of wall with one undulation does not lead to corrugated behavior of Nu_l, while in the case of the wall with two undulations one can find a weak oscillation of the local Nusselt number along the wall. Further increase in κ results in essential oscillation of Nu_l.

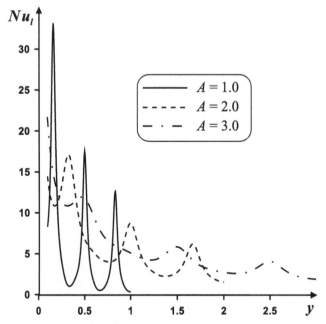

Fig. 8.12. Variation of the local Nusselt number with the aspect ratio
at $Ra = 500$, $\varphi = 0.04$, $\varepsilon = 0.5$, $\kappa = 3.0$, $a = 0.9$

It is worth noting that an increase in the undulation number leads
to a decrease in the local Nusselt number and average Nusselt number
$(\overline{Nu}_{l_{\kappa=1.0}} = 6.852 > \overline{Nu}_{l_{\kappa=2.0}} = 6.494 > \overline{Nu}_{l_{\kappa=5.0}} = 5.041)$.

The influence of the shape coefficient on the streamlines and
isotherms is presented in Fig. 8.15. Low value of this parameter
($a = 0.6$) reflects an essential blocking for the nanofluid motion.

Therefore, one can find a formation of recirculations in concave
zones of the porous cavity. Taking into account a weak motion in these
zones, the dominating heat transfer mechanism is heat conduction.

An increase in a leads to an increase in the length between the
undulations' crests and the right vertical wall. Such changes result in
intensification of convective flow inside the cavity.

Figure 8.16 presents the effect of the form coefficient on the local
Nusselt number along the wavy wall. An increase in a from 0.6 to 1.1
leads to an increase in Nu_l, and as a result, in the average Nusselt number. Further increase in a up to 1.4 leads to a formation of significant
oscillations with low average value.

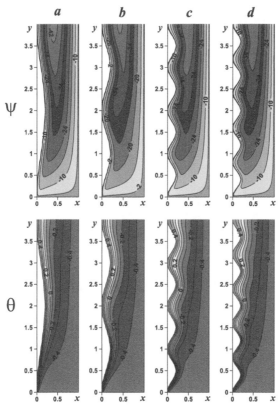

Fig. 8.13. Streamlines ψ and isotherms θ for $Ra = 500$, $\varphi = 0.04$, $\varepsilon = 0.5$, $A = 4.0$, $a = 0.9$: $\kappa = 1.0$ (a), $\kappa = 2.0$ (b), $\kappa = 4.0$ (c), $\kappa = 5.0$ (d)

The main findings can be listed as:

(1) An insertion of solid nanoparticles suppresses the convective flow.

(2) The average Nusselt number is a non-monotonic function of the aspect ratio and shape coefficient.

(3) The flow is enhanced with the Rayleigh number, aspect ratio, and shape coefficient, and is suppressed with nanoparticles' volume fraction and undulation number.

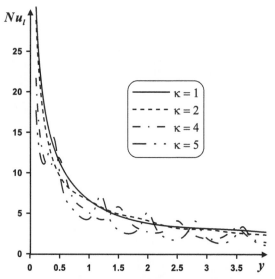

Fig. 8.14. Variation of the local Nusselt number with the undulation
number at $Ra = 500$, $\varphi = 0.04$, $\varepsilon = 0.5$, $A = 4.0$, $a = 0.9$

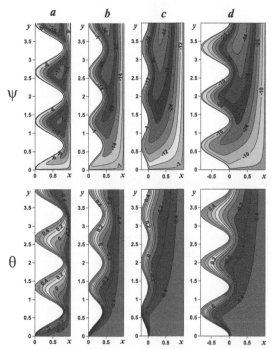

Fig. 8.15. Streamlines ψ and isotherms θ for $Ra = 500$, $\varphi = 0.04$,
$\varepsilon = 0.5$, $A = 4.0$, $\kappa = 3.0$: $a = 0.6$ (a), $a = 0.8$ (b), $a = 1.1$ (c),
$a = 1.4$ (d)

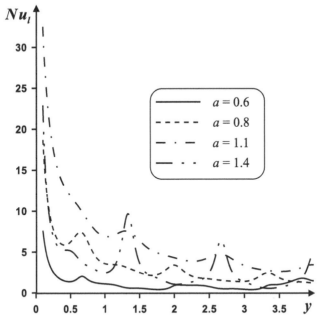

Fig. 8.16. Variation of the local Nusselt number with the form coefficient at $Ra = 500$, $\varphi = 0.04$, $\varepsilon = 0.5$, $A = 4.0$, $\kappa = 3.0$

C. Unsteady Free Convection in a Porous Open Wavy Cavity Filled with a Nanofluid Using Buongiorno's Mathematical Model

1. Introduction

Examples of unsteady flows are many. In fact, there is no actual flow situation, natural or artificial, that does not involve some unsteadiness. In many practical industrial applications as well as in geophysical processes, however, fluid flows and heat transfer are inherently time-dependent or unsteady in nature. For the sake of simplicity, the flows in all engineering applications were arbitrarily assumed to be steady. For example, the lifting characteristics of an airfoil or the drag characteristics of a blunt body were problems attacked both analytically and experimentally as steady problems. It is well-known, however, that in practice such devices encounter smooth or sudden changes in their aerodynamic environment. In many other engineering applications, unsteadiness is an integral part of the problem. The helicopter rotor, the cascades of blades of turbo machinery, and the ship propeller normally operate in an unsteady regime (Telionis, 1981).

The great majority of studies on convection in an enclosure are concerned primarily with steady-state solutions. In spite of frequent occurrences of time-dependent fluid and heat transport phenomena, comparatively scanty attention has been given to this time-varying heat transfer problem, particularly the subject of unsteady convection in an enclosure. This is possibly due to the added complexity and relatively poor state of understanding of the underlying physics of time-dependent convective flows (Pop, 1983; Joshi, 1990; Hyun, 1994; Hadjisophocleous et al., 1998; Dinarvand et al., 2015). It is worth saying that convection heat transfer in cavities is of great practical interest in many engineering applications, including electronic device cooling, chemical processing equipment, lubrication systems, food processing, solar energy collectors, and so on. Most of the studies on convective flow of viscous fluids and flows in porous media have used the base fluid with a low thermal conductivity, which, in turn, limits the heat transfer enhancement. An innovative technique to enhance heat transfer is by using nano-scale particles in the base fluid. Nanotechnology has been widely used in industry since materials with sizes of nanometers possess unique physical and chemical properties. Nano-scale particle-added fluids are called nanofluids (Choi, 1995), and show promise in significantly increasing heat transfer rate. In an illuminating paper by Buongiorno (2006), he mentions that nanofluids have higher thermal conductivity and single-phase heat transfer coefficients than their base fluids. In particular, the heat transfer coefficient increases appear to go beyond the mere thermal-conductivity effect, and cannot be predicted by traditional purefluid correlations such as Dittus–Boelter's. In the nanofluid literature, this behavior is generally attributed to thermal dispersion and intensified turbulence, brought about by nanoparticle motion. To test the validity of this assumption, Buongiorno (2006) has considered seven slip mechanisms that can produce a relative velocity between the nanoparticles and the base fluid. These are inertia, Brownian diffusion, thermophoresis, diffusiophoresis, Magnus effect, fluid drainage, and gravity. However, he concluded that, of these seven, only Brownian diffusion and thermophoresis are important slip mechanisms in nanofluids. Based on this finding, he developed a two-component, four-equation nonhomogeneous equilibrium model for mass, momentum, and heat transport in nanofluids. Very good literature reviews on convective flow and applications of nanofluids have been done by Buongiorno (2006), Nield and Bejan (2013), and Sakai et al. (2014).

Convection in fluid-saturated porous media has been of growing interest during the last several decades because of its great practical applications in modern industry, such as the design of building components for energy consideration, control of pollutant spread in groundwater, compact heat exchangers, solar power collectors, food industries, etc. (see Nield and Bejan, 2013). It is stated by Cho et al. (2013a,b) that wavy geometries are used in many engineering systems as a means of enhancing the transport performance. On the other hand, the heat transfer characteristics of nanofluids confined in cavities with wavy surfaces have attracted increasing attention in recent times. Several authors, such as Abu-Nada and Öztop (2011), Nikfar and Mahmoodi (2012), and Cho et al. (2012a–d) have performed numerical studies on natural convection heat transfer characteristics of Al_2O_3–water nanofluid in a cavity with wavy walls, while Esmaeilpour and Abdollahzadeh (2012) examined the natural convection heat transfer behavior and entropy generation of Cu–water nanofluid confined in a cavity with vertical wavy walls.

This section deals with unsteady convective heat transfer in a differentially heated wavy-walled open porous cavity saturated with a nanofluid using the nanofluid model proposed by Buongiorno.

2. Governing Equations

We will analyze the unsteady natural convection in a wavy-walled porous open cavity filled with a nanofluid. The domain of interest is presented in Fig. 8.17, where $\overline{x}$ and $\overline{y}$ are the Cartesian coordinates, D is the length of the bottom adiabatic wall, and H is the height of the cavity. The two vertical walls of the open cavity are assumed to be impermeable and isothermal while the bottom wall is adiabatic. It is assumed that the left wavy wall and right flat wall of the open cavity are described by the relations

$$\overline{x}_1 = D - D[a + b\cos(2\pi\kappa\overline{y}/H)] \text{ and } \overline{x}_2 = D, \text{ respectively, while}$$

$$\overline{\Delta} = \overline{x}_2 - \overline{x}_1 = D[a + b\cos(2\pi\kappa\overline{y}/H)]$$

is the distance between vertical walls.

The governing equations for the mass, momentum, and energy in a porous medium saturated with a nanofluid have been formulated using the Darcy–Boussinesq approximation and the local thermal equilibrium approach (Buongiorno, 2006; Kuznetsov and Nield, 2013; Nield and Kuznetsov, 2014).

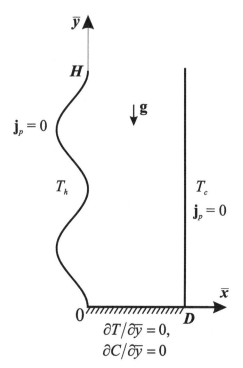

Fig. 8.17. Physical model and coordinate system [reprinted from Sheremet et al. (2015g) with permission from Elsevier]

Introducing the following dimensionless variables:

$$x = \overline{x}/D, \ y = \overline{y}/D, \ \psi = \overline{\psi}/\alpha_m,$$
$$\theta = (T - T_0)/(T_h - T_c), \ \phi = C/C_0, \ \tau = t\alpha_m/(\sigma D^2) \tag{8.34}$$

we obtain the governing equations (Sheremet et al., 2014; Sheremet and Pop, 2014a–c, 2015a–e)

$$\frac{\partial^2 \psi}{\partial x^2} + \frac{\partial^2 \psi}{\partial y^2} = -Ra\frac{\partial \theta}{\partial x} + Ra \cdot Nr\frac{\partial \phi}{\partial x} \tag{8.35}$$

$$\frac{\partial \theta}{\partial \tau} + \frac{\partial \psi}{\partial y}\frac{\partial \theta}{\partial x} - \frac{\partial \psi}{\partial x}\frac{\partial \theta}{\partial y} = \frac{\partial^2 \theta}{\partial x^2} + \frac{\partial^2 \theta}{\partial y^2}$$

$$+Nb\left(\frac{\partial \phi}{\partial x}\frac{\partial \theta}{\partial x} + \frac{\partial \phi}{\partial y}\frac{\partial \theta}{\partial y}\right) + Nt\left[\left(\frac{\partial \theta}{\partial x}\right)^2 + \left(\frac{\partial \theta}{\partial y}\right)^2\right] \tag{8.36}$$

$$\frac{\partial \phi}{\partial \tau} + \frac{1}{\varepsilon}\left(\frac{\partial \psi}{\partial y}\frac{\partial \phi}{\partial x} - \frac{\partial \psi}{\partial x}\frac{\partial \phi}{\partial y}\right)$$

$$= \frac{1}{Le}\left(\frac{\partial^2 \phi}{\partial x^2} + \frac{\partial^2 \phi}{\partial y^2}\right) + \frac{1}{Le}\frac{Nt}{Nb}\left(\frac{\partial^2 \theta}{\partial x^2} + \frac{\partial^2 \theta}{\partial y^2}\right). \tag{8.37}$$

It should be noted that the derivation of these equations has been described in detail by Sheremet and Pop (2014a–c, 2015a–e) and Sheremet et al. (2014).

Taking into account the considered dimensionless variables (8.34), the vertical walls of the cavity are described by the following relations:

$$x_1 = 1 - a - b\cos(2\pi\kappa y/A) \text{ and } x_2 = 1, \text{ respectively,}$$

where $\Delta = x_2 - x_1 = a + b\cos(2\pi\kappa y/A)$ is the distance between the vertical walls.

The corresponding boundary conditions for these equations are given by

$$\psi = 0, \qquad \theta = 0.5, \ \tilde{\mathbf{j}}_p = 0$$

$$\left(\text{or } Nb\frac{\partial\phi}{\partial n} + Nt\frac{\partial\theta}{\partial n} = 0\right) \text{ on } x = x_1$$

$$\psi = 0, \qquad \theta = -0.5, \ \tilde{\mathbf{j}}_p = 0$$

$$\left(\text{or } Nb\frac{\partial\phi}{\partial x} + Nt\frac{\partial\theta}{\partial x} = 0\right) \text{ on } x = x_2$$

$$\psi = 0, \qquad \partial\theta/\partial y = 0, \ \partial\phi/\partial y = 0 \text{ on } y = 0$$

$$\partial\psi/\partial y = 0, \quad \partial\theta/\partial y = 0, \ \partial\phi/\partial y = 0 \text{ on } y = A.$$

$$\tag{8.38}$$

3. Numerical Method

The open cavity in the x and y plane, i.e., physical domain, is transformed into a rectangular geometry in the computational domain using an algebraic coordinate transformation that introduces new independent variables ξ and η. The vertical walls of the cavity become coordinate lines having constant values of ξ. The cavity geometry is mapped into a rectangle on the basis of the following transformation:

$$\xi = \frac{x - x_1}{\Delta} = \frac{x - 1 + a + b\cos(2\pi\kappa y/A)}{a + b\cos(2\pi\kappa y/A)}, \ \eta = y. \tag{8.39}$$

Taking into account transformation (8.39), the governing equations (8.35)–(8.37) are rewritten in the following form:

$$\left[\left(\frac{\partial\xi}{\partial x}\right)^2 + \left(\frac{\partial\xi}{\partial y}\right)^2\right]\frac{\partial^2\psi}{\partial\xi^2} + 2\frac{\partial\xi}{\partial y}\frac{\partial^2\psi}{\partial\xi\partial\eta} + \frac{\partial^2\psi}{\partial\eta^2} + \frac{\partial^2\xi}{\partial y^2}\frac{\partial\psi}{\partial\xi}$$

$$= -Ra\frac{\partial \xi}{\partial x}\frac{\partial \theta}{\partial \xi} + Ra \cdot Nr\frac{\partial \xi}{\partial x}\frac{\partial \phi}{\partial \xi} \tag{8.40}$$

$$\frac{\partial \theta}{\partial \tau} + \frac{\partial \xi}{\partial x}\frac{\partial \psi}{\partial \eta}\frac{\partial \theta}{\partial \xi} - \frac{\partial \xi}{\partial x}\frac{\partial \psi}{\partial \xi}\frac{\partial \theta}{\partial \eta}$$

$$= \left[\left(\frac{\partial \xi}{\partial x}\right)^2 + \left(\frac{\partial \xi}{\partial y}\right)^2\right]\frac{\partial^2 \theta}{\partial \xi^2} + 2\frac{\partial \xi}{\partial y}\frac{\partial^2 \theta}{\partial \xi \partial \eta} + \frac{\partial^2 \theta}{\partial \eta^2} + \frac{\partial^2 \xi}{\partial y^2}\frac{\partial \theta}{\partial \xi}$$

$$+Nb\left\{\left[\left(\frac{\partial \xi}{\partial x}\right)^2 + \left(\frac{\partial \xi}{\partial y}\right)^2\right]\frac{\partial \phi}{\partial \xi}\frac{\partial \theta}{\partial \xi} + \frac{\partial \xi}{\partial y}\left(\frac{\partial \phi}{\partial \xi}\frac{\partial \theta}{\partial \eta} + \frac{\partial \theta}{\partial \xi}\frac{\partial \phi}{\partial \eta}\right) + \frac{\partial \theta}{\partial \eta}\frac{\partial \phi}{\partial \eta} + \right\}$$

$$+Nt\left[\left(\frac{\partial \xi}{\partial x}\right)^2\left(\frac{\partial \theta}{\partial \xi}\right)^2 + \left(\frac{\partial \xi}{\partial y}\right)^2\left(\frac{\partial \theta}{\partial \xi}\right)^2 + 2\frac{\partial \xi}{\partial y}\frac{\partial \theta}{\partial \xi}\frac{\partial \theta}{\partial \eta} + \left(\frac{\partial \theta}{\partial \eta}\right)^2\right] \tag{8.41}$$

$$\frac{\partial \phi}{\partial \tau} + \frac{1}{\varepsilon}\left(\frac{\partial \xi}{\partial x}\frac{\partial \psi}{\partial \eta}\frac{\partial \phi}{\partial \xi} - \frac{\partial \xi}{\partial x}\frac{\partial \psi}{\partial \xi}\frac{\partial \phi}{\partial \eta}\right)$$

$$= \frac{1}{Le}\left\{\left[\left(\frac{\partial \xi}{\partial x}\right)^2 + \left(\frac{\partial \xi}{\partial y}\right)^2\right]\frac{\partial^2 \phi}{\partial \xi^2} + 2\frac{\partial \xi}{\partial y}\frac{\partial^2 \phi}{\partial \xi \partial \eta} + \frac{\partial^2 \phi}{\partial \eta^2} + \frac{\partial^2 \xi}{\partial y^2}\frac{\partial \phi}{\partial \xi}\right\}$$

$$+\frac{1}{Le}\frac{Nt}{Nb}\left\{\left[\left(\frac{\partial \xi}{\partial x}\right)^2 + \left(\frac{\partial \xi}{\partial y}\right)^2\right]\frac{\partial^2 \theta}{\partial \xi^2} + 2\frac{\partial \xi}{\partial y}\frac{\partial^2 \theta}{\partial \xi \partial \eta} + \frac{\partial^2 \theta}{\partial \eta^2} + \frac{\partial^2 \xi}{\partial y^2}\frac{\partial \theta}{\partial \xi}\right\}. \tag{8.42}$$

The corresponding initial and boundary conditions for these equations are given by

$$\psi = 0,\ \theta = 0,\ \phi = 1 \text{ for } \tau = 0$$
$$\psi = 0,\ \theta = 0.5,\ Nb \cdot \partial\phi/\partial\xi + Nt \cdot \partial\theta/\partial\xi = 0 \text{ on } \xi = 0$$
$$\psi = 0,\ \theta = -0.5,\ Nb \cdot \partial\phi/\partial\xi + Nt \cdot \partial\theta/\partial\xi = 0 \text{ on } \xi = 1 \quad (8.43)$$
$$\psi = 0,\ \partial\theta/\partial\eta = 0,\ \partial\phi/\partial\eta = 0 \text{ on } \eta = 0$$
$$\partial\psi/\partial\eta = 0,\ \partial\theta/\partial\eta = 0,\ \partial\phi/\partial\eta = 0 \text{ on } \eta = A.$$

It should be noted here that

$$\frac{\partial \xi}{\partial x} = \frac{1}{a + b\cos(2\pi\kappa y/A)}, \quad \frac{\partial \xi}{\partial y} = \frac{2\pi\kappa b(x-1)\sin(2\pi\kappa y/A)}{A[a + b\cos(2\pi\kappa y/A)]^2}, \quad \frac{\partial^2 \xi}{\partial x^2} = 0,$$

$$\frac{\partial^2 \xi}{\partial y^2} = \frac{4\pi^2\kappa^2 b(x-1)[a\cos(2\pi\kappa y/A) + b + b\sin^2(2\pi\kappa y/A)]}{A^2[a + b\cos(2\pi\kappa y/A)]^3}.$$

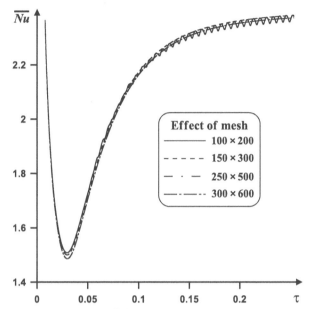

Fig. 8.18. Variation of the average Nusselt number versus
the dimensionless time and the mesh parameters [reprinted from
Sheremet et al. (2015g) with permission from Elsevier]

It is worth noting that for an analysis of the Sherwood number
along the wavy wall, it is possible to study only the Nusselt number
along this wall, because at the wavy wall we have

$$\frac{\partial \phi}{\partial \xi} = -\frac{Nt}{Nb}\frac{\partial \theta}{\partial \xi},$$

taking into account boundary conditions for ϕ (Eqs. (8.43)). Therefore,
the further analysis concerning integral parameters will be conducted
only for the average Nusselt number, because

$$Sh = -\frac{Nt}{Nb}Nu \text{ and } \overline{Sh} = -\frac{Nt}{Nb}\overline{Nu}.$$

The partial differential equations (8.40)–(8.42) with corresponding boundary conditions (8.43) were solved using the finite difference
method with the second order differencing schemes. Detailed description of the numerical technique used is presented by Aleshkova and
Sheremet (2010), Sheremet and Trifonova (2013), and Sheremet et al.
(2014, 2015a–g, 2016a,b). The present models, in the form of an in-
house computational fluid dynamics (CFD) code, have been validated
successfully (Sheremet et al., 2014, 2015a–g, 2016a,b).

For the purpose of obtaining a grid independent solution, a grid sensitivity analysis is performed. The grid independent solution was performed by preparing the solution for unsteady free convection in a wavy-wall open-porous cavity filled with a nanofluid at $Ra = 100$, $Le = 100$, $Nr = Nb = Nt = 0.1$, $\varepsilon = 0.8$, $A = 2$, $\kappa = 3$, $a = 0.9$. Four cases of the uniform grid are tested: a grid of 100×200 points, a grid of 150×300 points, a grid of 250×500 points, and a much finer grid of 300×600 points. Figure 8.18 shows an effect of the mesh on the average Nusselt number of the wavy wall.

On the basis of the conducted verifications the uniform grid of 250×500 points has been selected for the following analysis.

4. Results and Discussion

Numerical investigations of the boundary value problem (8.40)– (8.43) have been carried out at the following values of key parameters: $Ra = 100$, $Le = 100$, $Nr = Nb = Nt = 0.1$, $A = 2$, $\kappa = 1 - 4$, $a = 0.6 - 1.4$, $0 < \tau \le 0.25$. Figures 8.19–8.22 present isocontours of the stream function, temperature and nanoparticle volume fraction, and the variation of the average Nusselt number in dependence on the undulations number, shape parameter, and dimensionless time. Regardless of the governing parameters' values in the considered cavity, one can find a formation of a descending flow along the cooled right flat wall and an ascending flow along the heated left wavy wall. At the same time, a circulate flow is formed at the top open boundary of the cavity.

Figure 8.19 illustrates isocontours of the sought functions in the cavity for $a = 0.6$, $\tau = 0.25$ and different values of the undulations number. In the case of one undulation, a single convective cell is formed under the wave crest while, as has been mentioned above, the circulate flow is formed along the top open boundary over the wave crest. Such partitioning of the single descending/ascending flow inside the cavity is essentially due to the amplitude of the undulation that deforms the flow. At the same time, this partitioning can also be explained by the formation of the thermal boundary layer from the bottom part of the heated wall, where along the wave crest one can find a significant increase in the temperature gradient due to a narrowing of the flow.

Distortion of the isotherms of low temperature close to the bottom wall occurs due to the above-mentioned convective cell. Nanoparticle volume fraction reflects an increase in the nanoparticles' concentration in the bottom part of the cavity, namely, along the bottom wall, while in the top part close to the open border, the nanoparticles' concentration decreases. It is worth noting that the form of the isoconcentrations describes the flow motion with the formed convective cells. Also, one can find a decrease in ϕ in the cores of the convective cells and along the upper part of the wavy wall over the wave crest ($y > 1$). An increase in the undulations number leads to more essential heating of the area along the wave troughs, and as a result, one can find a decrease in the average Nusselt number at the wavy wall, presented in Figure 8.19.

Also, a single convective cell forms in each wave trough and an increase in κ leads to a decrease in the convective intensity, taking into account the maximum absolute values of the stream function.

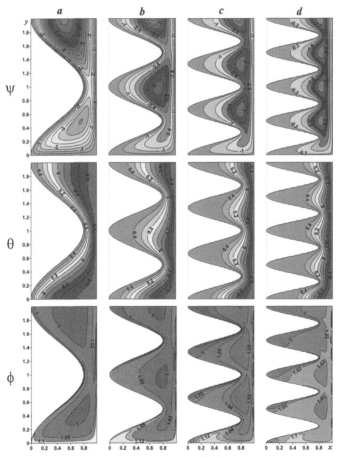

Fig. 8.19. Streamlines ψ, isotherms θ, isoconcentrations ϕ for $a = 0.6$, $\tau = 0.25$: $\kappa = 1.0$ (a), $\kappa = 2.0$ (b), $\kappa = 3.0$ (c), $\kappa = 4.0$ (d) [reprinted from Sheremet et al. (2015g) with permission from Elsevier]

It should be noted that in comparison with the case $\kappa = 1$ for high values of the undulations number, the first single convective cell is formed in the second wave trough. Distributions of the nanoparticles' concentration define an increase in ϕ along the bottom wall and a decrease in the nanoparticle volume fraction in the cores of the convective cells and along the flow-ward side of each undulation besides the last one, where one can find a decrease in ϕ along the whole undulation.

Figure 8.20 presents the dependencies of the average Nusselt number at the wavy wall on the undulations number and dimensionless time. First of all, it is necessary to note that an increase in κ leads to a decrease in $\overline{Nu}$ due to the significant heating of the wave troughs. An increase in dimensionless time reflects a formation of two main time levels, such as a decrease in $\overline{Nu}$ at initial time level due to heating of the surrounding nanofluid by the heat conduction mechanism, and the maintenance of the constant value of $\overline{Nu}$ due to convective flow of hot and cold nanofluid. It should be noted that for $\kappa = 1.0$, we have an additional time level that characterizes an increase in the average Nusselt number after the initial time level, due to more intensive convective flow in each wave trough.

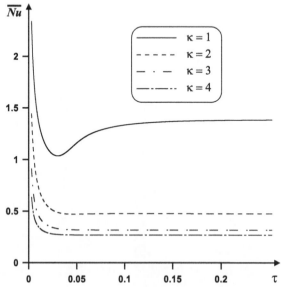

Fig. 8.20. Variation of the average Nusselt number versus the dimensionless time and undulation number [reprinted from Sheremet et al. (2015g) with permission from Elsevier]

Figure 8.21 demonstrates the contours of the stream function, temperature, and nanoparticle volume fraction for $\kappa = 1.0$, $\tau = 0.25$, and different values of the shape coefficient. It should be noted that values $a < 1.0$ characterize the presence of the concave wall and for $a > 1.0$ we have the convex wall. Also at $\kappa = 1.0$ the concave wall characterizes the presence of the wave crest inside the cavity while the convex wall characterizes the presence of the wave trough inside the cavity. An approach of the shape parameter to unity reflects a decrease in the amplitude of the undulations.

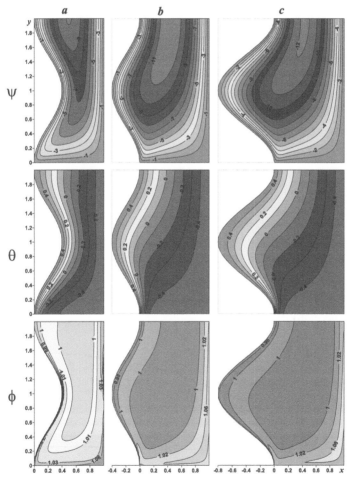

Fig. 8.21. Streamlines ψ, isotherms θ, isoconcentrations ϕ for $\kappa = 1.0$, $\tau = 0.25$: $a = 0.8$ (a), $a = 1.2$ (b), $a = 1.4$ (c) [reprinted from Sheremet et al. (2015g) with permission from Elsevier]

An increase in a leads to an essential intensification of convective flow inside the cavity. Also, an increase in a (> 1.0) leads to more essential heating of the wave trough that reflects a decrease in the temperature gradient along the wave wall. Behavior of the streamlines, isotherms, and isoconcentrations in the cases of the concave and convex walls has been described in detail above.

Variations of the average Nusselt number with the dimensionless time and shape parameter are presented in Fig. 8.22. It is worth noting that an increase in a (< 1.0) leads to heat transfer enhancement, while an increase in a (> 1.0) leads to non-monotonic changes in $\overline{Nu}$ due to more essential heating of the wave trough, as mentioned above. At the same time an increase in the shape coefficient leads to an increase in dimensionless time of an approach to the thermal steady-state.

The obtained results showed the following:

(1) An increase in the undulations number leads to a decrease in the average Nusselt number at the wavy wall due to the significant heating of the wave troughs.

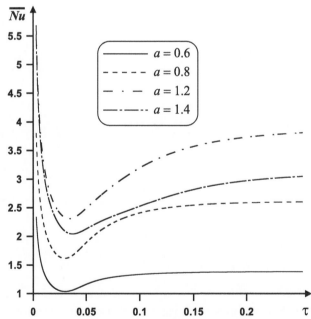

Fig. 8.22. Variation of the average Nusselt number versus the dimensionless time and shape coefficient [reprinted from Sheremet et al. (2015g) with permission from Elsevier]

(2) Distributions of the nanoparticles concentration define an increase in ϕ along the bottom wall and a decrease in the nanoparticle volume fraction in the cores of the convective cells and along the flowward side of each undulation besides the last one where one can find a decrease in ϕ along the whole undulation.

(3) An increase in the shape coefficient (< 1.0) leads to the heat transfer enhancement, while an increase in this parameter (> 1.0) leads to non-monotonic changes in the average Nusselt number due to more essential heating of the wave trough.

(4) An increase in the shape coefficient also leads to an increase in dimensionless time of an approach to the thermal steady-state.

D. Effect of Thermal Dispersion on Transient Natural Convection in a Wavy-Walled Porous Cavity Filled with a Nanofluid: Tiwari and Das' Nanofluid Model

1. Introduction

The study of convective flow and heat transfer in fluid-saturated porous media has a number of important applications in contemporary technologies such as, for example, thermal insulation of buildings, nuclear energy systems, geothermal energy systems, solar power collectors, regenerative heat exchangers, storage of grain, fruits, and vegetables, burying of drums containing heat generating chemicals in the earth, underground spread of pollutants, petroleum reservoirs, beds of fossil fuels, usage of porous conical bearings in lubrication technology, and many others (Nield and Bejan, 2013).

Murthy and Singh (1997) have pointed out, in a paper about the effect of viscous dissipation on a non-Darcy natural convection boundary layer along an isothermal vertical wall embedded in a fluid-saturated porous medium, that in the case when inertia terms are prevalent, the transverse thermal dispersion effects will become important. These effects were studied in detail by Plumb and Huenefeld (1981), Cheng (1981), Hong and Tien (1987), Hong et al. (1987), Cheng and Vortmeyer (1988), Amiri and Vafai (1994), and they have confirmed the importance of the thermal dispersion effects studied in these papers. It should be mentioned that except for Cheng and Vortmeyer (1988), all these boundary layer papers use the linear dependence of dispersion diffusivity on streamwise velocity. In order to correlate the available experimental data concerning the packed beds, Cheng and Vortmeyer

(1988) introduced a wall function term into the term of dispersion diffusivity.

This section focusses on the effects of thermal dispersion on transient free convection in a wavy-walled porous cavity filled with a nanofluid, using the mathematical nanofluid model proposed by Tiwari and Das (2007) with new physical properties (heat capacitance, thermal conductivity, and thermal diffusivity) of the nanofluid-saturated porous medium, published by Sheremet et al. (2015a).

2. Basic Equations

Consider the unsteady free convective heat transfer with thermal dispersion in a porous medium filled with a nanofluid (water and nanoparticles) located in a wavy-walled cavity. Following Nield and Kuznetsov (2009a,b, 2011, 2014), it is assumed that nanoparticles are suspended in the nanofluid using either surfactant or surface charge technology (Masuda et al., 2006). This prevents nanoparticles from agglomeration and deposition on the porous matrix. A schematic geometry of the problem under investigation is shown in Fig. 8.23, where $\bar{x}$ and $\bar{y}$ are the Cartesian coordinates measured in the horizontal and vertical directions, respectively.

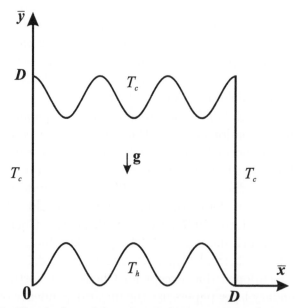

Fig. 8.23. Physical model and coordinate system [reprinted from Sheremet et al. (2016a) with permission from Elsevier]

It should be noted that the horizontal wavy walls are described by the following relations:

$\overline{y}_1 = D - D[a + b\cos(2\pi\kappa\overline{x}/D)]$ is the bottom wavy wall;

$\overline{y}_2 = D[a + b\cos(2\pi\kappa\overline{x}/D)]$ is the top wavy wall;

$\overline{\Delta} = \overline{y}_2 - \overline{y}_1 = 2D[a + b\cos(2\pi\kappa\overline{x}/D)] - D$ is the distance between these wavy walls.

Here a and b are constants that determine the shape and the wavy contraction ratio of the wavy wall $(a + b = 1)$.

The Darcy–Boussinesq approximation is employed. Homogeneity and local thermal equilibrium in the porous medium is assumed. For a description of nanofluid flow and heat transfer we use the Tiwari and Das' model (Tiwari and Das, 2007). It should be noted that this nanofluid approach is a single-phase effective model where the nanoparticles are assumed to have a uniform shape and size. The governing equations are (Murthy and Singh, 1997; Nield and Bejan, 2013):

$$\nabla \cdot \overline{\mathbf{V}} = 0 \tag{8.44}$$

$$0 = -\nabla p - \frac{\mu_{nf}}{K}\overline{\mathbf{V}} - (\rho\beta)_{nf}(T - T_0)\mathbf{g} \tag{8.45}$$

$$(\rho C_p)_{nf}\left(\frac{\partial T}{\partial t} + (\overline{\mathbf{V}} \cdot \nabla)T\right) = \frac{\partial}{\partial \overline{x}}\left[(k_{mnf} + k_d)\frac{\partial T}{\partial \overline{x}}\right]$$
$$+ \frac{\partial}{\partial \overline{y}}\left[(k_{mnf} + k_d)\frac{\partial T}{\partial \overline{y}}\right]. \tag{8.46}$$

Here $k_d = (\rho C_p)_{nf}|\overline{\mathbf{V}}|d_p\varphi$ is the enhancement in the thermal conductivity due to the thermal dispersion (Khanafer et al., 2003; Sakai et al., 2014).

The physical properties (heat capacitance, thermal conductivity, and thermal diffusivity) of the nanofluid-saturated porous medium have been described in detail by Sheremet et al. (2015a).

Equations (8.44)–(8.46) can be written in Cartesian coordinates as

$$\frac{\partial \overline{u}}{\partial \overline{x}} + \frac{\partial \overline{v}}{\partial \overline{y}} = 0 \tag{8.47}$$

$$\frac{\mu_{nf}}{K}\left(\frac{\partial \overline{u}}{\partial \overline{y}} - \frac{\partial \overline{v}}{\partial \overline{x}}\right) = -g(\rho\beta)_{nf}\frac{\partial T}{\partial \overline{x}} \tag{8.48}$$

$$\frac{\partial T}{\partial t} + \bar{u}\frac{\partial T}{\partial \bar{x}} + \bar{v}\frac{\partial T}{\partial \bar{y}} = \frac{\partial}{\partial \bar{x}}\left[\left(\alpha_{mnf} + d_p\varphi\sqrt{\bar{u}^2 + \bar{v}^2}\right)\frac{\partial T}{\partial \bar{x}}\right]$$

$$+ \frac{\partial}{\partial \bar{y}}\left[\left(\alpha_{mnf} + d_p\varphi\sqrt{\bar{u}^2 + \bar{v}^2}\right)\frac{\partial T}{\partial \bar{y}}\right]. \qquad (8.49)$$

Further, we introduce the following dimensionless variables:

$$x = \frac{\bar{x}}{D}, \; y = \frac{\bar{y}}{D}, \; u = \frac{\bar{u}}{\sqrt{g\beta(T_h - T_c)D}}, \; v = \frac{\bar{v}}{\sqrt{g\beta(T_h - T_c)D}},$$

$$\theta = \frac{T - T_0}{T_h - T_c}, \; \tau = t\sqrt{\frac{g\beta(T_h - T_c)}{D}}$$

$$(8.50)$$

where $T_0 = (T_h + T_c)/2$ is the mean temperature of heated and cooled walls.

One can introduce a dimensionless stream function ψ defined by

$$u = \frac{\partial \psi}{\partial y}, \; v = -\frac{\partial \psi}{\partial x} \qquad (8.51)$$

so that Eq. (8.47) is satisfied identically. We are then left with the following equations

$$\frac{\partial^2 \psi}{\partial x^2} + \frac{\partial^2 \psi}{\partial y^2} = -\sqrt{\frac{Ra \cdot Da}{Pr}} \cdot H_1(\varphi)\frac{\partial \theta}{\partial x} \qquad (8.52)$$

$$\frac{\partial \theta}{\partial \tau} + \frac{\partial \psi}{\partial y}\frac{\partial \theta}{\partial x} - \frac{\partial \psi}{\partial x}\frac{\partial \theta}{\partial y} = \sqrt{\frac{Da}{Ra \cdot Pr}} \cdot H_2(\varphi)$$

$$\times \frac{\partial}{\partial x}\left[\left(1 + \frac{Ds}{H_2(\varphi)}\sqrt{\frac{Ra \cdot Pr}{Da}}\sqrt{\left(\frac{\partial \psi}{\partial x}\right)^2 + \left(\frac{\partial \psi}{\partial y}\right)^2}\right)\frac{\partial \theta}{\partial x}\right] + \sqrt{\frac{Da}{Ra \cdot Pr}}$$

$$\times H_2(\varphi)\frac{\partial}{\partial y}\left[\left(1 + \frac{Ds}{H_2(\varphi)}\sqrt{\frac{Ra \cdot Pr}{Da}}\sqrt{\left(\frac{\partial \psi}{\partial x}\right)^2 + \left(\frac{\partial \psi}{\partial y}\right)^2}\right)\frac{\partial \theta}{\partial y}\right]$$

$$(8.53)$$

where $Ds = \varphi d_p/D$ is the dispersion parameter. We notice that when $Ds = 0.0$, Eqs. (8.52) and (8.53) reduce to equations presented in the paper by Sheremet et al. (2015a).

Taking into account the considered dimensionless variables (8.50), the bottom and top wavy walls of the cavity are described by the following relations:

$y_1 = 1 - a - b\cos(2\pi\kappa x)$ is the bottom wavy wall;

$y_2 = a + b\cos(2\pi\kappa x)$ is the top wavy wall;

$\Delta = y_2 - y_1 = 2a + 2b\cos(2\pi\kappa x) - 1$ is the distance between the horizontal walls.

The corresponding initial and boundary conditions for these equations are given by

$$\psi = 0, \quad \theta = 0 \qquad \text{at} \quad \tau = 0$$
$$\psi = 0, \quad \theta = -0.5 \quad \text{at} \quad x = 0$$
$$\psi = 0, \quad \theta = -0.5 \quad \text{at} \quad x = 1 \tag{8.54}$$
$$\psi = 0, \quad \theta = 0.5 \quad \text{at} \quad y = y_1$$
$$\psi = 0, \quad \theta = -0.5 \quad \text{at} \quad y = y_2.$$

Here $Ra = gK(\rho\beta)_f(T_h - T_c)D/(\alpha_m\mu_f)$ is the Rayleigh number for the porous medium, $Da = K/D^2$ is the Darcy number, $Pr = v_f/\alpha_m$ is the Prandtl number, and the functions $H_1(\varphi)$ and $H_2(\varphi)$ are given by

$$H_1(\varphi) = [1 - \varphi + \varphi(\rho\beta)_p/(\rho\beta)_f](1 - \varphi)^{2.5}$$

$$H_2(\varphi) = \frac{1 - \dfrac{3\varepsilon\varphi k_f(k_f - k_p)}{k_m[k_p + 2k_f + \varphi(k_f - k_p)]}}{[1 - \varphi + \varphi(\rho C_p)_p/(\rho C_p)_f]}. \tag{8.55}$$

These functions depend on the nanoparticles concentration φ and physical properties of the fluid, the nanoparticles and the solid structure of the porous medium.

3. Numerical Method and Validation

The cavity in the x and y plane, i.e., physical domain, is transformed into a rectangular geometry in the computational domain using an algebraic coordinate transformation by introducing new independent variables ξ and η. The left and right walls of the cavity become coordinate lines having constant values of ξ. The independent variables in the physical domain are transformed to independent variables in the computational domain by the following equations:

$$x = x(\xi, \eta), \quad y = y(\xi, \eta). \tag{8.56}$$

The cavity geometry is mapped into a rectangle on the basis of the following transformation:

$$
\begin{cases}
\xi = x, \\
\eta = \dfrac{y - y_1}{\Delta} = \dfrac{y - 1 + a + b\cos(2\pi\kappa x)}{2a + 2b\cos(2\pi\kappa x) - 1}.
\end{cases}
\tag{8.57}
$$

Taking into account transformation (8.57), the governing equations (8.52) and (8.53) will be rewritten in the following form:

$$
\frac{\partial^2\psi}{\partial\xi^2} + 2\frac{\partial\eta}{\partial x}\frac{\partial^2\psi}{\partial\xi\partial\eta} + \left[\left(\frac{\partial\eta}{\partial x}\right)^2 + \left(\frac{\partial\eta}{\partial y}\right)^2\right]\frac{\partial^2\psi}{\partial\eta^2} + \frac{\partial^2\eta}{\partial x^2}\frac{\partial\psi}{\partial\eta}
$$

$$
= -\sqrt{\frac{Ra \cdot Da}{Pr}} \cdot H_1(\varphi)\left(\frac{\partial\theta}{\partial\xi} + \frac{\partial\eta}{\partial x}\frac{\partial\theta}{\partial\eta}\right)
\tag{8.58}
$$

$$
\frac{\partial\theta}{\partial\tau} + \frac{\partial\eta}{\partial y}\frac{\partial\psi}{\partial\eta}\frac{\partial\theta}{\partial\xi} - \frac{\partial\eta}{\partial y}\frac{\partial\psi}{\partial\xi}\frac{\partial\theta}{\partial\eta} = \sqrt{\frac{Da}{Ra \cdot Pr}} \cdot H_2(\varphi)
$$

$$
\times\left\{\frac{\partial}{\partial\xi}\left[\left(1 + \frac{Ds}{H_2(\varphi)}\sqrt{\frac{Ra \cdot Pr}{Da}}\sqrt{\left(\frac{\partial\psi}{\partial\xi} + \frac{\partial\eta}{\partial x}\frac{\partial\psi}{\partial\eta}\right)^2 + \left(\frac{\partial\eta}{\partial y}\frac{\partial\psi}{\partial\eta}\right)^2}\right)\right.\right.
$$

$$
\times\left.\left(\frac{\partial\theta}{\partial\xi} + \frac{\partial\eta}{\partial x}\frac{\partial\theta}{\partial\eta}\right)\right] + \frac{\partial\eta}{\partial x}\frac{\partial}{\partial\eta}\left[\left(1 + \frac{Ds}{H_2(\varphi)}\sqrt{\frac{Ra \cdot Pr}{Da}}\right.\right.
$$

$$
\times\left.\left.\sqrt{\left(\frac{\partial\psi}{\partial\xi} + \frac{\partial\eta}{\partial x}\frac{\partial\psi}{\partial\eta}\right)^2 + \left(\frac{\partial\eta}{\partial y}\frac{\partial\psi}{\partial\eta}\right)^2}\right)\left(\frac{\partial\theta}{\partial\xi} + \frac{\partial\eta}{\partial x}\frac{\partial\theta}{\partial\eta}\right)\right]
$$

$$
+ \frac{\partial\eta}{\partial y}\frac{\partial}{\partial\eta}\left[\left(1 + \frac{Ds}{H_2(\varphi)}\sqrt{\frac{Ra \cdot Pr}{Da}}\right.\right.
$$

$$
\times\left.\left.\left.\sqrt{\left(\frac{\partial\psi}{\partial\xi} + \frac{\partial\eta}{\partial x}\frac{\partial\psi}{\partial\eta}\right)^2 + \left(\frac{\partial\eta}{\partial y}\frac{\partial\psi}{\partial\eta}\right)^2}\right)\left(\frac{\partial\eta}{\partial y}\frac{\partial\theta}{\partial\eta}\right)\right]\right\}.
\tag{8.59}
$$

The corresponding initial and boundary conditions of these equations are given by

$$
\begin{aligned}
\psi = 0, \quad &\theta = 0 \qquad \text{at} \quad \tau = 0 \\
\psi = 0, \quad &\theta = -0.5 \quad \text{at} \quad \xi = 0 \\
\psi = 0, \quad &\theta = -0.5 \quad \text{at} \quad \xi = 1 \\
\psi = 0, \quad &\theta = 0.5 \qquad \text{at} \quad \eta = 0 \\
\psi = 0, \quad &\theta = -0.5 \quad \text{at} \quad \eta = 1.
\end{aligned}
\tag{8.60}
$$

It should be noted here that

$$\frac{\partial \eta}{\partial x} = \frac{2\pi\kappa b(2y-1)\sin(2\pi\kappa x)}{[2a + 2b\cos(2\pi\kappa x) - 1]^2},$$

$$\frac{\partial \eta}{\partial y} = \frac{1}{2a + 2b\cos(2\pi\kappa x) - 1}, \quad \frac{\partial^2 \eta}{\partial y^2} = 0,$$

$$\frac{\partial^2 \eta}{\partial x^2} = \frac{4\pi^2\kappa^2 b(2y-1)[(2a-1)\cos(2\pi\kappa x) + 2b + 2b\sin^2(2\pi\kappa x)]}{[2a + 2b\cos(2\pi\kappa x) - 1]^3}.$$

The physical quantities of interest are the local Nusselt number along the bottom wavy wall Nu, which is defined as

$$Nu = -\frac{k_{mnf} + k_d}{k_m}\left(\frac{\partial \theta}{\partial \eta}\right)_{\eta=0}$$

$$= -\frac{k_{mnf}}{k_m}\left(1 + \frac{Ds}{H_2(\varphi)}\sqrt{\frac{Ra \cdot Pr}{Da}}\left|\frac{\partial \psi}{\partial \eta}\right|_{\eta=0}\sqrt{\left(\frac{\partial \eta}{\partial x}\right)^2 + \left(\frac{\partial \eta}{\partial y}\right)^2}\right)\left(\frac{\partial \theta}{\partial \eta}\right)_{\eta=0},$$

(8.61)

and the average Nusselt numbers $\overline{Nu}$, which is given by

$$\overline{Nu} = \int_0^1 Nu \, d\eta. \tag{8.62}$$

The unsteady partial differential equations (8.58) and (8.59) with corresponding boundary conditions (8.60) were solved using the finite difference method with the second order differencing schemes. A detailed description of the numerical technique used is presented by Aleshkova and Sheremet (2010), Sheremet and Trifonova (2013), Sheremet and Pop (2014a–c, 2015a–e), and Sheremet et al. (2014, 2015a–g). The present models, in the form of an in-house computational fluid dynamics code, have been validated successfully (Aleshkova and Sheremet, 2010; Sheremet and Trifonova, 2013; Sheremet and Pop, 2014a–c, 2015a–e; Sheremet et al., 2014, 2015a–g, 2016a,b).

We have conducted the grid independent test, analyzing the unsteady free convection in a wavy-walled porous cavity filled with a Cu–water nanofluid at $Ra = 300$, $Da = 10^{-4}$, $\varphi = 0.04$, $\varepsilon = 0.5$, $Ds = 0.1$, $\kappa = 3$, $a = 0.9$, where the solid matrix of the porous medium is the aluminum foam. Three cases of the uniform grid are tested: a grid of 100×100 points, a grid of 200×200 points, and a much finer grid of 300×300 points. Figure 8.24 shows the effect of the mesh parameters on the stream function and temperature profiles at cross-section $y = 0.5$. Figure 8.25 presents profiles of the local Nusselt number along the bottom wavy wall for different mesh parameters.

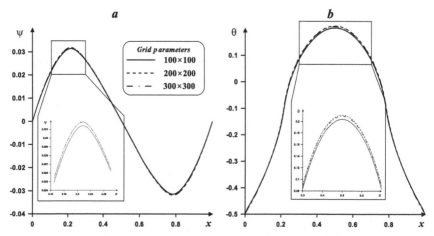

Fig. 8.24. Grid selection test: comparison of ψ (a) and θ (b) along $y = 0.5$ on three different grid systems [reprinted from Sheremet et al. (2016a) with permission from Elsevier]

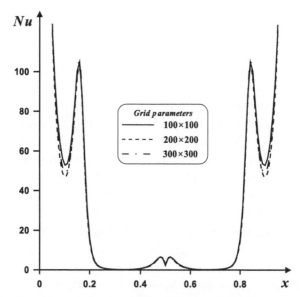

Fig. 8.25. Profiles of the local Nusselt number along the bottom wavy wall for three different grid systems [reprinted from Sheremet et al. (2016a) with permission from Elsevier]

Taking into account the results presented in Figs. 8.24 and 8.25, and also knowing that for fine grid we have essential computational time, the uniform grid of 200×200 points has been selected for the following analysis.

4. Results and Discussion

Numerical study has been conducted at the following values of key parameters: Rayleigh number ($Ra = 300$), Darcy number ($Da = 10^{-4}$), Prandtl number ($Pr = 0.041$), the porosity of the porous medium ($\varepsilon = 0.5$), undulation number ($\kappa = 3$), shape coefficient ($a = 0.9$), the solid volume fraction parameter of nanoparticles ($\varphi = 0.0 - 0.05$), the dispersion parameter ($Ds = 0.0 - 0.3$), the dimensionless time ($\tau = 0.0 - 100.0$), the solid matrix of the porous medium (aluminum foam). Particular efforts have been focused on the effects of the dimensionless time, thermal dispersion parameter, and nanoparticles' volume fraction on the fluid flow and heat transfer. Streamlines, isotherms, and average Nusselt number at the bottom wavy wall for different values of the key parameters mentioned above are illustrated in Figs. 8.26–8.29.

Figure 8.26 shows contours of the stream function and temperature for $\varphi = 0.04$, $Ds = 0.1$ and different values of the dimensionless time.

At initial time ($\tau < 0.1$) one can find isotherms of low temperature close to the vertical walls and top wavy wall and isotherms of high temperature close to the wavy bottom wall. The dominant heat transfer mechanism is heat conduction. There are twelve convective cells inside the cavity at $\tau = 0.02$.

The main reason for the formation of these circulations is the temperature differences between isothermal walls and initial temperature of the domain of interest. Therefore all these circulations are located near the walls. It should be noted that two elongated vortices are formed close to cold vertical walls. An increase in the dimensionless time ($\tau = 0.1$) leads to an intensification of two main circulations near the vertical walls due to an absorption of several small vortices. Also, one can find the cooling of the domain of interest from cold walls and the heating of the bottom part of the cavity. There are two main convective cells inside the cavity, with small circulations near the top central wave crest at $\tau = 0.5$. Distortion of isotherms in the upper part of the cavity is due to not only the undulations of the top wall but also an intensification of convective heat transfer.

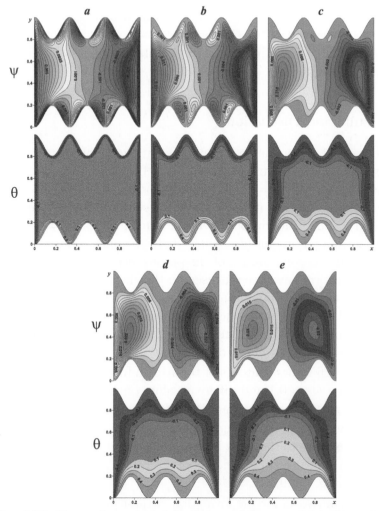

Fig. 8.26. Streamlines ψ and isotherms θ for $\varphi = 0.04$, $Ds = 0.1$:
$\tau = 0.02$ (a), $\tau = 0.1$ (b), $\tau = 0.5$ (c), $\tau = 1.0$ (d), $\tau = 100.0$ (e)
[reprinted from Sheremet et al. (2016a) with permission from Elsevier]

Further increase in the dimensionless time leads to a formation of
steady fluid flow and heat transfer regime with two convective cells and
thermal plume above the isothermal bottom wavy wall. Taking into
account the isotherms presented in Fig. 8.26e, it is possible to conclude
that high values of the local Nusselt number along the bottom wall are
at the left wall of the first left undulation and at the right wall of the
last undulation. Such a conclusion is due to an essential density of the
isotherms in these zones.

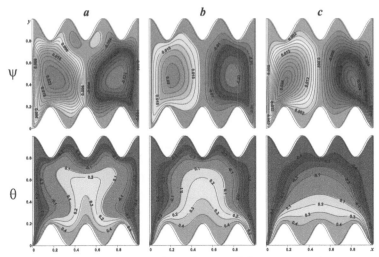

Fig. 8.27. Streamlines ψ and isotherms θ for $\varphi = 0.04$, $\tau = 100.0$: $Ds = 0.0$ (a), $Ds = 0.05$ (b), $Ds = 0.3$ (c) [reprinted from Sheremet et al. (2016a) with permission from Elsevier]

At the same time one can find an essential increase in the temperature in the bottom part of the cavity between the first and last undulations. Observed heating leads to significant decrease in the local Nusselt number along the bottom wavy wall between these undulations.

Figure 8.27 presents the streamlines and isotherms for $\varphi = 0.04$, $\tau = 100.0$, and different values of the dispersion parameter. It should be noted that value $Ds = 0.0$ characterizes the lack of the thermal dispersion where $k_d = 0$. In this case there are two large-scale convective cells and two small recirculations in the upper central wave troughs. An interaction of central hot thermal plume and cold thermal wave leads to a formation of two thermal plumes in the right and left parts of the cavity (see the shape of the isotherm $\theta = 0.1$). An increase in the dispersion parameter ($Ds = 0.05$ in Fig. 8.27b) leads to an intensification of convective flow and a dissipation of small recirculations in the upper part of the cavity. At the same time, the thermal plume is not so essential in comparison with the case of $Ds = 0.0$. Further increase in the dispersion parameter (Fig. 8.27c) leads to an attenuation of convective flow ($|\psi|_{\max}^{Ds=0.0} = 0.024 < |\psi|_{\max}^{Ds=0.05} = 0.034 > |\psi|_{\max}^{Ds=0.1} = 0.032 > |\psi|_{\max}^{Ds=0.3} = 0.027$) inside the cavity, with more essential cooling of the domain of interest where the thermal plume vanishes. Also, it should be noted that an increase in the dispersion parameter leads to a change of the convective cores' orientation.

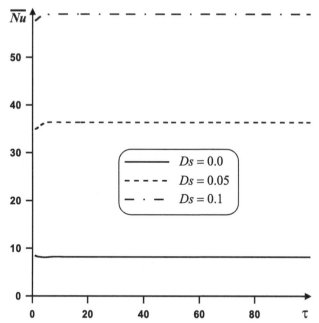

Fig. 8.28. Variation of the average Nusselt number versus
the dimensionless time and dispersion parameter for $\varphi = 0.04$
[reprinted from Sheremet et al. (2016a) with permission from Elsevier]

One can find an elongation of convective cores along the bisectors of
the first half of the bottom wave troughs for the left and right convective
cells, while in the case of $Ds = 0.0$, the convective cores are elongated
along the bisectors of the first half of the upper wave troughs.

Figure 8.28 demonstrates the dependencies of the average Nusselt
number at the bottom wavy wall on the dispersion parameter and dimensionless time. First of all, it is necessary to note that an increase
in Ds leads to an essential increase in $\overline{Nu}$ due to the inertia effect. An
increase in dimensionless time reflects a rapid approach of the steady-
state heat transfer mode. Therefore, the time moment $\tau = 100.0$ characterizes the steady-state regime.

The effect of the nanoparticles' volume fraction on the average Nusselt number is presented in Fig. 8.29. Taking into account the effect of
φ on the thermal dispersion conductivity in Eq. (8.46), Fig. 8.29 also
reflects an essential influence of this parameter on the heat transfer
rate.

An increase in the solid volume fraction parameter of nanoparticles
leads to a decrease in the heat transfer rate.

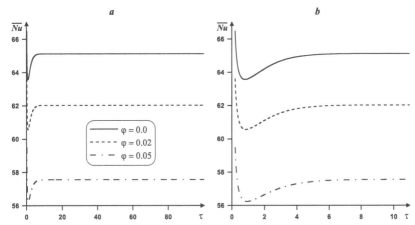

Fig. 8.29. Variation of the average Nusselt number versus
the dimensionless time and solid volume fraction parameter
of nanoparticles for $Ds = 0.1$, for the entire time interval (a)
and for the initial time interval (b) [reprinted from Sheremet et al.
(2016a) with permission from Elsevier]

It should be noted that in the case of $Ds = 0.0$, an increase in
the nanoparticles' volume fraction leads to not-so-significant decrease
in the average Nusselt number (Sheremet et al., 2015a).

The obtained results in Fig. 8.29 can be explained by an increase in
the thermal conductivity due to the thermal dispersion. An insertion
of nanoparticles in water leads to an increase in the thermal conductivity due to the thermal dispersion, and also characterizes the observed
variations of the average Nusselt number.

The obtained results showed the following:

(1) At the initial time level, several convective cells are formed close
to the isothermal walls due to the temperature differences between
these walls and initial temperature of the cavity.

(2) An increase in the dimensionless time leads to a rapid approach
of the steady-state regime.

(3) Regardless of the values of the dispersion parameter and
nanoparticles' volume fraction in the considered range, the convective
flow is characterized by double cell flow regime.

(4) An increase in the dispersion parameter leads to an essential
increase in the heat transfer rate.

(5) An increase in the solid volume fraction parameter of nanoparticles leads to a decrease in the heat transfer rate, while in the case of

$Ds = 0.0$ an increase in φ leads to not-so-significant decrease in the average Nusselt number (Sheremet et al., 2015a).

E. Free Convection in a Porous Wavy Cavity Filled with a Nanofluid using Buongiorno's Mathematical Model with Thermal Dispersion Effect

1. Introduction

Free convection within enclosures filled with porous media has been studied extensively during the last several decades due to its widespread engineering applications, including geothermal systems, underground spread of pollutants, storage of nuclear waste materials, solidification of casting, thermal insulation, electronic cooling, petroleum reservoir modeling, burying of drums containing heat generating chemicals in the earth, design of chemical catalytic reactors, powder metallurgy, ceramic engineering, the food and medical industries, and many others (Chen and Hsiao, 1998). Geothermal systems are an important energy source in many countries, both for direct use of hot groundwater for cooking, bathing, and chemical processes as well as usage of higher-enthalpy fluids for electricity generation (McKibbin, 1998).

Most of the early theoretical studies on the convective flow in porous cavities were based on Darcy's law with the assumption of neglecting a number of effects, such as inertia, non-Darcian, variable porosity, anisotropy, thermal dispersion, waviness of the walls, nanofluids, etc. The dispersion concept helps to explain the differences often observed between transport parameters measured along and across the principal direction of fluid flow in simple geometries. Among the first fundamental work on this subject, it is important to mention the papers by Taylor (1953) and Saffman (1960). Hydrodynamic dispersion in porous media is modeled in the literature as a tensorial quantity, with its components being either parallel or orthogonal to the main flow direction (Bear, 1972).

The effects of thermal dispersion on the boundary layer past flat surfaces embedded in fluid-saturated porous media has been studied by many researchers. Cheng (1981) considered the free convection boundary layer over a vertical surface embedded in a porous medium by assuming that the dispersion coefficients were proportional to the velocity components and to the Forchheimer coefficient. With this formulation, Cheng (1981) found that the effect of thermal dispersion was to decrease the surface heat flux. On the other hand, Plumb (1983) assumed that the longitudinal coefficient was negligible and the transverse coefficient was proportional to the streamwise velocity component. The effect of thermal dispersion on the free convection boundary layer past a vertical plate embedded in a porous medium has also been analyzed by Hong and Tien (1987). From a perturbation analysis carried on a Brinkman flow model, the authors concluded that dispersion tends to increase heat transfer, while boundary and inertia effects tends to act contrarily. Further, following the basic equations proposed by Kvernvold and Tyvand (1979, 1980), Telles and Trevisan (1993) extended the paper by Hong and Tien (1987) by considering the convective flows promoted by the density variation due to the combination of temperature and concentration gradients (double-diffusive convection). It is found that in the asymptotic limits, dispersion in boundary layer flows is described by simple-form equations. Wavy geometries are used in many engineering systems as a means of enhancing the transport performance (Wang and Chen, 2002; Mahmud and Fraser, 2004; Chen and Cho, 2007; Al-Amiri et al., 2007). The heat transfer performance of the traditional working fluids can also be enhanced through a careful design of the cavity geometry. For example, in many engineering applications (e.g., solar collectors, condensers in refrigerators, and so on), the heat transfer performance is improved by means of wavy surfaces. Overall, the results show that the heat transfer performance depends strongly on both the geometry parameters of the wavy surface (e.g., the wave amplitude and wavelength) and the flow parameters (e.g., the Grashof number, Rayleigh number, cavity length, etc.).

It is already well-known that traditional working fluids such as water, oil, or ethylene glycol have a low thermal conductivity. As a result, their heat transfer performance is inevitably limited. In an attempt to improve the heat transfer performance, Choi (1995) dispersed metallic nanoparticles with a high thermal conductivity in traditional working fluids to form so-called nanofluids. Many subsequent studies have shown that nanofluids yield a significant improvement in the heat transfer performance in a diverse range of engineering fields. Good literature on convective flow and applications of nanofluids can be found in the books by Das et al. (2007), Nield and Bejan (2013), and in the review papers by Buongiorno (2006), Das et al. (2006), Kakaç and Pramuanjaroenkij (2009), Wen et al. (2011), Haddad et al. (2012a,b), Mahian et al. (2013), and many others.

This section deals with the effect of thermal dispersion in steady free convection in a porous wavy cavity filled with a nanofluid, using Buongiorno's mathematical model. The effect of different governing parameters on the average Nusselt and Sherwood numbers, stream function, isotherms, etc., is investigated.

2. Basic Equations

Figure 8.30 shows the considered free convective flow, heat, and mass transfer in a porous square cavity filled with a water-based nanofluid. It is assumed that nanoparticles are suspended in the nanofluid using either surfactant or surface charge technology. This prevents nanoparticles from agglomeration and deposition on the porous matrix (see Nield and Kuznetsov, 2011; Kuznetsov and Nield, 2011a,b,c). The porous medium is assumed to be homogeneous and isotropic and the local thermal equilibrium is valid.

The flow in the cavity is considered to be steady, laminar, and incompressible. It is assumed that the bottom horizontal wall is maintained at temperature T_h, while the rest are kept at a temperature T_c, where we assume that $T_h > T_c$.

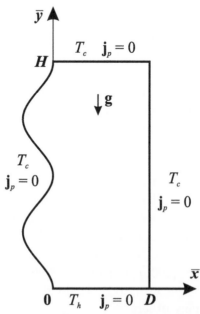

Fig. 8.30. Physical model and coordinate system

It is assumed the left wavy wall and right flat wall of the cavity are described by the relations

$$\overline{x}_1 = D - D[a + b\cos(2\pi\kappa\overline{y}/H)] \text{ and } \overline{x}_2 = D, \text{ respectively, and}$$

$$\overline{\Delta} = \overline{x}_2 - \overline{x}_1 = D[a + b\cos(2\pi\kappa\overline{y}/H)]$$

is the distance between vertical walls. Here a and b are constants that determine the shape and the wavy contraction ratio of the wavy wall $(a + b = 1)$.

Except for the density, the properties of the fluid are taken to be constant. It is further assumed that the effect of buoyancy is included through the Boussinesq approximation. The viscous, radiation, and Joule heating effects are neglected.

Under the above assumptions, the conservation equations for mass, momentum, thermal energy, and nanoparticles can be written as follows (see Buongiorno, 2006; Kuznetsov and Nield, 2011a,b,c; Nield and Kuznetsov, 2014):

$$\nabla \cdot \mathbf{V} = 0 \qquad (8.63)$$

$$\frac{\rho_{f0}}{\varepsilon}\frac{\partial \mathbf{V}}{\partial t} = -\nabla p - \frac{\mu}{K}\mathbf{V} + [\overline{\varphi}\rho_p + (1 - \overline{\varphi})\rho_{f0}(1 - \beta(T - T_c))]\mathbf{g} \quad (8.64)$$

$$(\rho C_p)_m\frac{\partial T}{\partial t} + (\rho C_p)_f(\mathbf{V} \cdot \nabla)T = \nabla[(k_m + k_d)\nabla T]$$
$$+ \varepsilon(\rho C_p)_p[D_B\nabla\overline{\varphi} \cdot \nabla T + (D_T/T_c)\nabla T \cdot \nabla T] \qquad (8.65)$$

$$\rho_p\left(\frac{\partial\overline{\varphi}}{\partial t} + \frac{1}{\varepsilon}(\mathbf{V} \cdot \nabla)\overline{\varphi}\right) = -\nabla \cdot \overline{j}_p \qquad (8.66)$$

where $\mathbf{V}$ is the Darcy velocity vector, T is the fluid temperature, $\overline{\varphi}$ is the nanoparticle volume fraction, t is the time, p is the fluid pressure, $\mathbf{g}$ is the gravity vector, ε is the porosity of the porous medium, K is the permeability of the porous medium, ρ_{f0} is the reference density of the fluid, μ is the viscosity of the fluid, β is the thermal expansion coefficient, ρ_p is the nanoparticle mass density, $(\rho C_p)_f$ is the volume heat capacity of the fluid, $(\rho C_p)_m$ is the effective volume heat capacity of the porous medium, k_m is the effective thermal conductivity of the porous medium, $k_d = (\rho C_p)_f|\overline{\mathbf{V}}|d_p\overline{\varphi}$ is the enhancement in the thermal conductivity due to the thermal dispersion (see Nakayama, 1995; Khanafer et al., 2003; Sakai et al., 2014), d_p is the pore or particle diameter, $(\rho C_p)_p$ is the effective volume heat capacity of the nanoparticle material, D_B is the Brownian diffusion coefficient, D_T is the thermophoretic diffusion coefficient, and $\mathbf{j}_p = -\rho_p[D_B\nabla\overline{\varphi} + (D_T/T_c)\nabla T]$ is the nanoparticles mass flux.

The flow is assumed to be slow so that an advective term and a Forchheimer quadratic drag term do not appear in the momentum equation. In keeping with the Boussinesq approximation and an assumption that the nanoparticle concentration is dilute, and with a suitable choice for the reference pressure, we can linearize the momentum equation and write Eq. (8.64) as

$$0 = -\nabla p - \frac{\mu}{K}\mathbf{V} + \{\overline{\varphi}(\rho_p - \rho_{f0}) + \rho_{f0}[1 - (1 - \overline{\varphi}_0)\beta(T - T_c)]\}\mathbf{g}. \quad (8.67)$$

Equations (8.63) and (8.65)–(8.67) for the steady problem under consideration can be written in dimensional coordinates $\overline{x}, \overline{y}$ as follows:

$$\frac{\partial\overline{u}}{p\overline{x}} + \frac{\partial\overline{v}}{\partial\overline{y}} = 0 \quad (8.68)$$

$$\frac{\partial p}{\partial\overline{x}} = -\frac{\mu}{K}\overline{u} \quad (8.69)$$

$$\frac{\partial p}{\partial\overline{y}} = -\frac{\mu}{K}\overline{v} - \{\overline{\varphi}(\rho_p - \rho_{f0}) + \rho_{f0}[1 - (1 - \overline{\varphi}_0)\beta(T - T_c)]\}g \quad (8.70)$$

$$(\rho C_p)_f \left(\overline{u}\frac{\partial T}{\partial\overline{x}} + \overline{v}\frac{\partial T}{\partial\overline{y}}\right) = \frac{\partial}{\partial\overline{x}}\left[\left(k_m + (\rho C_p)_f d_p\overline{\varphi}\sqrt{\overline{u}^2 + \overline{v}^2}\right)\frac{\partial T}{\partial\overline{x}}\right]$$

$$+ \frac{\partial}{\partial\overline{y}}\left[\left(k_m + (\rho C_p)_f d_p\overline{\varphi}\sqrt{\overline{u} + \overline{v}^2}\right)\frac{\partial T}{\partial\overline{y}}\right]$$

$$+ \varepsilon(\rho C_p)_p \left\{D_B\left(\frac{\partial\overline{\varphi}}{\partial\overline{x}}\frac{\partial T}{\partial\overline{x}} + \frac{\partial\overline{\varphi}}{\partial\overline{y}}\frac{\partial T}{\partial\overline{y}}\right) + \frac{D_T}{T_c}\left[\left(\frac{\partial T}{\partial\overline{x}}\right)^2 + \left(\frac{\partial T}{\partial\overline{y}}\right)^2\right]\right\} \quad (8.71)$$

$$\frac{1}{\varepsilon}\left(\overline{u}\frac{\partial\overline{\varphi}}{\partial\overline{x}} + \overline{v}\frac{\partial\overline{\varphi}}{\partial\overline{y}}\right) = D_B\left(\frac{\partial^2\overline{\varphi}^2}{\partial\overline{x}} + \frac{\partial^2\overline{\varphi}}{\partial\overline{y}^2}\right)$$

$$+ \frac{D_T}{T_c}\left(\frac{\partial^2 T}{\partial\overline{x}^2} + \frac{\partial^2 T}{\partial\overline{y}^2}\right) \quad (8.72)$$

where $\overline{u}, \overline{v}$ are the velocity components along the $\overline{x}, \overline{y}$ directions, respectively.

One can introduce a stream function $\overline{\psi}$ defined by

$$\overline{u} = \frac{\partial\overline{\psi}}{\partial\overline{y}}, \quad v = -\frac{\partial\overline{\psi}}{\partial\overline{x}} \quad (8.73)$$

so that Eq. (8.68) is satisfied identically. We are then left with the following equations:

$$\frac{\partial^2 \overline{\psi}}{\partial \overline{x}^2} + \frac{\partial^2 \overline{\psi}}{\partial \overline{y}^2} = \frac{Kg}{\mu} \left[(\rho_p - \rho_{f0}) \frac{\partial \overline{\varphi}}{\partial \overline{x}} - \rho_{f0}\beta(1 - \overline{\varphi}_0) \frac{\partial T}{\partial \overline{x}} \right] \qquad (8.74)$$

$$\frac{\partial \overline{\psi}}{\partial \overline{y}} \frac{\partial T}{\partial \overline{x}} - \frac{\partial \overline{\psi}}{\partial \overline{x}} \frac{\partial T}{\partial \overline{y}} = \frac{\partial}{\partial \overline{x}} \left[\left(\alpha_m + d_p \overline{\varphi} \sqrt{\left(\frac{\partial \overline{\psi}}{\partial \overline{x}}\right)^2 + \left(\frac{\partial \overline{\psi}}{\partial \overline{y}}\right)^2} \right) \frac{\partial T}{\partial \overline{x}} \right]$$

$$+ \frac{\partial}{\partial \overline{y}} \left[\left(\alpha_m + d_p \overline{\varphi} \sqrt{\left(\frac{\partial \overline{\psi}}{\partial \overline{x}}\right)^2 + \left(\frac{\partial \overline{\psi}}{\partial \overline{y}}\right)^2} \right) \frac{\partial T}{\partial \overline{y}} \right]$$

$$+ \delta \left\{ D_B \left(\frac{\partial \overline{\varphi}}{\partial \overline{x}} \frac{\partial T}{\partial \overline{x}} + \frac{\partial \overline{\varphi}}{\partial \overline{y}} \frac{\partial T}{\partial \overline{y}} \right) + \frac{D_T}{T_c} \left[\left(\frac{\partial T}{\partial \overline{x}}\right)^2 + \left(\frac{\partial T}{\partial \overline{y}}\right)^2 \right] \right\} \qquad (8.75)$$

$$\frac{1}{\varepsilon} \left(\frac{\partial \overline{\psi}}{\partial \overline{y}} \frac{\partial \overline{\varphi}}{\partial \overline{x}} - \frac{\partial \overline{\psi}}{\partial \overline{x}} \frac{\partial \overline{\varphi}}{\partial \overline{y}} \right) = D_B \left(\frac{\partial^2 \overline{\varphi}}{\partial \overline{x}^2} + \frac{\partial^2 \overline{\varphi}}{\partial \overline{y}^2} \right)$$

$$+ \frac{D_T}{T_c} \left(\frac{\partial^2 T}{\partial \overline{x}^2} + \frac{\partial^2 T}{\partial \overline{y}^2} \right) \qquad (8.76)$$

where $\alpha_m = k_m/(\rho C_p)_f$ is effective thermal diffusivity of the porous medium.

Introducing the following dimensionless variables:

$$x = \overline{x}/D, \ y = \overline{y}/D,$$

$$\psi = \overline{\psi}/\alpha_m, \ \theta = (T - T_c)/(T_h - T_c), \ \varphi = \overline{\varphi}/\overline{\varphi}_0 \qquad (8.77)$$

and substituting (8.77) into Eqs. (8.74)–(8.76), we obtain

$$\frac{\partial^2 \psi}{\partial x^2} + \frac{\partial^2 \psi}{\partial y^2} = -Ra \left(\frac{\partial \theta}{\partial x} - Nr \frac{\partial \varphi}{\partial x} \right) \qquad (8.78)$$

$$\frac{\partial \psi}{\partial y} \frac{\partial \theta}{\partial x} - \frac{\partial \psi}{\partial x} \frac{\partial \theta}{\partial y} = \frac{\partial}{\partial x} \left[\left(1 + Ds \cdot \varphi \sqrt{\left(\frac{\partial \psi}{\partial x}\right)^2 + \left(\frac{\partial \psi}{\partial y}\right)^2} \right) \frac{\partial \theta}{\partial x} \right]$$

$$+ \frac{\partial}{\partial y} \left[\left(1 + Ds \cdot \varphi \sqrt{\left(\frac{\partial \psi}{\partial x}\right)^2 + \left(\frac{\partial \psi}{\partial y}\right)^2} \right) \frac{\partial \theta}{\partial y} \right]$$

$$+Nb\left(\frac{\partial\varphi}{\partial x}\frac{\partial\theta}{\partial x}+\frac{\partial\varphi}{\partial y}\frac{\partial\theta}{\partial y}\right)+Nt\left[\left(\frac{\partial\theta}{\partial x}\right)^2+\left(\frac{\partial\theta}{\partial y}\right)^2\right] \tag{8.79}$$

$$\frac{\partial\psi}{\partial y}\frac{\partial\varphi}{\partial x}-\frac{\partial\psi}{\partial x}\frac{\partial\varphi}{\partial y}=\frac{1}{Ln}\left(\frac{\partial^2\varphi}{\partial x^2}+\frac{\partial^2\varphi}{\partial y^2}\right)$$

$$+\frac{Nt}{Ln\cdot Lb}\left(\frac{\partial^2\theta}{\partial x^2}+\frac{\partial^2\theta}{\partial y^2}\right). \tag{8.80}$$

Taking into account the considered dimensionless variables (8.77), the left wavy wall of the cavity is described by the following relation:

$$x_1=1-a-b\cos(2\pi\kappa y/A);$$

$$\Delta=x_2-x_1=a+b\cos(2\pi\kappa y/A)$$

is the distance between vertical walls.

The corresponding boundary conditions for these equations are as follows (see Kuznetsov and Nield, 2013):

$$\psi=0,\quad\theta=0,\quad\tilde{\mathbf{j}}_p=0$$

$$\left(\text{or } Nb\frac{\partial\phi}{\partial n}+Nt\frac{\partial\theta}{\partial n}=0\right)\text{ on }x=x_1$$

$$\psi=0,\quad\theta=0,\quad\tilde{\mathbf{j}}_p=0$$

$$\left(\text{or } Nb\frac{\partial\phi}{\partial x}+Nt\frac{\partial\theta}{\partial x}=0\right)\text{ on }x=x_2$$

$$\psi=0,\quad\theta=1,\quad\tilde{\mathbf{j}}_p=0 \tag{8.81}$$

$$\left(\text{or } Nb\frac{\partial\phi}{\partial y}+Nt\frac{\partial\theta}{\partial y}=0\right)\text{ on }y=0$$

$$\psi=0,\quad\theta=0,\quad\tilde{\mathbf{j}}_p=0$$

$$\left(\text{or } Nb\frac{\partial\phi}{\partial y}+Nt\frac{\partial\theta}{\partial y}=0\right)\text{ on }y=A.$$

Here Ra is the Rayleigh number for the porous media, Nr is the nanofluid buoyancy ratio, Nb is the Brownian motion parameter, Nt is the thermophoresis parameter, Ln is the nanofluid Lewis number, Ds is the thermal dispersion parameter, and A is the aspect ratio, which are defined as

$$Ra=\frac{gK\rho_{f0}\beta(T_h-T_c)(1-\overline{\varphi}_0)D}{\alpha_m\mu},\quad Nr=\frac{(\rho_p-\rho_{f0})\overline{\varphi}_0}{\rho_{f0}\beta(T_h-T_c)(1-\overline{\varphi}_0)},$$

$$Nb = \frac{\delta D_B \overline{\varphi}_0}{\alpha_m}, \quad Nt = \frac{\delta D_T (T_h - T_c)}{\alpha_m T_c},$$

$$Ln = \frac{\alpha_m}{\varepsilon D_B}, \quad Ds = \frac{d_p \overline{\varphi}_0}{D}, \quad A = \frac{H}{D}. \tag{8.82}$$

The physical quantities of interest are the local Nusselt Nu and the Sherwood Sh numbers along the bottom horizontal wall, which are defined as

$$Nu = -\left(\frac{\partial \theta}{\partial y}\right)_{y=0}, \quad Sh = -\left(\frac{\partial \varphi}{\partial y}\right)_{y=0} \tag{8.83}$$

and the average Nusselt $\overline{Nu}$ and Sherwood $\overline{Sh}$ numbers, which are given by

$$\overline{Nu} = \int_0^1 Nu \, dx, \quad \overline{Sh} = \int_0^1 Sh \, dx. \tag{8.84}$$

3. Numerical Method and Validation

The cavity in the x and y plane, i.e., physical domain, is transformed into a rectangular geometry in the computational domain using an algebraic coordinate transformation, by introducing new independent variables ξ and η. The left and right walls of the cavity become coordinate lines having constant values of ξ. The independent variables in the physical domain are transformed to independent variables in the computational domain by the following equations:

$$x = x(\xi, \eta), \quad y = y(\xi, \eta). \tag{8.85}$$

The cavity geometry is mapped into a rectangle on the basis of the following transformation:

$$\begin{cases} \xi = \dfrac{x - x_1}{\Delta} = \dfrac{x - 1 + a + b\cos(2\pi\kappa y/A)}{a + b\cos(2\pi\kappa y/A)}, \\[2ex] \eta = y. \end{cases} \tag{8.86}$$

Taking into account transformation (8.86), the governing equations (8.78)–(8.80) are rewritten in the following form:

$$\left[\left(\frac{\partial \xi}{\partial x}\right)^2 + \left(\frac{\partial \xi}{\partial y}\right)^2\right]\frac{\partial^2 \psi}{\partial \xi^2} + 2\frac{\partial \xi}{\partial y}\frac{\partial^2 \psi}{\partial \xi \partial \eta} + \frac{\partial^2 \psi}{\partial \eta^2} + \frac{\partial^2 \xi}{\partial y^2}\frac{\partial \psi}{\partial \xi}$$

$$= -Ra\frac{\partial \xi}{\partial x}\frac{\partial \theta}{\partial \xi} + Ra \cdot Nr\frac{\partial \xi}{\partial x}\frac{\partial \varphi}{\partial \xi} \tag{8.87}$$

$$\frac{\partial \xi}{\partial x}\frac{\partial \psi}{\partial \eta}\frac{\partial \theta}{\partial \xi} - \frac{\partial \xi}{\partial x}\frac{\partial \psi}{\partial \xi}\frac{\partial \theta}{\partial \eta}$$

$$= \frac{\partial \xi}{\partial x}\frac{\partial}{\partial \xi}\left[\left(1 + Ds \cdot \varphi\sqrt{\left(\frac{\partial \xi}{\partial x}\frac{\partial \psi}{\partial \xi}\right)^2 + \left(\frac{\partial \xi}{\partial y}\frac{\partial \psi}{\partial \xi} + \frac{\partial \psi}{\partial \eta}\right)^2}\right)\frac{\partial \xi}{\partial x}\frac{\partial \theta}{\partial \xi}\right]$$

$$+ \frac{\partial \xi}{\partial y}\frac{\partial}{\partial \xi}\left[\left(1 + Ds \cdot \varphi\sqrt{\left(\frac{\partial \xi}{\partial x}\frac{\partial \psi}{\partial \xi}\right)^2 + \left(\frac{\partial \xi}{\partial y}\frac{\partial \psi}{\partial \xi} + \frac{\partial \psi}{\partial \eta}\right)^2}\right)\left(\frac{\partial \xi}{\partial y}\frac{\partial \theta}{\partial \xi} + \frac{\partial \theta}{\partial \eta}\right)\right]$$

$$+ \frac{\partial}{\partial \eta}\left[\left(1 + Ds \cdot \varphi\sqrt{\left(\frac{\partial \xi}{\partial x}\frac{\partial \psi}{\partial \xi}\right)^2 + \left(\frac{\partial \xi}{\partial y}\frac{\partial \psi}{\partial \xi} + \frac{\partial \psi}{\partial \eta}\right)^2}\right)\left(\frac{\partial \xi}{\partial y}\frac{\partial \theta}{\partial \xi} + \frac{\partial \theta}{\partial \eta}\right)\right]$$

$$+ Nb\left[\left(\frac{\partial \xi}{\partial x}\right)^2\frac{\partial \varphi}{\partial \xi}\frac{\partial \theta}{\partial \xi} + \left(\frac{\partial \xi}{\partial y}\frac{\partial \varphi}{\partial \xi} + \frac{\partial \varphi}{\partial \eta}\right)\left(\frac{\partial \xi}{\partial y}\frac{\partial \theta}{\partial \xi} + \frac{\partial \theta}{\partial \eta}\right)\right]$$

$$+ Nt\left[\left(\frac{\partial \xi}{\partial x}\frac{\partial \theta}{\partial \xi}\right)^2 + \left(\frac{\partial \xi}{\partial y}\frac{\partial \theta}{\partial \xi} + \frac{\partial \theta}{\partial \eta}\right)^2\right] \qquad (8.88)$$

$$\frac{\partial \xi}{\partial x}\frac{\partial \psi}{\partial \eta}\frac{\partial \varphi}{\partial \xi} - \frac{\partial \xi}{\partial x}\frac{\partial \psi}{\partial \xi}\frac{\partial \varphi}{\partial \eta}$$

$$= \frac{1}{Ln}\left\{\left[\left(\frac{\partial \xi}{\partial x}\right)^2 + \left(\frac{\partial \xi}{\partial y}\right)^2\right]\frac{\partial^2 \varphi}{\partial \xi^2} + 2\frac{\partial \xi}{\partial y}\frac{\partial^2 \varphi}{\partial \xi \partial \eta} + \frac{\partial^2 \varphi}{\partial \eta^2} + \frac{\partial^2 \xi}{\partial y^2}\frac{\partial \varphi}{\partial \xi}\right\}$$

$$+ \frac{Nt}{Ln \cdot Nb}\left\{\left[\left(\frac{\partial \xi}{\partial x}\right)^2 + \left(\frac{\partial \xi}{\partial y}\right)^2\right]\frac{\partial^2 \theta}{\partial \xi^2} + 2\frac{\partial \xi}{\partial y}\frac{\partial^2 \theta}{\partial \xi \partial \eta} + \frac{\partial^2 \theta}{\partial \eta^2} + \frac{\partial^2 \xi}{\partial y^2}\frac{\partial \theta}{\partial \xi}\right\}.$$

$$(8.89)$$

The corresponding boundary conditions for these equations are given by

$$\psi = 0, \quad \theta = 0, \quad Nb\frac{\partial \varphi}{\partial \xi} + Nt\frac{\partial \theta}{\partial \xi} = 0 \text{ on } \xi = 0$$

$$\psi = 0, \quad \theta = 0, \quad Nb\frac{\partial \varphi}{\partial \xi} + Nt\frac{\partial \theta}{\partial \xi} = 0 \text{ on } \xi = 1$$

$$\psi = 0, \quad \theta = 1, \quad Nb\frac{\partial \varphi}{\partial \eta} + Nt\frac{\partial \theta}{\partial \eta} = 0 \text{ on } \eta = 0$$

$$\psi = 0, \quad \theta = 0, \quad Nb\frac{\partial \varphi}{\partial \eta} + Nt\frac{\partial \theta}{\partial \eta} = 0 \text{ on } \eta = A.$$

$$(8.90)$$

It should be noted that

$$\frac{\partial \xi}{\partial x} = \frac{1}{a + b\cos(2\pi\kappa y/A)}, \quad \frac{\partial \xi}{\partial y} = \frac{2\pi\kappa b(x-1)\sin(2\pi\kappa y/A)}{A[a + b\cos(2\pi\kappa y/A)]^2}, \quad \frac{\partial^2 \xi}{\partial x^2} = 0,$$

$$\frac{\partial^2 \xi}{\partial y^2} = \frac{4\pi^2\kappa^2 b(x-1)[a\cos(2\pi\kappa y/A) + b + b\sin^2(2\pi\kappa y/A)]}{A^2[a + b\cos(2\pi\kappa y/A)]^3}.$$

The physical quantities of interest are the local Nusselt number along the hot bottom wall Nu, which is defined as

$$Nu = -\frac{k_m + k_d}{k_m}\left(\frac{\partial\theta}{\partial\eta}\right)_{\eta=0}$$

$$= -\left(1 + Ds\cdot\varphi\Big|_{\eta=0}\left|\frac{\partial\psi}{\partial\eta}\right|_{\eta=0}\right)\left(\frac{\partial\theta}{\partial\eta}\right)_{\eta=0} \tag{8.91}$$

and the average Nusselt numbers $\overline{Nu}$, which is given by

$$\overline{Nu} = \int_0^1 Nu\,d\xi. \tag{8.92}$$

The steady partial differential equations (8.87)–(8.89) with corresponding boundary conditions (8.90) were solved using the finite difference method with the second order differencing schemes. Detailed descriptions of the utilized numerical technique have been presented earlier (see Aleshkova and Sheremet, 2010; Sheremet and Trifonova, 2013; Sheremet and Pop, 2014a-c; Sheremet et al., 2014). The present models, in the form of an in-house computational fluid dynamics (CFD) code, have been validated successfully (see Aleshkova and Sheremet, 2010; Sheremet and Trifonova, 2013; Sheremet and Pop, 2014a-c; Sheremet et al, 2014).

We have conducted the grid independent test, analyzing the steady free convection in a wavy-walled porous cavity filled with a nanofluid at $Ra = 100$, $Ln = 10.0$, $Ds = 0.1$, $Nr = 0.1$, $Nb = 0.1$, $Nt = 0.1$, $\kappa = 2$, $a = 0.8$. Three cases of the uniform grid are tested: a grid of 100×100 points, a grid of 200×200 points, and a much finer grid of 300×300 points. Figures 8.31 and 8.32 shows the effect of the mesh parameters on the stream function and temperature profiles at cross-sections $x = 0.5$ and $y = 0.5$.

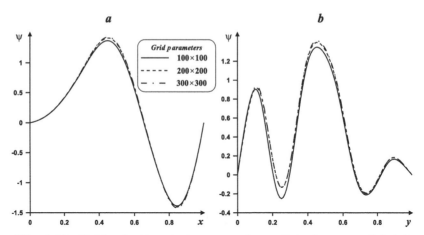

Fig. 8.31. Grid selection test: comparison of ψ along $y = 0.5$ (a) and $x = 0.5$ (b) on three different grid systems

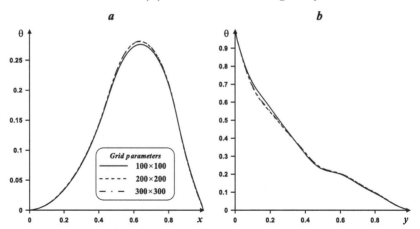

Fig. 8.32. Grid selection test: comparison of θ along $y = 0.5$ (a) and $x = 0.5$; (b) on three different grid systems

Taking into account the results presented in Figs. 8.31 and 8.32 and also knowing that for fine grid we have essential computational time, the uniform grid of 200×200 points has been selected for the following analysis.

4. Results and Discussion

Numerical investigations of the boundary value problem (8.87)–(8.90) have been carried out at the following values of key parameters: $Ra = 50 - 300$, $\kappa = 1 - 3$, $Ds = 0.0 - 0.3$, $Nr = Nb = Nt = 0.1$, $Ln = 10$, $A = 1$.

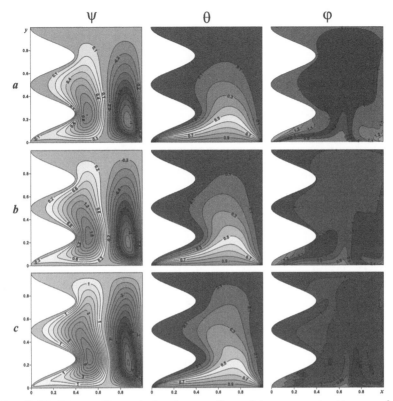

Fig. 8.33. Streamlines ψ, isotherms θ, and isoconcentrations φ for $\kappa = 2.0$, $Ds = 0.1$: $Ra = 50$ (a), $Ra = 100$ (b), $Ra = 300$ (c)

Figures 8.33–8.36 present isocontours of the stream function, temperature, and nanoparticle volume fraction, and the variation of the average Nusselt number in dependence on the Rayleigh number, undulation number, and thermal dispersion parameter.

Figure 8.33 shows the effect of the Rayleigh number on distributions of stream function, temperature, and nanoparticle volume fraction in the wavy cavity. Regardless of the Rayleigh number values, two convective cells appear inside the cavity illustrating a formation of ascending flows in a central part of the cavity, and two descending flows near the cold vertical walls. The sizes of the clockwise vortices are less than the sizes of the counterclockwise vortices, which is due to the presence of the wavy wall with crests and troughs. Such differences in the sizes lead to a formation of more intensive motion inside the small vortices. It should be noted that ascending flows are caused by an evolution of a thermal plume over the heat source of constant temperature.

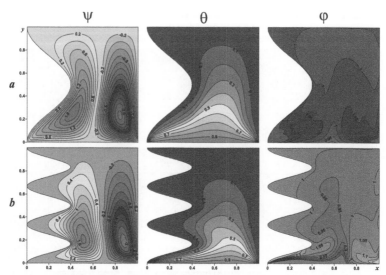

Fig. 8.34. Streamlines ψ, isotherms θ, and isoconcentrations φ for
$Ra = 100$, $Ds = 0.1$: $\kappa = 1.0$ (a), $\kappa = 3.0$ (b)

At the same time, cores of these two convective cells are located
close to the bottom hot wall due to the large hydrodynamic resistance from the solid matrix of the porous medium. Distortion of the
border between two convective cells is due to the effect of the wavy
wall, namely, one can find both a convex border in front of the wavy
crests and a concave border in front of the central wavy trough. Taking into account the distributions of the nanoparticle volume fraction, one can find high concentration of nanoparticles near the bottom right and left corners, while inside the thermal plume zone this
concentration decreases due to the inertia effect. An increase in the
Rayleigh number leads to both an intensification of convective flow
($|\psi|_{\max}^{Ra=50} = 1.08 < |\psi|_{\max}^{Ra=100} = 2.29 < |\psi|_{\max}^{Ra=300} = 7.05$) and the
weak displacement of the convective cells' cores along the vertical axis.
The boundary layers' thickness decreases with Ra, which is confirmed
by the high density of isolines close to the vertical walls. Also an increase in the Rayleigh number leads to more intensive heating of the
cavity, taking into account the position of isotherm $\theta = 0.1$. Such behavior is due to an intensification of convective flow in comparison with
low values of Ra where heat conduction is the dominating heat transfer mechanism. An intensification of convective heat transfer leads to
more homogeneous distributions of the nanoparticles inside the cavity,
while one can find the bottom corner zones with high concentration of
nanoparticles due to the geometrical features.

An intensification of convective heat transfer with Ra is presented in Fig. 8.36a for the average Nusselt number along the bottom hot wall (see Eqs. (8.91) and (8.92)). An increase in the Rayleigh number leads to an increase in the heat transfer rate.

Figures 8.33b and 8.34 present streamlines, isotherms, and isoconcentrations of the nanoparticles for different values of the undulation number. Regardless of the undulation number values, there are two convective cells of different shapes inside the cavity. It should be noted that the form of the vortices depends on the wavy wall shape. An increase in the undulation number leads to a displacement of the convective cells' cores along the vertical axis to the bottom hot wall. Also, the form of the wavy wall defines the shape of the border between two convective cells. One can find that an increase in κ leads to an attenuation of the convective flow ($|\psi|_{\max}^{\kappa=1.0} = 2.85 > |\psi|_{\max}^{\kappa=2.0} = 2.29 > |\psi|_{\max}^{\kappa=3.0} = 2.12$), and as a result, to less intensive heating of the domain of interest. At the same time, an increase in the undulation number characterizes a formation of more heterogeneous distribution of the nanoparticles inside the cavity, namely, close to the bottom hot wall. An intensification of convective heat transfer with κ is presented in Fig. 8.36b.

The effect of the thermal dispersion parameter on streamlines, isotherms, and isoconcentrations is demonstrated in Fig. 8.35. It should be noted that thermal dispersion characterizes heat transfer due to hydrodynamic mixing of the interstitial fluid at the pore scale. In addition to the molecular diffusion of heat, there is mixing due to the nature of the porous medium (see Nield and Bejan, 2013). An increase in Ds leads to both an attenuation of the convective flow ($|\psi|_{\max}^{Ds=0.0} = 2.7 > |\psi|_{\max}^{Ds=1.0} = 2.29 > |\psi|_{\max}^{Ds=2.0} = 2.08 > |\psi|_{\max}^{Ds=3.0} = 1.99$) and less intensive heating of the cavity, taking into account the position of isotherm $\theta = 0.1$. It should be noted that at the presence of thermal dispersion effect one can find a displacement of the right convective cell core to the right bottom corner, and of the left convective cell core, to the bottom hot wall. At the same time, an influence of this parameter on the distribution of the nanoparticle volume fraction is inessential.

Figure 8.36 shows variations of the average Nusselt number along the bottom hot wall with Ds, Ra, and κ. An increase in the thermal dispersion parameter leads to an increase in the heat transfer rate. It is worth noting that more intensive increment of $\overline{Nu}$ occurs for small values of the Rayleigh number, where diffusion heat transfer is a dominating heat transfer mechanism.

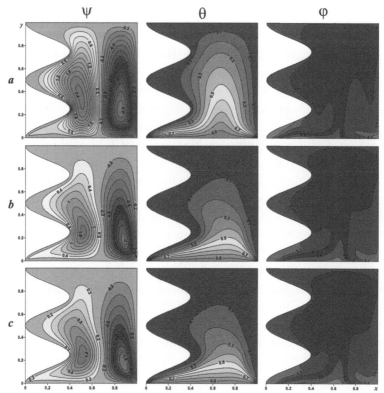

Fig. 8.35. Streamlines ψ, isotherms θ, and isoconcentrations φ for $Ra = 100$, $\kappa = 2.0$: $Ds = 0.0$ (a), $Ds = 0.2$ (b), $Ds = 0.3$ (c)

Also, one can find that in the case of $Ds = 0$, an increase in the Rayleigh number from 100 to 300 leads to a more intensive increment of the average Nusselt number in comparison with other values of the thermal dispersion parameter. Rate of the average Nusselt number growth with Ds also depends on the undulation number. For high values of κ the heat transfer rate increases intensively with the thermal dispersion parameter.

Based on the findings in this study, we conclude the following:

(1) An increase in the Rayleigh number leads to an intensification of convective flow and heat transfer. Regardless of the Rayleigh number value, there are two convective cells inside the cavity. An increase in Ra leads to a homogenization of the nanoparticles distribution inside the cavity.

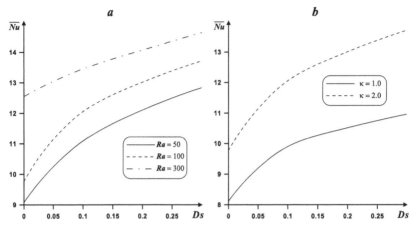

Fig. 8.36. Variation of the average Nusselt number versus
the thermal dispersion parameter and Rayleigh number for
$\kappa = 2.0$ (a); the thermal dispersion parameter and
undulation number for $Ra = 100$ (b)

(2) An increase in the undulation number leads to an attenuation
of the convective flow and an intensification of the convective heat
transfer. Also, the form of the wavy wall defines the shape of the border
between two formed convective cells inside the cavity. An increase in
the undulation number characterizes a formation of less homogeneous
distribution of the nanoparticles inside the cavity, namely, close to the
bottom hot wall.

(3) An increase in the thermal dispersion parameter leads to both an
attenuation of the convective flow and an increase in the heat transfer
rate. At the same time, an influence of Ds on the nanoparticle volume
fraction distribution is insignificant.

Concluding Remarks

These concluding remarks are meant for recapitulating the various aspects that were dealt with in the foregoing chapters, and to provide a forum to suggest some possible research ideas for advancing this important area of convective flow and heat transfer from wavy surfaces. Irregular surfaces are encountered in a number of heat transfer devices, such as solar collectors, condensers in refrigerators, double-walled thermal insulators, underground cable systems, electric machinery, cooling systems of micro-electronic devices, etc., and in most cases, the surfaces are intentionally made irregular or wavy in order to enhance heat transfer. It is obvious that wavy surfaces do bring in an added level of mathematical complexity, which is greater than when dealing with smooth flat surfaces. However, it is seen from the treatise in this book that commonsensical mathematical simplifications can result in reducing the complexity and providing elegant analyzes and solutions of the problems.

Chapter 1 was dedicated to the governing equations for convective flow and heat transfer from wavy surfaces. It was shown that the governing equations for flow and heat transfer from wavy surfaces are not truly different from those that are used when dealing with flat and smooth surfaces. The only difference is the use of an additional equation $y = \sigma(x)$, where the function $\sigma(x)$ takes various forms depending upon the profile of the wavy surface. Physical models have been shown for the horizontal wavy wall and vertical wavy wall for the case of external flow. In the case of internal flows, physical models have been described for flow between a wavy wall and a parallel wall, flow between two symmetrically configured wavy walls, and flow between two asymmetrically configured wavy walls. In all cases, simple coordinate transformations were used to change the complex wavy geometry into a smooth one so that the governing equations could be solved by well-known analytical and numerical methods. Although this procedure brought about considerable simplifications, the effort required to solve the transformed equations numerically was often just about the same as that for the original equations. However, the advantage of the transformation lay in the fact that the profile of the wavy surface could be changed during the numerical calculations without having to resort to realterations in the governing equations.

Chapter 2 dealt with steady natural and mixed convection flow in viscous fluids over a wavy vertical wall. It was pointed out that for the natural convection case, the increase in heat transfer relative to the flat plate of equal projected area occurred due to an increase in surface area for the wavy surface and not due to any improvement in heat transfer mechanism. In other words, a flat plate with a surface area equal to that of the wavy surface would transfer more heat. For the mixed convection case, under the assumption of the term Gr/Re^2 being finite, but the Reynolds number being large enough so that the motion is confined to a thin boundary layer along the wavy surface, a numerical solution of the boundary layer equations was obtained using a finite difference scheme and results presented for a smooth flat plate and the wavy plate for a value of $Pr = 1$.

Chapter 3 considered steady natural convection flow in fluid-saturated porous media induced by constant temperature and constant heat flux over a wavy vertical wall. The natural convection Darcy flow along a wavy vertical wall placed in an isotropic porous medium saturated with a Newtonian fluid was presented. It was assumed that the Rayleigh number was large such that natural convection took place within a boundary layer whose cross-stream width was substantially smaller than the amplitude of the surface waves. The governing equations were a little different from those in Chapter 2 for viscous fluids because the presence of the porous media brought in the equation for Darcy's law. It was indicated that flow past a wavy vertical surface placed in a porous medium depended only on the local slope of the surface wave, which is in contrast to an identical configuration in viscous fluids without porous medium discussed in Chapter 2. It was shown that the values of local Nusselt numbers are less than or equal to those corresponding to a smooth flat plate at constant temperature, embedded in a porous medium. For the constant heat flux case, the transformed boundary layer equations were non-similar, and they were solved numerically using the Keller box method. It was found that the presence of the boundary layer flow ensured that the positions of maximum temperature were shifted downstream of the positions of maximum slope.

Chapter 4 shifted the focus from external flows that were studied in Chapters 2 and 3 to internal flows, namely, the wavy vertical channel. Three types of channel flows were considered, namely, (1) flow between a long vertical wall and a parallel wall, (2) flow between two symmetrically configured wavy walls, and (3) flow between two asymmetrically configured wavy walls. Only the natural convection flow within these three configurations of internal flow were presented, and a perturbation solution of the boundary layer equations was given based on the long-wave approximation accounting for the contribution from the gradual waviness of the wavy wall.

Chapter 5 also dealt with internal flow, similar to Chapter 4, but the wavy channel was considered to be horizontal instead of vertical. Since the published work relating to fluid flow confined between two horizontal walls has been concentrated on predominantly forced convection flow, this chapter's focus was on forced convection. Two types of channels were considered, each having frequent changes in cross-section: (1) symmetric and (2) asymmetric. As is customary, from a pragmatic viewpoint, an expression for the skin friction coefficient at the walls of the channel was given. Convective flow in fluid-saturated porous media confined between two wavy walls of mean horizontal disposition, with the medium being heated from below, was also presented. Analytical and numerical solutions of the coupled full Darcy and energy equations were given.

Chapter 6 paid attention to the flow in periodically constricted tubes (PCT). Five types of geometries were considered, namely, square wave, conical, corrugated, parabolic, and sinusoidal. Pressure drops, flow patterns, and velocity profiles were studied. It was found that the heat transfer deviations between PCT and straight tubes are rather small. It was shown that for PCT, the Nusselt numbers depend on Prandtl number value. For $Pr = 0.7$, Nusselt numbers for the PCT generally fall below those for the straight tube; for $Pr = 2.5$, their values are equivalent in value to those for the straight tube; while for $Pr = 5$, their values are larger than those for the straight tube. However, the moderate enhancements in heat transfer are accompanied by a substantial increase in pressure drop. For wavy tubes with wall corrugations, it was found that effect of the corrugations is minimal for stagnation or increased resistance. However, the friction loss in a corrugated tube was found to be larger than that in a smooth one.

Chapter 7 focused on the natural convection flow saturated with nanoparticles in wavy-walled cavities. Two distinct cases were presented: (1) the magnetic field effect on the unsteady natural convection, and (2) entropy generation in natural convection with non-uniform heating. Single-phase and two-phase models were utilized for mathematical simulation of nanofluid flow and heat transfer. In the case of a two-phase model, two dominant particle transfer mechanisms, such as

Brownian motion and thermophoresis, were taken into account. Governing equations were formulated in dimensionless "stream function-vorticity" variables and solved on the basis of the finite difference method of the second order accuracy. It was found in the first case that an increase in the wavy contraction ratio led to an increase in the wave amplitude and an attenuation of the convective flow, with more intensive heating of the wavy troughs where heat conduction was a dominating heat transfer regime. In the second case, it was found that an increase in the wavy contraction ratio led to an attenuation of the convective flow due to an intensification of the secondary vortices located in the bottom and upper wavy troughs.

Chapter 8 looked at the effect of the simultaneous presence of porous media and nanoparticles on flow and heat transfer inside wavy cavities. Five different cases were presented in this chapter. The first case dealt with natural convection in a wavy porous cavity, with sinusoidal temperature distributions on both side walls filled with a nanofluid. For a description of nanofluid transport processes, Brownian motion and thermophoresis were taken into account. It was found that non-homogenous distribution of nanoparticles inside the cavity occurred at low Rayleigh numbers, low Lewis numbers, low values of Brownian motion parameter, and high values of thermophoresis. These obtained results characterize a presence of essential differences in the case of a single-phase model where it is impossible to analyze non-homogeneous distribution of nanoparticles. The second case focused on the natural convection in a wavy open porous cavity filled with a nanofluid. In this case, a single-phase nanofluid model was utilized, and it was seen that the presence of solid nanoparticles suppressed the convective flow, the average Nusselt number was a non-monotonic function of the aspect ratio, and shape coefficient, and the flow was enhanced with the Rayleigh number, aspect ratio, and shape coefficient, but suppressed by increasing nanoparticles volume fraction and undulation number. The third case described was unsteady free convection in a porous open wavy cavity filled with a nanofluid. In this case, a Buongiorno two-phase nanofluid model was utilized and it was found that an increase in the undulations number led to a decrease in the average Nusselt number at the wavy wall due to the significant heating of the wave troughs; so also, an increase in the shape coefficient (< 1.0) led to the heat transfer enhancement, while an increase in this parameter (> 1.0) led to non-monotonic changes in the average Nusselt number due to more essential heating of the wave troughs. The fourth case showed the effect of thermal dispersion on transient natural convection in a wavy-walled porous cavity filled with a nanofluid. It was concluded that in the case of a single-phase nanofluid model and regardless of the values of the dispersion parameter and nanoparticles volume fraction in the considered range, the convective flow was characterized by a double cell flow regime, and an increase in the dispersion parameter led to an essential increase in the heat transfer rate. The fifth case was free convection in a porous wavy cavity filled with a nanofluid with thermal dispersion effect in the case of a two-phase nanofluid model. It was found that an increase in the Rayleigh number led to homogenization of the nanoparticles' distribution inside the cavity and an intensification of convective flow and heat transfer; so also, an increase in the thermal dispersion parameter led to both an attenuation of the convective flow and an increase in the heat transfer rate.

In the book, an attempt was made to include some of the basic topics and treatises related to the convective flow and heat transfer from wavy surfaces. The important fundamental issues were treated in detail and cursory mention was made of all the other available research efforts in the literature. There are a number of areas where more research work is needed to elucidate the subject matter further, and some of these perceived areas are given below for the benefit of those who would want to contribute to this exciting field of convective flow and heat transfer from wavy surfaces.

It should be mentioned that there is only very little work published on convective flow and heat transfer of non-Newtonian fluids over wavy surfaces. In fact, the governing equations for such fluids can be set up easily using cues from those given in this book combined with those that are available elsewhere, such as in the Handbook chapter (Shenoy, 1986) for viscous fluids, or the review article (Shenoy, 1994) for porous media. Extensions of Darcy's law through inclusion of the non-linear porous inertia term known as the Forchheimer term (Nakayama and Pop, 1991; Shenoy, 1992; Nakayama and Shenoy, 1992a,b, 1993), and the viscous effects term known as the Brinkman term (Shenoy, 1993a,b), can be considered for additional investigations of the various classes of problems tackled in this book.

References

Abu-Nada, E., Chamkha, A.J., 2010. Mixed convection flow in a lid-driven inclined square enclosure filled with a nanofluid. *European Journal of Mechanics, B/Fluids* 29, 472–482.

Abu-Nada, E., Chamkha, A.J., 2014. Mixed convection flow of a nanofluid in a lid-driven cavity with a wavy wall. *International Communications in Heat and Mass Transfer* 57, 36–47.

Abu-Nada, E., Öztop, H.F., 2009. Effects of inclination angle on natural convection in enclosures filled with Cu–water nanofluid. *International Journal of Heat and Fluid Flow* 30, 669–678.

Abu-Nada, E., Öztop, H.F., 2011. Numerical analysis of Al_2O_3-water nanofluids natural convection in a wavy walled cavity. *Numerical Heat Transfer, Part A* 59, 403–419.

Adjlout, L., Imine, O., Azzi, A., Belkadi, M., 2002. Laminar natural convection in an inclined cavity with a wavy wall. *International Journal of Heat and Mass Transfer* 45, 2141–2152.

Ahmed, S.E., Abd El-Aziz, M.M., 2013. Effect of local thermal non-equilibrium on unsteady heat transfer by natural convection of a nanofluid over a vertical wavy surface. *Meccanica* 48, 33–43.

Aiboud, S., Saouli, S., 2010. Entropy analysis for viscoelastic magneto hydrodynamic flow over a stretching surface. *International Journal of Non-Linear Mechanics* 45, 482–489.

Al-Amiri, A., Khanafer, K., Bull, J., Pop, I., 2007. Effect of sinusoidal wavy bottom surface on mixed convection heat transfer in a lid-driven cavity. *International Journal of Heat and Mass Transfer* 50, 1771–1780.

Aleshkova, I.A., Sheremet, M. A., 2010. Unsteady conjugate natural convection in a square enclosure filled with a porous medium. *International Journal of Heat and Mass Transfer* 53, 5308–5320.

Ali, M.M., Ramadhyani, S., 1992. Experimental convective heat transfer in corrugated channels. *Experimental Heat Transfer* 5, 175–193.

Alinia, M., Ganji, D.D., Gorji-Bandpy, M., 2011. Numerical study of mixed convection in an inclined two sided lid driven cavity filled with nanofluid using two phase mixture model. *International Communications in Heat and Mass Transfer* 38, 1428–1435.

Al-Najem, N.M., Khanafer, K.M., El-Rafaee, M.M., 1998. Numerical study of laminar natural convection in tilted enclosure with transverse magnetic field, *International Journal of Numerical Methods for Heat and Fluid Flow* 8 (1998) 651–672.

Amano, R.S., 1985. A numerical study of laminar and turbulent heat transfer in a periodically corrugated wall channel. *ASME Journal of Heat Transfer* 107, 564–569.

Amano, R.S., Bagherlee, A., Smith, R.J., Niess, T.G., 1987. Turbulent heat transfer in corrugated wall channels with and without fins. *ASME Journal of Heat Transfer* 109, 62–67.

Aminossadati, S.M., Ghasemi, B., 2009. Natural convection cooling of a localised heat source at the bottom of a nanofluid-filled enclosure. *European Journal of Mechanics, B/Fluids* 28, 630–640.

Aminossadati, S.M., Kargar, A., Ghasemi, B., 2012. Adaptive network-based fuzzy influence system analysis of mixed convection in a two-side lid-driven cavity filled with a nanofluid. *International Journal of Thermal Science* 52, 102–111.

Amiri, A., Vafai, K., 1994. Analysis of dispersion effects and non-thermal equilibrium, non-Darcian variable porosity incompressible flow through porous media. *International Journal of Heat and Mass Transfer* 37, 939–954.

Andreev, V.K., Gaponenko, Y.A., Goncharova, O.N., Pukhnachev, V.N., 2011. *Mathematical Models of Convection*. De Gruyer, Berlin, Germany.

Apelblat, A., 1969. Application of the Laplace transformation to the solution boundary layer equations. III. Magnetohydrodynamic Falkner-Skan problem. *Journal of the Physical Society of Japan* 27, 235–239.

Arakawa, A., 1966. Computational design for long-term numerical integration of the equations of fluid motion: Two-dimensional incompressible flow. Part I. *Journal of Computation Physics* 1, 119–143.

Arefmanesh, A., Mahmoodi, M., 2011. Effects of uncertainties of viscosity models for Al_2O_3-water nanofluid on mixed convection numerical simulations. *International Journal of Thermal Science* 50, 1706–1719.

Arikoglu, A., Ozkol, I., Komurgoz, G., 2008. Effect of slip on entropy generation in a single rotating disk in MHD flow. *Applied Energy* 85, 1225–1236.

Asako, Y., Faghri, M., 1987. Finite-volume solutions for laminar flow and heat transfer in a corrugated duct. *ASME Journal of Heat Transfer* 109, 627–634.

Asako, Y., Nakamura, H., Faghri, M., 1988. Heat transfer and pressure drop characteristics in a corrugated duct with rounded corners. *International Journal of Heat and Mass Transfer* 31, 1237–1245.

Aydin, O., Unal, A., Ayhan, T., 1999. Natural convection in rectangular enclosures heated from one side and cooled from the ceiling. *International Journal of Heat and Mass Transfer* 42, 2345–2355.

Aziz, A., Na, T.Y., 1984. *Perturbation Methods in Heat Transfer.* Hemisphere, Washington DC.

Azzam, I.S., Dullien, F.A.L., 1977. Flow in tubes with periodic step changes in diameter: a numerical solution. *Chemical Engineering Science* 32, 1445–1455.

Bagchi, A., Kulacki, F.A., 2014. *Natural Convection in Superposed Fluid-Porous Layers.* Springer, New York.

Bahaidarah, H.M.S., Anand, N.K., Chen, H.C., 2005. Numerical study of heat and momentum transfer in channels with wavy walls. *Numerical Heat Transfer, Part A* 47, 417–439.

Bahiraei, M., Hangi, M., 2015. Flow and heat transfer characteristics of magnetic nanofluids: A review. *Journal of Magnetism & Magnetic Materials* 374, 125–138.

Bang, I.C., Chang, S.H., 2005. Boiling heat transfer performance and phenomena of Al_2O_3-water nanofluids from a plain surface in a pool. *International Journal of Heat and Mass Transfer* 48, 2407–2419.

Baytas, A.C., 2000. Entropy generation for natural convection in an inclined porous cavity, *International Journal of Heat and Mass Transfer* 43, 2089–2099.

Baytas, A.C., Pop, I., 1999. Free convection in oblique enclosures filled with a porous medium. *International Journal of Heat and Mass Transfer* 42, 1047–1057.

Bear, J., 1972. *Dynamics of Fluids in Porous Media.* Elsevier, New York.

Bejan, A., 1980. Second law analysis in heat transfer. *Energy* 5, 721–732.

Bejan, A., 1982. *Entropy Generation Through Heat and Fluid Flow.*

Wiley, New York.

Bejan, A., 1994. *Entropy Generation through Heat and Fluid Flow* (2nd *edition*). Wiley, New York.

Bejan, A., 1996. *Entropy Generation Minimization.* CRC Press, Boca Raton.

Bejan, A., 2014. *Convective Heat Transfer* (4th *edition*). John Wiley & Sons, New York.

Bejan, A., Dincer, I., Lorente, S., Miguel, A.F., Reis, A.H., 2004. *Porous and Complex Flow Structures in Modern Technologies.* Springer, New York.

Benjamin, T.Br., 1959. Shearing flow over a wavy boundary. *Journal of Fluid Mechanics* 6, 161–205.

Bhardwaj, S., Dalal, A., Pati, S., 2015. Influence of wavy wall and non-uniform heating on natural convection heat transfer and entropy generation inside porous complex enclosure. *Energy* 79, 467–481.

Bhavnani, S.H., Bergles, A.E., 1991. Natural convection heat transfer from sinusoidal wavy surfaces. *Wärme- und Stoffübertragung* 26, 341–349.

Bilgen, E., Wang, X., Vasseur, P., Meng, F., Rabillard, L., 1995. On the periodic conditions to simulate mixed convection heat transfer in horizontal channels. *Numerical Heat Transfer, Part A* 27, 461–472.

Bird, R.B., Stewart, W.E., Lightfoot, E.N., 1960. *Transport Phenomena.* Wiley, New York.

Blancher, S., Batina, J., Creff, R., Andre, P., 1990. Analysis of convective heat transfer by a spectral method for an expanding vortex in a wavy-walled channel. In: *Proceedings of IX International Heat Transfer Conference*, Jerusalem, Israel.

Blazek, J., 2001. *Computational Fluid Dynamics: Principles and Applications.* Elsevier, Oxford, UK.

Blottner, F.G., 1970. Finite-difference methods of solution of the boundary layer equations. *AIAA Journal* 8, 193–205.

Bondareva, N.S., Sheremet, M.A., Pop, I., 2015. Magnetic field effect on the unsteady natural convection in a right-angle trapezoidal cavity filled with a nanofluid: Buongiorno's. *International Journal of Numerical Methods for Heat and Fluid Flow* 25, 1924–1946.

Bordner, G.L., 1978. Nonlinear analysis of laminar boundary layer flow over a periodic wavy surface. *Physics of Fluids* 21, 1414–1474.

Branover, G.G., Tinober, A.B., 1970. *Magnetohydrodynamics of Incompressible Media* (in Russian). Nauka, Moscow.

Brust, M., Walker, M., Bethell, D., Schriffrin, D.J., Whyman, R.J., 1994. Synthesis of thiol-derivated gold nanoparticles in a 2-phase liquid-liquid system. *Journal of the Chemical Society: Chemical Communications* no. 7, 801–802.

Buongiorno, J., 2006. Convective transport in nanofluids. *ASME Journal of Heat Transfer* 128, 240–250.

Buongiorno, J., Hu, W., 2005. Nanofluid coolant for advanced nuclear power plants. Paper No. 5705. In: *Proceedings of ICAPP'05*, Seoul, South Korea, May 15–19.

Burley, D.M., Moult, A., Rawson, H., 1978. Application of the finite element method to calculate flow patterns in glass tank furnaces. *Glass Technology* 19, 86–91.

Burns, J.C., Parkes, T., 1967. Peristaltic motion. *Journal of Fluid Mechanics* 29, 731–743.

Butler, G.A., 1979. Blood flow in arteries with and without prosthetic inserts. PhD Thesis, University of Manchester, UK.

Butt, A.S., Ali, A., 2012. Effects of magnetic field on entropy generation in flow and heat transfer due to a radially stretching surface. *Chinese Physics Letter* 30, 024704–024708.

Butt, A.S., Ali, A., 2014a. Entropy analysis of flow and heat transfer caused by a moving plate with thermal radiation. *Journal of Mechanical Science and Technology* 28, 343–348.

Butt, A.S., Ali, A., 2014b. A computational study of entropy generation in magnetohydrodynamic flow and heat transfer over an unsteady stretching permeable sheet. *European Physics Journal Plus* 129, 1–13.

Butt, A.S., Munawar, S., Ali, A., Mehmood, A., 2012a. Entropy generation in the Blasius flow under thermal radiation. *Physica Scripta* 85, 035008.

Butt, A.S., Munawar, S., Ali, A., Mehmood, A., 2012b. Effect of viscoelasticity on entropy generation in a porous medium over a stretching plate. *World Applied Sciences Journal* 17, 516–523.

Butt, A.S., Munawar, S., Ali, A., Mehmood, A., 2012c. Entropy generation in hydrodynamic slip flow over a vertical plate with convective boundary. *Journal of Mechanical Science and Technology* 26, 2977–2984.

Caponi, E.A., Fornberg, B., Knight, D.D., McLean, J.W., Saffman, P.G., Yuen, H.C., 1982. Calculations of laminar viscous flow over a moving wavy surface. *Journal of Fluid Mechanics* 124, 347–362.

CEA, 2007. *Nanofluids for heat transfer applications.* France: Marketing Study Unit.

Celli, M., 2013. Non-homogeneous model for a side heated square cavity filled with a nanofluid. *International Journal of Heat and Fluid Flow* 44, 327–335.

Chamkha, A.J., Abu-Nada, E., 2012. Mixed convection flow in single- and double-lid driven square cavities filled with water-Al2O3 nanofluid: Effect of viscosity models. *European Journal of Mechanics - B/Fluids* 36, 82–96.

Chandran, P., Sacheti, N.C., Singh, A.K., 1996. Hydromagnetic flow and heat transfer past a continuously moving porous boundary. *International Communications in Heat and Mass Transfer* 23, 889–898.

Chen, C.K., Cho, C.C., 2007. Electrokinetically-driven flow mixing in microchannels with wavy surface. *Journal of Colloid and Interface Science* 312, 470–480.

Chen, C.K., Hsiao, S.W., 1998. Transport phenomena in enclosed porous cavities. In: *Transport Phenomena in Porous Media* (Ingham, D.B. and Pop, I., eds). Pergamon, Oxford, 31–56.

Chen, C.K., Lai, H.-Y., Liu, C.-C., 2011. Numerical analysis of entropy generation in mixed convection flow with viscous dissipation effects in vertical channel. *International Communications in Heat and Mass Transfer* 38, 285–290.

Chen, H.S., Yang, W., He, Y.R., Ding, Y.L., Zhang, L.L., Tan, C.Q., 2008. Heat transfer and flow behaviour of aqueous suspensions of titanate nanotubes (nanofluids). *Powder Technology* 183, 63–72.

Cheng, P., 1981. Thermal dispersion effects in non-Darcian convective flows in a saturated porous medium. *Letters in Heat and Mass Transfer* 8, 267–270.

Cheng, P., 1985. Natural convection in a porous medium: external flows. *Natural Convection: Fundamentals and Applications* (Kakaç, S., Aung, W., Viskanta, R., eds.). Hemisphere, Washington DC.

Cheng, P., Minkowycz, W.J., 1977. Free convection about a vertical flat plate embedded in a porous medium with application to heat transfer from a dike. *Journal of Geophysics Research* 82, 2040–2044.

Cheng, P., Vortmeyer, D., 1988. Transverse thermal dispersion and wall channeling in a packed bed with forced convective flow. *Chemical Engineering Science* 43, 2523–2532.

Chengara, A., Nikolov, A.D., Wasan, D.T., 2004. Spreading of nanofluids driven by the structural disjoining pressure gradient. *Journal of Colloid Interface Science* 280, 192–201.

Chiu, C.P., Chou, H.M., 1994. Transient analysis of natural convection along a vertical wavy surface in micropolar fluids. *International Journal of Engineering Science* 32, 19–33.

Cho, C.C., 2014. Heat transfer and entropy generation of natural convection in nanofluid-filled square cavity with partially-heated wavy surface. *International Journal of Heat and Mass Transfer* 77, 818–827.

Cho, C.C., Chen, C.L., Chen, C.K., 2012a. Mixing of non-Newtonian fluids in wavy serpentine microchannel using electrokinetically-driven flow. *Electrophoresis* 33, 743–750.

Cho, C.C., Chen, C.L., Chen, C.K., 2012b. Natural convection heat transfer performance in complex-wavy wall enclosed cavity filled with nanofluid. *International Journal of Thermal Science* 60, 255–263.

Cho, C.C., Chen, C.L., Chen, C.K., 2012c. Characteristics of combined electroosmotic flow and pressure-driven flow in microchannels with complex-wavy surfaces. *International Journal of Thermal Science* 61, 94–105.

Cho, C.C., Chen, C.L., Chen, C.K., 2012d. Electrokinetically-driven non-Newtonian fluid flow in rough microchannel with complex-wavy surface. *Journal of Non-Newtonian Fluid Mechanics* 173-174, 13–20.

Cho, C.C., Chen, C.L., Chen, C.K., 2013a. Mixed convection heat transfer performance of water-based nanofluids in lid-driven cavity with wavy surfaces. *International Journal of Thermal Science* 68, 181–190.

Cho, C.C., Chen, C.L., Chen, C.K., 2013b. Natural convection heat transfer and entropy generation in wavy-wall enclosure containing water-based nanofluid. *International Journal of Heat and Mass Transfer* 61, 749–758.

Cho, C.C., Chen, C.L., Hwang, J.J., Chen, C.K., 2014. Natural convection heat transfer performance of non-Newtonian power-law fluids enclosed in cavity with complex-wavy surfaces. *ASME Journal of Heat Transfer* 136, 014502.

Choi, S.U.S., 1995. Enhancing thermal conductivity of fluids with nanoparticles. In: Siginer, D.A., Wang, H.P. (eds.). Developments and Applications of Non-Newtonian Flows. *FED ASME*, Vol.

231/MD-66. New York, 99–105.

Choi, S.U.S., 2009. Nanofluids: From vision to reality through research. *ASME Journal of Heat Transfer* 131, Art. No. 033106.

Choi, S.U.S., Zhang, Z.G., Yu, W., Lockwood, F.E., Grulke, E.A., 2001. Anomalously thermal conductivity enhancement in nanotube suspensions. *Applied Physics Letters* 79, 2252–2254.

Chopkar, M., Das, A.K., Manna, I., Das, P.K., 2008. Pool boiling heat transfer characteristics of ZrO_2-water nanofluids from a flat surface in a pool. *Heat and Mass Transfer* 44, 999–1004.

Chow, J.C.F., Soda, K., 1972. Laminar flow in tubes with constriction. *Physics of Fluids* 15, 1700–1706.

Chow, J.C.F., Soda, K., 1973. Heat or mass transfer in laminar flow in conduits with constriction. *ASME Journal of Heat Transfer* 95, 352–356.

Chu, Y.-H., Hsu, P.-T., Liu, Y.-H., Chiu, L.-H., 2002. Mixed convection of micropolar fluids along a vertical wavy surface with a discontinuous temperature profile. *Numerical Heat Transfer, Part A* 42, 733–755.

Cîmpean, D., Lungu, N., Pop, I., 2008. A problem of entropy generation in a channel filled with a porous medium. *Creative Mathematics and Informatics* 17, 357–362.

Cramer, K.R. Pai, S.-I., 1973. *Magneto-fluid dynamics for engineers and applied physicists*. McGraw-Hill Book Company, Washington DC.

Dalal, A., Das, M.K., 2006. Natural convection in a cavity with a wavy wall heated from below and uniformly cooled from the top and both sides. *ASME Journal of Heat Transfer* 128, 717–725.

Das, P.K., Mahmud, S., 2003. Numerical investigation of natural convection inside a wavy enclosure. *International Journal of Thermal Sciences* 42, 397–406.

Das, U.N., Ahmed, N., 1992. Free convective MHD flow and heat transfer in a viscous incompressible fluid confined between a long vertical wavy wall and a parallel flat wall. *Indian Journal of Pure and Applied Mathematics* 23, 295–304.

Das, S.K., Choi, U.S., 2009. A review of heat transfer in nanofluids. In: Greene, G.A., Cho, Y.I., and Bar-Cohen, A. (eds.). *Advances in Heat Transfer* 41, 81–197.

Das, S.K., Choi, S.U.S., Patel, H.E., 2006. Heat transfer in nanofluids — a review. *ASME Heat Transfer Engineering* 27, 3–19.

Das, S.K., Choi, S.U.S., Yu, W., Pradeep, T., 2007. Nanofluids: Science and Technology. Wiley, New York.

Das, S.K., Putra, N., Thiesen, P., Roetzel, W., 2003. Temperature dependence of thermal conductivity enhancement for nanofluids. *ASME Journal of Heat Transfer* 125, 567–574.

Daungthongsuk, W., Wongwises, S., 2007. A critical review of convective heat transfer of nanofluids. *Renewable and Sustainable Energy Reviews* 11, 797–817.

Deiber, J.A., Schowalter, R., 1979. Flow through tubes with sinusoidal axial variations in diameter. *AIChE Journal* 25, 638–644.

de Lemos, M.J.S., 2012. *Turbulence in Porous Media: Modeling and Applications (2nd edition)*. Elsevier, Oxford.

Deng, Q.-H., Chang, J.-J., 2008. Natural convection in a rectangular enclosure with sinusoidal temperature distributions on both side walls. *Numerical Heat Transfer, Part A* 54, 507–524.

Dhanai, R., Rana, P., Kumar, L., 2015. Critical values in slip flow and heat transfer analysis of non-Newtonian nanofluid utilizing heat source/sink and variable magnetic field: Multiple solutions. *Journal of the Taiwan Institute of Chemical Engineering* 47, 18–27.

Dinarvand, S., Hosseini, R., Pop, I., 2015. Unsteady convective heat and mass transfer of a nanofluid in Howarth's stagnation point by Buongiorno's model. *International Journal of Numerical Methods for Heat and Fluid Flow* 25, 1176–1197.

Ding, Y., Chen, H., Wang, L., Yang, C.-Y., He, Y., Yang, W., Lee, W.P., Zhang, L., Huo, R., 2007. Heat transfer intensification using nanofluids. *KONA* 25, 23–38.

Dodson, A.G., Torwnsend, P., Walters, K., 1971. On the flow of Newtonian and non-Newtonian liquids through corrugated pipes. *Rheologica Acta* 10, 508–516.

Dullien, F.A.L., Azzam, M.I.S., 1973. Flow rate–pressure gradient measurements in periodically nonuniform capillary tubes. *AIChE Journal* 19, 222–229.

Eastman, J.A., Choi, S.U.S., Yu, W., Thompson, L.J., 2001. Anomalously increased effective thermal conductivities of ethylene glycol-based nanofluids containing copper nanoparticles. *Applied Physics Letters* 78, 718–720.

Edel, J.B., deMello, A.J. (eds.), 2008. *Nanofluidics: Nanoscience and Nanotechnology*. Published by The Royal Society of Chemistry, Thomas Graham House, Science Park, Milton Road, Cambridge,

UK.

Elshehabey, H.M., Ahmed, S.E., 2015. MHD mixed convection in a lid-driven cavity filled by a nanofluid with sinusoidal temperature distribution on the both vertical walls using Buongiorno's nanofluid model. *International Journal of Heat and Mass Transfer* 88, 181–202.

Esfahani, J.A., Jafarian, M.M., 2005, Entropy generation analysis of a flat plate boundary layer with various solution methods. *Scientia Iranica* 12, 233–240.

Esmaeilpour, M., Abdollahzadeh, M., 2012. Free convection and entropy generation of nanofluid inside an enclosure with different patterns of vertical wavy walls. *International Journal of Thermal Science* 52, 127–136.

Faghri, M., Sparrow, E.M., Prata, A.T., 1984. Finite difference solutions of convection-diffusion problems in irregular domains using a non-orthogonal coordinate transformation. *Numerical Heat Transfer, Part A* 7, 183–209.

Farr, W.W., Gabitto, J.F., Luss, D., Balakotaiah, V., 1991. Reaction-driven convection in a porous medium. *AIChE Journal* 37, 963–985.

Fedkiw, P.S., Newman, J., 1977. Mass transfer at high Péclet numbers for creeping flow in a packed-bed reactor. *AIChE Journal* 23, 255–263.

Fedkiw, P.S., Newman, J., 1978. Numerical calculations for the asymptotic, diffusion dominated mass-transfer coefficient in packed bed reactors. *Chemical Engineering Science* 33, 1563–1566.

Fedkiw, P.S., Newman, J., 1979. Entrance region (Leveque like) mass transfer coefficients in packed-bed reactors. *AIChE Journal* 25, 1077–1080.

Fedkiw, P.S., Newman, J., 1987. Friction factors for creeping flow in sinusoidal periodically constricted tubes. *Chemical Engineering Science* 42, 2962–2963.

Forrester, J.H., 1968. Flow through a converging-diverging tube and its implications in occlusive vascular disease. Ph.D. Thesis, Iowa State University, Ames, Iowa.

Forrester, J.H., Young, D.F., 1970a. Flow through a converging-diverging tube and its implications in occlusive vascular disease – I. Theoretical development. *Journal of Biomechanics* 3, 297–306.

Forrester, J.H., Young, D.F., 1970b. Flow through a converging-diverging tube and its implications in occlusive vascular disease

– II. Theoretical and experimental results and their implications. *Journal of Biomechanics* 3, 307–316.

Franzen, P., 1977. Strömungskanal mit Veranderlichem Kreisquerschnitt als Modell für Zufallsschuttungen gleich grosser Kugeln. *Rheologica Acta* 16, 548–552.

Garg, V.K., Maji, P.K., 1988. Laminar flow and heat transfer in a periodically converging-diverging channel. *International Journal of Numerical Methods in Fluids* 8, 579–597.

Gebhart, B., Jaluria, Y., Mahajan, R.L., Sammakia, B., 1988. *Buoyancy Induced Flows and Transport.* Hemisphere, Washington DC.

Gerrard, J.H., 1971. The stability of unsteady axisymmetric incompressible pipe flow close to a piston. *Journal of Fluid Mechanics* 50, 625–644.

Ghalambaz, M., Sheremet, M.A., Pop, I., 2015. Free convection in a parallelogrammic porous cavity filled with a nanofluid using Tiwari and Das' nanofluid model. *PLOSE ONE* 10(5): e0126486. doi:10.1371/journal.pone.0126486.

Ghasemi, B., Aminossadati, S.M., 2009. Natural convection heat transfer in an inclined enclosure filled with a water–CuO nanofluid. *Numerical Heat Transfer, Part A* 55, 807–823.

Gillani, N.V., Swanson, W.M., 1976. Time dependent laminar incompressible flow through a spherical cavity. *Journal of Fluid Mechanics* 78, 99–127.

Goldstein, L., Sparrow, E.M., 1977. Heat/mass transfer characteristics for flow in a corrugated wall channel. *ASME Journal of Heat Transfer* 99, 187–195.

Goldsworthy, F.A., 1961. Magnetohydrodynamic flow of a perfectly conducting, viscous fluid. *Journal of Fluid Mechanics* 11, 519–528.

Gopalan, N.P., Ponnalagarsamy, N., 1992. Investigation of laminar flow of a suspension in corrugated straight tubes. *International Journal of Engineering Science* 30, 631–644.

Gosman, A.D., Pun, W.M., Runchal, A.K., Spalding, D.B., Wolfstein, M., 1969. *Heat and Mass Transfer in Recirculating Flows.* Academic Press, London, UK.

Groşan, T., 2011. Thermal dispersion effect on fully developed free convection of nanofluids in a vertical channel. *Sains Malaysiana* 40, 1429–1435.

Groşan, T., Pop, I., 2011a. Axisymmetric mixed convection boundary layer flow past a vertical cylinder in a nanofluid. *International Jour-*

nal of Heat and Mass Transfer 54, 3139–3145.

Groşan, T., Pop, I., 2011b. Forced convection boundary layer flow past nonisothermal thin needles in nanofluids. *ASME Journal of Heat Transfer* 133, 054503 (four pages).

Groşan, T., Pop, I., 2012. Fully developed mixed convection in a vertical channel filled by a nanofluid. *ASME Journal of Heat Transfer* 134, 082501 (five pages).

Groşan, T., Revnic, C., Pop, I., Ingham, D.B., 2009. Magnetic field and internal heat generation effects on the free convection in a rectangular cavity filled with a porous medium. *International Journal of Heat and Mass Transfer* 52, 1525–1533.

Groşan, T., Revnic, C., Pop, I., Ingham, D.B., 2015. Free convection heat transfer in a square cavity filled with a porous medium saturated by a nanofluid. *International Journal of Heat and Mass Transfer* 87, 36–41.

Habibzadeh, S., Kazemi-Beydokhti, A., Khodadadi, A.A., Mortazavi, Y., Omanovic, S., Shariat-Niassar, M., 2010. Stability and thermal conductivity of nanofluids of tin dioxide synthesized via microwave-induced combustion route. *Chemical Engineering Journal* 156, 471–478.

Haddad, Z., Abu-Nada, E., Öztop, H.F., Mataoui, A., 2012a. Natural convection in nanofluids: Are the thermophoresis and Brownian motion effects significant in nanofluid heat transfer enhancement? *International Journal of Thermal Sciences* 57, 152–162.

Haddad, Z., Öztop, H.F., Abu-Nada, E., Mataoui, A., 2012b. A review on natural convective heat transfer of nanofluids. *Renewable and Sustainable Energy Reviews* 16, 5363–5378.

Hadjadj, A., Kyal, M.E., 1999. Effect of two sinusoidal protuberances on natural convection in a vertical annulus. *Numerical Heat Transfer, Part A* 36, 273–289.

Hadjisophocleous, G.V., Sousa, A.C.M., Venart, J.E.S., 1998. Prediction of transient natural convection in enclosures of arbitrary geometry using a nonorthogonal numerical model. *Numerical Heat Transfer, Part A* 13, 373–392.

Hickox, C.E., Gartling, D.K., McVey, D.F., Russo, A.J., 1980. Analysis of heat and mass transfer in sub-seabed disposal of nuclear waste. *ASME Winter Meeting*, Chicago, Illinois.

Hong, J.T., Tien, C.L., 1987. Analysis of thermal dispersion effect on vertical plate natural convection in porous media. *International*

Journal of Heat and Mass Transfer 30, 143–150.

Hong, J.T., Yamada, Y., Tien, C.L., 1987. Effects of non-Darcian and non-uniform porosity on vertical plate natural convection in porous media. *ASME Journal of Heat Transfer* 109, 356–361.

Horton, C.W., Rogers, C.T., 1945. Convection currents in a porous medium. *Journal of Applied Physics* 16, 367–370.

Hossain, M.S., Alim, M.A., 2014. MHD free convection within trapezoidal cavity with non-uniformly heated bottom wall. *International Journal of Heat and Mass Transfer* 69, 327–336.

Hossain, M.A., Munir, M.S., Pop, I., 2001. Natural convection with variable viscosity and thermal conductivity from a vertical wavy cone. *International Journal of Thermal Sciences* 40, 437–443.

Hossain, M.A., Pop, I., 1996. Magnetohydrodynamic boundary layer flow and heat transfer on a continuous moving wavy surface. *Archives of Mechanics* 48, 813–823.

Hossain, M.A., Rees, D.A.S., 1999. Combined heat and mass transfer in natural convection flow from a vertical wavy surface. *Acta Mechanica* 136, 133–141.

Houpeurt, A., 1959. Sur l'ecoulsnent des gaz dans les milieux poreux. *Revue de l'Institut français du pétrole et Annales des combustibles liquides* 14, 1468.

Hsu, P.-T., Chen, C.-K., Wang, C.-C., 2000. Mixed convection of micropolar fluids along a vertical wavy surface. *Acta Mechanica* 144, 231–247.

Hyun, J.M., 1994. Unsteady buoyant convection in an enclosure. *Advances in Heat Transfer* 24, 277–320.

Ilis, G.G., Mobedi, M., Sunden, B., 2008. Effect of aspect ratio on entropy generation in a rectangular cavity with differentially heated vertical walls. *International Communications in Heat and Mass Transfer* 35, 696–703.

Inger, G.R. 1971. Compressible boundary layer flow past a swept wavy wall with heat transfer and ablation. *Astronautica Acta* 16, 325–338.

Ingham, D.B., 1973. Impulsively started viscous flows past a finite flat plate with and without an applied magnetic field. *International Journal for Numerical Methods in Engineering* 6, 521–527.

Ingham, D.B., Pop, I. (eds.), 1998. *Transport Phenomena in Porous Media*. Pergamon, Oxford, UK.

Ingham, D.B., Pop, I. (eds.), 2005. *Transport Phenomena in Porous Media III*, Elsevier, Oxford, UK.

Jaluria, Y., 1980. *Natural Convection Heat and Mass Transfer*. Pergamon Press, Oxford, UK.

Jaluria, Y., Manca, O., Poulikakos, D., Vafai, K., Wang, L., 2012. Heat transfer in nanofluids. *Advances in Mechanical Engineering*. Article ID 972973 (two pages).

Jang, J.H., Yan, W.M., 2004. Mixed convection heat and mass transfer along a vertical wavy surface. *International Journal of Heat and Mass Transfer* 47, 419–428.

Jang, J.H., Yan, W.M., Liu, H.C., 2003. Natural convection heat and mass transfer along a vertical wavy surface. *International Journal of Heat and Mass Transfer* 46, 1075–1083.

Jansen, J.D., 2013. A Systems Description of Flow Through Porous Media. Springer, New York.

Joshi, Y., 1990. Transient Natural Convection Flows. In: *Encyclopedia of Fluid Mechanics* (Cheremisinoff, N.P. ed.), Vol. 8, Ch. 15, 477–533.

Jou, R.Y., Tzeng, S.C., 2006. Numerical research of nature convective heat transfer enhancement filled with nanofluids in rectangular enclosures. *International Communications in Heat and Mass Transfer* 33, 727–736.

Junqi, D., Jiangping, C., Zhijiu, C., Yimin, Z., Wenfeng, Z., 2007. Heat transfer and pressure drop correlations for the wavy fin and flat tube heat exchangers. *Applied Thermal Engineering* 27, 2066–2073.

Kakaç, S., Pramuanjaroenkij, A., 2009. Review of convective heat transfer enhancement with nanofluid. *International Journal of Heat and Mass Transfer* 52, 3187–3196.

Kakaç, S., Shah, R.K., Aung, W., 1987. *Handbook of Single-Phase Convective Heat Transfer*. John Wiley & Sons, New York.

Kang, H.U., Kim, S.H., Oh, J.M., 2006. Estimation of thermal conductivity of nanofluid using experimental effective particle volume. *Experimental Heat Transfer* 19, 181–191.

Karami, A., Yousefi, T., Harsini, I., Maleki, E., Mahmoudinezha, S., 2015. Neuro-Fuzzy modeling of the free convection heat transfer from a wavy surface. *Heat Transfer Engineering* 36, 847–855.

Kasaeian, A., Eshghi, A.T., Sameti, M., 2015. A review on the applications of nanofluids in solar energy systems. *Renewable and Sustainable Energy Reviews* 43, 584–598.

Kashani, S., Ranjbar, A.A., Mastiani, M., Mirzaei, H., 2014. Entropy generation and natural convection of nanoparticle-water mixture (nanofluid) near water density inversion in an enclosure with various patterns of vertical wavy walls. *Applied Mathematics and Computation* 226, 180–193.

Kaviany, M., 1991. *Principles of Heat Transfer in Porous Media.* Springer, New York.

Keller, H.B., 1978. Numerical methods in boundary layer theory. *Annual Review of Fluid Mechanics* 10, 417–433.

Keller, H.B., Cebecci, T., 1971. Accurate numerical methods for boundary layer flows. Part-I: two dimensional laminar flows In: *Proceedings of the International Conference on Numerical Methods in Fluid Dynamics.* Lecture Notes in Physics. Springer, Berlin, Germany.

Kelly, R.E., Pal, D., 1978. Thermal convection with spatially periodic boundary conditions: resonant wavelength excitation. *Journal of Fluid Mechanics* 86, 433–456.

Khanafer, K., 2014. Comparison of flow and heat transfer characteristics in a lid-driven cavity between flexible and modified geometry of a heated bottom wall. *International Journal of Heat and Mass Transfer* 78, 1032–1041.

Khanafer, K., Chamkha, A.J., 1999. Mixed convection flow in a lid-driven enclosure filled with a fluid-saturated porous medium. *International Journal of Heat and Mass Transfer* 42, 2465–2481.

Khanafer, K., Vafai, K., Lightstone, M., 2003. Buoyancy-driven heat transfer enhancement in a two-dimensional enclosure utilizing nanofluids. *International Journal of Heat and Mass Transfer* 46, 3639–3653.

Khanafer, K., Vafai, K., 2011. A critical synthesis of thermophysical characteristics of nanofluids. *International Journal of Heat and Mass Transfer* 54, 4410–4428.

Kim, E., Chen, J.L.S., 1991. Natural convection of non-Newtonian fluids along a wavy vertical plate. *ASME, Heat Transfer Division* – Vol. 174, 45–49.

Kimura, S., Bejan, A., 1983. The "heatline" visualization of convective heat transfer. *ASME Journal of Heat Transfer* 105, 916–919.

Kimura, S., Schubert, G., Straus, J.M., 1987. Instabilities of steady and periodic, and quasi-periodic modes of convection in porous media. *ASME Journal of Heat Transfer* 109, 350–355.

Kishinami, K., Saito, H., Tokura, I., 1990. An experimental study on natural convective heat transfer from a vertical wavy surface heated at convex/concave elements. *Experimental Thermal and Fluid Sci-*

ence 3, 305–315.

Kleinstreuer, C., 2013. *Microfluidics and Nanofluidics. Theory and Selected Applications.* Wiley, Hoboken, New Jersey.

Kleinstreuer, C., Li, J., Koo, J., 2008. Microfluidics of nano-drug delivery. *International Journal of Heat and Mass Transfer* 51, 5590–5597.

Krishna, R., Calero, S., Smit, B., 2002, Investigation of entropy effects during sorption of mixtures of alkanes in MFI zeolite. *Chemical Engineering Journal* 88, 81–94.

Kumar, B.V.R., 2000. A study of free convection induced by a vertical wavy surface with heat flux in a porous enclosure. *Numerical Heat Transfer*, Part A 37, 493–510.

Kumar, B.V.R., Gupta, S., 2005. Combined influence of mass and thermal stratification on double-diffusion non-Darcian natural convection from a wavy vertical wall to porous media. *ASME Journal of Heat Transfer* 127, 637–647.

Kumar, B.V.R., Singh, P., Murthy, P.V.S.N., 1997. Effect of surface undulations on natural convection in a porous square cavity. *ASME Journal of Heat Transfer* 119, 848–851.

Kuznetsov, A.V., Nield, D.A., 2010a. Natural convective boundary-layer flow of a nanofluid past a vertical plate. *International Journal of Thermal Sciences* 49, 243–247.

Kuznetsov, A.V., Nield, D.A., 2010b. Thermal instability in a porous medium layer saturated by a nanofluid: Brinkman model. *Transport in Porous Media* 81, 409–422.

Kuznetsov, A.V., Nield, D.A., 2010c. Effect of local thermal non-equilibrium on the onset of convection in a porous medium layer saturated by a nanofluid. *Transport in Porous Media* 83, 425–436.

Kuznetsov, A.V., Nield, D.A., 2010d. The onset of double-diffusive nanofluid convection in a layer of a saturated porous medium. *Transport in Porous Media* 85, 941–951.

Kuznetsov, A.V., Nield, D.A., 2011a. The Cheng-Minkowycz problem for the double-diffusive natural convective boundary layer flow in a porous medium saturated by a nanofluid. *International Journal of Heat and Mass Transfer* 54, 374–378.

Kuznetsov, A.V., Nield, D.A., 2011b. The effect of vertical through flow on thermal instability in a porous medium layer saturated by a nanofluid. *Transport in Porous Media* 87, 765–775.

Kuznetsov, A.V., Nield, D.A., 2011c. Double-diffusive natural convective boundary-layer flow of a nanofluid past a vertical plate. *International Journal of Thermal Science* 50, 712–717.

Kuznetsov, A.V., Nield, D.A., 2013. The Cheng-Minkowycz problem for natural convective boundary layer flow in a porous medium saturated by a nanofluid: A revised model. *International Journal of Heat and Mass Transfer* 65, 682–685.

Kuznetsov, G.V., Sheremet, M.A., 2008. New approach to the mathematical modeling of thermal regimes for electronic equipment. *Russian Microelectronics* 37, 131–138.

Kuznetsov, G.V., Sheremet, M.A., 2010. On the possibility of controlling thermal conditions of a typical element of electronic equipment with a local heat source via natural convection. *Russian Microelectronics* 39, 427–442.

Kuznetsov, G.V., Sheremet, M.A., 2011. Unsteady natural convection of nanofluids in an enclosure having finite thickness walls. *Computational Thermal Sciences* 3, 427–443.

Kvernvold, O., Tyvand, P.A., 1979. Nonlinear thermal convection in anisotropic porous media. *Journal of Fluid Mechanics* 90, 609–624.

Kvernvold, O., Tyvand, P.A., 1980. Dispersion effects on thermal convection in porous media. *Journal of Fluid Mechanics* 99, 673–686.

Ladyzhenskaya, O.A., 1969. *The Mathematical Theory of Viscous Incompressible Fluid*. Gordon & Breach, New York.

Lage, J.L., Bejan, A., 1993. The resonance of natural convection in an enclosure heated periodically from the side. *International Journal of Heat and Mass Transfer* 36, 2027–2038.

Lahbabi, A., Chang, H.C., 1985. High Reynolds number flow through cubic arrays of spheres, steady state and transition to turbulence. *Chemical Engineering Science* 40, 435–447.

Lahbabi, A., Chang, H.C., 1986. Flow in periodically constricted tubes: transition to inertial and nonsteady flows. *Chemical Engineering Science* 41, 2481–2505.

Lai, F.C., 2000. Mixed convection in saturated porous media. In: *Handbook of Porous Media* (Vafai, K., ed.). Marcel Dekker, New York.

Lapwood, E.R., 1948. Convection of a fluid in a porous medium. *Proceedings of the Cambridge Philosophical Society* 44, 508–521.

Leal, Gary, L., 2007. *Advanced Transport Phenomena: Fluid Mechanics and Convective Transport Processes*. Cambridge University Press, New York.

Lekoudis, S.G., Nayfeh, A.H., Saric, W.S., 1970. Compressible boundary layers over wavy walls. *Physics of Fluids* 19, 514–519.

Lemcoff, N.O., Pereira Duarte, S.I., Martinez, O.M., 1990. Heat transfer in packed beds. *Reviews in Chemical Engineering* (Aumundson, N. R., Luss, D., eds.) 6, 229–292.

Lessen, H., Gangwani, S.T., 1976. Effect of small amplitude wall waviness upon the stability of the laminar boundary layer. *Physics of Fluids* 19, 510–513.

Liron, N., Wilhelm, H.E., 1974. Integration of the magnetohydrodynamic boundary-layer equations by Meksyn' method. *Journal of Applied Mathematics and Mechanics* (ZAMM) 54, 27–37.

Lomascolo, M., Colangelo, G., Milanese, M., de Risi, A., 2015. Review of heat transfer in nanofluids: conductive, convective and radiative experimental results. *Renewable and Sustainable Energy Reviews* 43, 1182–1198.

Lowe, M., Albert, B.S., Gollub, J.P., 1986. Convective flows with multiple spatial periodicities. *Journal of Fluid Mechanics* 173, 253–272.

Macagno, E.O., Hung, T.K., 1967. Computational and experimental study of a captive annular eddy. *Journal of Fluid Mechanics* 28, 43–64.

Mahdy, A., Ahmed, S.E., 2012. Laminar free convection over a vertical wavy surface embedded in a porous medium saturated with a nanofluid. *Transport in Porous Media* 91, 423–435.

Mahian, O., Kianifar, A., Kalogirou, S.A., Pop, I., Wongwises, S., 2013. A review of the applications of nanofluids in solar energy. *International Journal of Heat and Mass Transfer* 57, 582–594.

Mahmoudi, A.H., Shahi, M., Shahedin, A.M., Hemati, N., 2011. Numerical modeling of natural convection in an open cavity with two vertical thin heat sources subjected to a nanofluid. *International Communications in Heat and Mass Transfer* 38, 110–118.

Mahmud, S., Das, P.K., Hyder, N., Islam, A.K.M.S., 2002. Free convection in an enclosure with vertical wavy walls. *International Journal of Thermal Sciences* 41, 440–446.

Mahmud, S., Fraser, R.A., 2003. The second law analysis in fundamental convective heat transfer problems. *International Journal of Thermal Sciences* 42, 177–186.

Mahmud, S., Fraser, R.A., 2004. Free convection and entropy generation inside a vertical inphase wavy cavity. *International Communications in Heat and Mass Transfer* 31, 455–466.

Mahmud, S., Sadrul Islam, 2003. Laminar free convection and entropy generation inside an inclined wavy enclosure. *International Journal of Thermal Sciences* 42, 1003–1012.

Makinde, O.D., 2006. Irreversibility analysis for gravity driven non-Newtonian liquid film along an inclined isothermal plate. *Physica Scripta* 74, 642–645.

Makinde, O.D., 2010. Thermodynamic second law analysis for a gravity driven variable viscosity liquid film along an inclined heated plate with convective cooling. *Journal of Mechanical Science and Technology* 24, 899–908.

Makinde, O.D., 2011. Second law analysis for variable viscosity hydromagnetic boundary layer flow with thermal radiation and Newtonian heating. *Entropy* 13, 1446–1464.

Makinde, O.D., 2012. Entropy analysis for MHD boundary layer flow and heat transfer over a flat plate with a convective surface boundary condition. *International Journal of Exergy* 10, 142–154.

Malashetty, M.S., Umavathi, J.C., Leela, V., 2001. Magnetoconvective flow and heat transfer between vertical wavy wall and a parallel flat wall. *International Journal of Applied Mechanics and Engineering* 6, 437–456.

Maliga, S.E.B., Palm, S.M., Nguyen, C.T., Roy, G., Galanis, N., 2005. Heat transfer enhancement using nanofluid in forced convection flow. *International Journal of Heat and Fluid Flow* 26, 530–546.

Manglik, R.M., Zhang, J., Muley, A., 2005. Low Reynolds number forced convection in three-dimensional wavy-platefin compact channels: fin density effects. *International Journal of Heat and Mass Transfer* 48, 1439–1449.

Mansour, M.A., Mohamed, R.A., Abd-Elaziz, M.M., Ahmed, S.E., 2010. Numerical simulation of mixed convection flows in a square lid-driven cavity partially heated from below using nanofluid. *International Communications in Heat and Mass Transfer* 37, 1504–1512.

Martynenko, O.G., Khramtsov, P.P., 2005. *Free-Convective Heat Transfer*. Springer, New York.

Masuda, H., Ebata, A., Teramae, K., Hishinuma, N., 1993. Alteration of thermal conductivity and viscosity of liquid by dispersing ultra fine particles. *Netsu Bussei* 7, 227–233.

Masuda, H., Higashitani, K., Yoshida, H., 2006. *Powder Technology: Fundamentals of Particles, Powder Beds, and Particle Generation*. CRC Press, Tokyo.

McDonald, D.A., 1960. *Blood flow in arteries.* Williams and Wilkins, Baltimore, Maryland, 90.

McKibbin, R., 1998. Mathematical models for heat and mass transport in geothermal systems. In: *Transport Phenomena in Porous Media* (Ingham, D.B. and Pop, I., eds). Pergamon, Oxford, 131–154.

Mehrez, Z., El Cafsi, A., Belghith, A., Le Quere, P., 2015. MHD effects on heat transfer and entropy generation of nanofluid flow in an open cavity. *Journal of Magnetism and Magnetic Materials* 374, 214–224.

Merkin, J.H., 1980. Mixed convection boundary layer flow on a vertical surface in a saturated porous medium. *Journal of Engineering Mathematics* 14, 301–313.

Merrill, E.W., Gilliland, E.R., Lee, T.S., Salzman, E.W., 1966. Blood rheology: effect of fibrinogen deduced by addition. *Circulation Research* 13, 437–446.

Merrill, E.W., Pelletier, G.A., 1967. Viscosity of human blood: transition from Newtonian to non-Newtonian. *Journal of Applied Physiology* 23, 178–182.

Metwally, H.M., Manglik, R.M., 2000. A computational study of enhanced heat transfer in laminar flows of non-Newtonian fluids in corrugated-plate channels. In: Manglik, R.M. et al. (eds.). *Advances in Enhanced Heat Transfer*, HTD – vol. 365/PID – vol. 4, 41–48.

Michaels, A.S., 1959. Diffusion in a pore of irregular cross-section: a simplified treatment. *AIChE Journal* 5, 270–271.

Miles, J.W. 1957. On the generation of surface waves by shear flows. *Journal of Fluid Mechanics* 3, 185–204.

Misirlioglu, A., Cihat Baytas, A., Pop, I., 2005. Free convection in a wavy cavity filled with a porous medium. *International Journal of Heat and Mass Transfer* 48, 1840–1850.

Moffatt, H.K., 1964. Viscous and resistive eddies near a sharp corner. *Journal of Fluid Mechanics* 18, 1–18.

Moghaddam, M.B., Goharshadi, E.K., Entezari, M.H., Nancarrow, P., 2013. Preparation, characterization, and rheological properties of graphene-glycerol nanofluids. *Chemical Engineering Journal* 231, 365–372.

Molla, M.M., Gorla, R.S.R., 2009. Natural convection laminar flow with temperature dependent viscosity and thermal conductivity along a vertical wavy surface. *International Journal of Fluid Mechanics Research* 36, 272–288.

Molla, M.M., Hossain, M.A., 2007. Radiation effect on mixed convection laminar flow along a vertical wavy surface. *International Journal of Thermal Sciences* 46, 926–935.

Molla, M.M., Hossain, M.A., Yao, L.-S., 2007. Natural convection flow along a vertical complex wavy surface with uniform heat flux. *ASME Journal of Heat Transfer* 129, 1403–1407.

Moulic, S.G., Yao, L.S., 1989a. Mixed convection along a wavy surface. *ASME Journal of Heat Transfer* 111, 974–979.

Moulic, S.G., Yao, L.S., 1989b. Natural convection along a vertical wavy surface with uniform heat flux. *ASME Journal of Heat Transfer* 111, 1106–1108.

Muley, A., Borghese, J., Manglik, R.M., Kundu, J., 2002. Experimental and numerical investigation of thermal-hydraulic characteristics of wavy-channel compact heat exchanger. In: *Proceedings of* 12th *International Heat Transfer Conference*, France, vol. 4, 417–422.

Muley, A., Borghese, J.B., White, S.L., Manglik, R.M., 2006. Enhanced thermal-hydraulic performance of a wavy-platefin compact heat exchanger: effect of corrugation severity. In: *Proceedings 2006 ASME International Mechanical Engineering Congress and Exposition (IMECE2006)*, Chicago.

Murthy, P.V.S.N., Singh, P., 1997. Effect of viscous dissipation on a non-Darcy natural convection regime. *International Journal of Heat and Mass Transfer* 40, 1251–1260.

Muthtamilselvan, M., Kandaswamy, P., Lee, J., 2010. Heat transfer enhancement of copper-water nanofluids in a lid-driven enclosure. *Communications in Nonlinear Science and Numerical Simulation* 15, 1501–1510.

Nakayama, A., 1995. *PC-Aided Numerical Heat Transfer and Convective Flow*. CRC Press, Boca Raton.

Nakayama, A., Pop, I., 1991. A unified similarity transformation for free, forced and mixed convection in Darcy and non-Darcy porous media. *International Journal of Heat and Mass Transfer* 34, 357–367.

Nakayama, A., Shenoy, A. V., 1992a. Combined forced and free convection heat transfer in non-Newtonian fluid-saturated porous media, *Applied Scientific Research* 50, 83–95.

Nakayama, A., Shenoy, A. V., 1992b. A unified similarity transformation for Darcy and non-Darcy forced, free and mixed convection heat transfer in non-Newtonian inelastic fluid-saturated porous media, *Chemical Engineering Journal* 50, 33–45.

Nakayama, A., Shenoy, A.V., 1993. Non-Darcy forced convection heat transfer in a channel embedded in a non-Newtonian fluid-saturated porous medium. *Canadian Journal of Chemical Engineering* 71, 168–173.

Narasimhan, A., 2013. *Essentials of Heat and Fluid Flow in Porous Media*. CRC Press, Boca Raton.

Nasrin, R., Alim, M.A., Chamkha, A.J., 2012. Combined convection flow in triangular wavy chamber filled with water–CuO nanofluid: effect of viscosity models. *International Journal of Heat and Mass Transfer* 39, 1226–1236.

Nayak, R.K., Bhattacharyya, S., Pop, I., 2015. Numerical study on mixed convection and entropy generation of Cu–water nanofluid in a differentially heated skewed enclosure. *International Journal of Heat and Mass Transfer* 85, 620–634.

Nayak, R.K., Bhattacharyya, S., Pop, I., 2016. Numerical study on mixed convection and entropy generation of a nanofluid in a lid-driven square enclosure. *ASME Journal of Heat Transfer* 138, 012503-1-012503-11.

Neagu, M., 2011. Free convective heat and mass transfer induced by a constant heat and mass fluxes vertical wavy wall in a non-Darcy double stratified porous medium. *International Journal of Heat and Mass Transfer* 54, 2310–2318.

Neira, A., Payatakes, A.C., 1978. Collocation solution of creeping Newtonian flow through periodically constricted tubes with piecewise continuous wall profile. *AIChE Journal* 24, 43–54.

Neira, M.A., Payatakes, A.C., 1979. Collocation solution of creeping Newtonian flow through sinusoidal tubes. *AIChE Journal* 25, 725–730.

Nejad, M.M., Javaherdeh, K., Moslemi, M., 2015. MHD mixed convection flow of power law non-Newtonian fluids over an isothermal wavy plate. *Journal of Magnetism and Magnetic Materials* 389, 66–72.

Nemati, H., Farhadi, M., Sedighi, K., Fattahi, E., Darzi, A.A.R., 2010. Lattice Boltzmann simulation of nanofluid in lid-driven cavity. *International Communications in Heat and Mass Transfer* 37, 1528–1534.

Nield, D.A., Bejan, A., 2013. *Convection in Porous Media* (4th edition). Springer, New York.

Nield, D.A., Kuznetsov, A.V., 2009a. The Cheng-Minkowycz problem for natural convective boundary-layer flow in a porous medium saturated by a nanofluid. *International Journal of Heat and Mass Transfer* 52, 5792–5795.

Nield, D.A., Kuznetsov, A.V., 2009b. Thermal instability in a porous medium layer saturated by a nanofluid. *International Journal of Heat and Mass Transfer* 52, 5796–5801.

Nield, D.A., Kuznetsov, A.V., 2011. The onset of double-diffusive convection in a nanofluid layer. *International Journal of Heat and Fluid Flow* 32, 771–776.

Nield, D.A., Kuznetsov, A.V., 2014. Thermal instability in a porous medium layer saturated by a nanofluid: A revised model. *International Journal of Heat and Mass Transfer* 68, 211–214.

Nikfar, M., Mahmoodi, M., 2012. Meshless local Petrov-Galerkin analysis of free convection of nanofluid in a cavity with wavy side walls. *Engineering Analysis with Boundary Elements* 36, 433–455.

Nishimura, T., Kajimoto, Y., Kawamura, Y., 1986. Mass transfer enhancement in channels with a wavy wall. *Journal of Chemical Engineering Japan* 19, 142–144.

Nishimura, T., Murakami, S., Arakawa, S., Kawamura, Y., 1990. Flow observations and mass transfer characteristics in symmetrical wavy-walled channels at moderate Reynolds numbers for steady flow. *International Journal of Heat and Mass Transfer* 33, 835–845.

Nishimura, T., Ohori, Y., Kawamura, Y., 1984. Flow characteristics in a channel with symmetric wavy wall for steady flow. *Journal of Chemical Engineering Japan* 17, 446–471.

Nishimura, T., Ohori, Y., Kawamura, Y., 1985. Mass transfer characteristics in a channel with symmetric wavy wall for steady flow. *Journal of Chemical Engineering Japan* 18, 550–555.

Noghrehabadi, A., Behbahan, A.S., Pop, I., 2015. Natural convection of a nanofluid in a square enclosure with two pairs of heat sources/sinks. *International Journal of Numerical Methods for Heat and Fluid Flow* 25, 1030–1046.

O'Brien, J.E., Sparrow, E.M., 1982. Corrugated duct heat transfer, pressure drop, and flow visualization. *ASME Journal of Heat Transfer* 104, 410–416.

Odat, M.Q.A., Damseh, R.A., Nimr, M.A.A., 2004. Effect of magnetic field on entropy generation due to laminar forced convection past a horizontal flat plate. *Entropy* 4, 293–303.

Oldroyd, J.G., 1950. On the formulation of rheological equations of state. *Proceedings of the Royal Society London* 200A, 523–541.

Oldroyd, J.G., 1958. Non-Newtonian effects in steady motion of some idealized elastico-viscous liquids. *Proceedings of the Royal Society London* 245A, 275–297.

Oosthuizen, P.H., Naylor, D., 1996. Natural convective heat transfer from a cylinder in an enclosure partially filled with a porous medium. *International Journal of Numerical Methods for Heat and Fluid Flow* 6, 51–63.

Ostrach, S.,1952. Laminar natural convection flow and heat transfer of fluids with and without heat sources in channels with constant wall temperature. *NACA TN*, 2863.

Ostrach, S., 1972. Natural convection in enclosures. *Advanced Heat Transfer* 8, 161–227.

Ostrach, S., 1982. Natural convection heat transfer in cavities and cells. *Proceedings of 7th International Heat Transfer Conference*, Munich, Germany 6, 365–379.

Öztop, H.F., Abu-Nada, E., 2008. Numerical study of natural convection in partially heated rectangular enclosures filled with nanofluids. *International Journal of Heat and Fluid Flow* 29, 1326–1336.

Öztop, H.F., Abu-Nada, E., Varol, Y., Chamkha, A., 2011. Natural convection in wavy enclosures with volumetric heat sources. *International Journal of Thermal Sciences* 50, 502–514.

Öztop, H.F., Al-Salem, K., 2012. A review on entropy generation in natural and mixed convection heat transfer for energy systems. *Renewable Sustainable Energy Rev.* 16, 911–920.

Palm, E., Weber, J.E., Kvernvold, O., 1972. On steady convection in a porous medium. *Journal of Fluid Mechanics* 54, 153–161.

Papazoglou, E.S., Parthasarathy, A., 2007. *Bio Nanotehnology*. Morgan & Klaypool, London, UK.

Patankar, S.V., 1980. *Numerical Heat Transfer and Fluid Flow*. Hemisphere, Washington DC.

Patankar, S.V., Liu, C.H., Sparrow, E.M., 1977. Fully developed flow and heat transfer in ducts having streamwise-periodic variations of cross-sectional area. *ASME Journal of Heat Transfer* 99, 180–186.

Patel, V.C., Chon, J.T., Yoon, J.Y., 1991. Laminar flow over wavy walls. *ASME Journal of Heat Transfer* 113, 574–578.

Payatakes, A.C., Tien, C., Turian, R.H., 1973. A new model for granular porous media, Part I. Model formulation. *AIChE Journal* 19, 58-66; Part II. 1973. Numerical solution of steady state incompressible Newtonian flow through periodically constricted tubes. *AIChE Journal* 19, 67-76; 1973. Further work on the flow through periodically constricted tubes a reply. *AIChE Journal* 19, 1036–1039.

Pearson, C.E., 1965. A computational method for viscous flow problems. *Journal of Fluid Mechanics* 21, 611–622.

Petersen, E.E., 1958. Diffusion in a pore of varying cross-section. *AIChE Journal* 4, 343–345.

Peterson, S.D., 2010. Steady flow through a curved tube with wavy walls. *Physics of Fluids* 22, 023602.

Pirmohammadi, M., Ghassemi, M., 2009. Effect of magnetic field on convection heat transfer inside a tilted square enclosure. *International Communications in Heat and Mass Transfer* 36, 776–780.

Plumb, O.A., 1983. The effect of thermal dispersion on heat transfer in packed bed boundary layers. In: *Proceedings ASME JSME Thermal Engineering Joint Conference* 2, 17–22.

Plumb, O., Huenefeld, J.C., 1981. Non-Darcy natural convection from heated surfaces in saturated porous medium. *International Journal of Heat and Mass Transfer* 24, 765–768.

Pop, I. 1983. Theory of the Unsteady Laminar Boundary Layer (in Romanian). Editura Ştiinţifică şi Enciclopedică, Bucharest, Romania.

Pop, I., Ingham, D.B., 2001. *Convection Heat Transfer: Mathematical and Computational Modeling of Viscous Fluids and Porous Media.* Pergamon, Oxford, UK.

Pop, I., Ingham, D.B., 2000. Convective boundary layers in porous media: external flows. In: *Handbook of Porous Media* (Vafai, K., ed.). Marcel Dekker, New York.

Pop, I., Na, T.-Y., 1995. Natural convection over a frustum of a wavy cone in a porous medium. *Mechanics Research Communications* 22, 181–190.

Poulikakos, D., 1985. Natural convection in a confined fluid-filled space driven by a single vertical wall with warm and cold regions. *ASME Journal of Heat Transfer* 107, 867–876.

Pozrikidis, C., 2006. Stokes flow through a twisted tube. *Journal of Fluid Mechanics* 567, 261–280.

Prata, A.T., Sparrow, E.M., 1984. Heat transfer and fluid flow characteristics for an annulus of periodically varying cross section. *Numerical Heat Transfer* 7, 285–304.

Prince, C., Gu, M., Peterson, S.D., 2013. A numerical study of the impact of wavy walls on steady fluid flow through a curved tube.

ASME Journal of Fluids Engineering 135, 07120 (13 pages).

Radko, T., 2013. *Double-Diffusive Convection.* Cambridge University Press, Cambridge, UK.

Rahman, S.U., 2001. Natural convection along vertical wavy surfaces: an experimental study. *Chemical Engineering Journal* 84, 587–591.

Ralph, M.E., 1987. Steady flow structures and pressure drops in wavy-walled tubes. *ASME Journal of Fluids Engineering* 109, 255–261.

Rao, Y., 2010. Nanofluids: Stability, phase diagram, rheology and applications. *Particuology* 8, 549–555.

Rao Prasad, D.R.V., Krishna, D.V., Louenath, D., 1983. Free convection in hydromagnetic flows in a vertical wavy channel. *International Journal of Engineering Science* 21, 1025–1039.

Rashidi, S., Abelman, N., Freidooni, M., 2013. Entropy generation in steady MHD flow due to a rotating porous disk in a nanofluid. *International Journal of Heat and Mass Transfer* 62, 515–525.

Rees, D.A.S., Bassom, A.P., 1991. Some exact solutions for free convective flow over heated semi-infinite surfaces in porous media. *International Journal of Heat and Mass Transfer* 31, 1564–1567.

Rees, D.A.S., Pop, I., 1994a. A note on free convection along a vertical wavy surface in a porous medium. *ASME Journal of Heat Transfer* 116, 505–508.

Rees, D.A.S., Pop, I., 1994b. Free convection induced by a horizontal wavy surface in a porous medium. *Fluid Dynamics Research* 14, 151–166.

Rees, D.A.S., Pop, I., 1995a. Free convection induced by a vertical wavy surface with uniform heat flux in a porous medium. *ASME Journal of Heat Transfer* 117, 547–550.

Rees, D.A.S., Pop, I., 1995b. Non-Darcy natural convection from a vertical wavy surface in a porous medium. *Transport in Porous Media* 20, 223–234.

Rees, D.A.S., Riley, D.S., 1986. Free convection in an undulating saturated porous layer: resonant wavelength excitation. *Journal of Fluid Mechanics* 166, 503–530.

Rees, A., Xu, H., Sun, Q., Pop, I., 2015. Mixed convection in gravity-driven nano-liquid film containing both nanoparticles and gyrostatic microorganisms. *Applied Mathematics and Mechanics (English Edition)* 36, 163–178.

Revnic, C., Groşan, T., Pop, I., Ingham, D.B., 2011. Magnetic field effect on the unsteady free convection flow in a square cavity filled with a porous medium with a constant heat generation. *International Journal of Heat and Mass Transfer* 54, 1734–1742.

Riley, D.S., 1988. Steady two-dimensional thermal convection in a vertical porous slot with spatially periodic boundary imperfections. *International Journal of Heat and Mass Transfer* 31, 2365–2380.

Roache, P.J., 1972. *Computational Fluid Dynamics*. Hermosa, Albuquerque, New Mexico.

Rubin, S.G., Himansu, A., 1989. Convergence properties of high-Reynolds-number separated flow calculations. *International Journal of Numerical Methods in Fluids* 9, 1395–1411.

Rush, T.A., Newell, T.A., Jacobi, A.M., 1999. An experimental study of flow and heat transfer in sinusoidal wavy passages. *International Journal of Heat and Mass Transfer* 42, 1541–1553.

Saffman, P.G., 1960. Dispersion due to molecular diffusion and macroscopic mixing in flow through a network of capillaries. *Journal of Fluid Mechanics* 7, 194–208.

Sahimi, M., 2011. *Flow and Transport in Porous Media and Fractured Rock*. Wiley-VCH, Weinheim, Germany.

Saidi, C., Legay-Desesquelles, F., Prunet-Foch, B., 1987. Laminar flow past a sinusoidal cavity. *International Journal of Heat and Mass Transfer* 30, 649–661.

Saidur, R., Leong, K.Y., Mohammad, H.A., 2011. A review on applications and challenges of nanofluids. *Renewable and Sustainable Energy Reviews* 15, 1646–1668.

Sakai, F., Li, W., Nakayama, A., 2014. A rigorous derivation and its applications of volume averaged transport equations for heat transfer in nanofluids saturated metal foam. In: *Proceedings of 15th International Heat Transfer Conference*. IHTC-15, August 10-15, Kyoto, Japan.

Salari, M., Tabar, M.M., Tabar, A.M., Danesh, H.A., 2012. Mixed convection of nanofluid flows in a square lid-driven cavity heated partially from both the bottom and side walls. *Numerical Heat Transfer*, Part A 62, 158–177.

Samarski, A.A., 1983. Theory of Difference Schemes, Nauka, Moscow.

Saniei, N., and Dini, S., 1993. Heat transfer characteristics in a wavy-walled channel. *ASME Journal of Heat Transfer* 115, 788–792.

Saouli, S., Aiboud-Saouli, S., 2004. Second law analysis of laminar falling liquid film along an inclined heated plate. *International Communications in Heat and Mass Transfer* 31, 879–886.

Sarkar, J., Ghosh, P., Adil, A., 2015. A review on hybrid nanofluids: Recent research, development and applications. *Renewable and Sustainable Energy Reviews* 43, 164–177.

Sarris, I.E., Leakakis, I., Vlachos, N.S., 2002. Natural convection in a 2D enclosure with sinusoidal upper wall temperature. *Numerical Heat Transfer*, Part A 42, 513–530.

Sarris, I.E., Zikos, G.K., Grecos, A.P., Vlachos, N.S., 2006. On the limits of validity of the low magnetic Reynolds number approximation in MHD natural-convection heat transfer. *Numerical Heat Transfer Part B* 50, 157–180.

Savvides, C.N., Gerrard, J.H., 1984. Numerical analysis of the flow through a corrugated tube with application to arterial prostheses. *Journal of Fluid Mechanics* 138, 129–160.

Sayegh, S.G., Vera, J.H., 1980. Lattice-model expressions for the combinatorial entropy of liquid mixtures: A critical discussion. *Chemical Engineering Journal* 19, 1–10.

Scarborough, J.B., 1958. *Numerical Mathematical Analysis* (4th edition). Johns Hopkins Press, Baltimore.

Schaefer, H.-E., 2010. *Nanoscience: The Science of the Small in Physics, Engineering, Chemistry, Biology and Medicine.* Springer, Berlin, Germany.

Schlichting, H., Gersten, K., 2000. *Boundary Layer Theory.* Springer, New York.

Selimefendigil, F., Öztop, H.F., 2015. Natural convection and entropy generation of nanofluid filled cavity having different shaped obstacles under the influence of magnetic field and internal heat generation. *Journal of the Taiwan Institute of Chemical Engineers*, 56, 42–56.

Selvan, M.M., 2010. Mixed convection in a lid-driven cavity utilizing nanofluids. *CFD Letters* 2, 163–175.

Sha, W.T., 2011. *Novel Porous Media Formulation for Multiphase Flow Conservation Equations.* Cambridge University Press, Cambridge.

Shankar, P.N., Sinha, U.N., 1976. The Rayleigh problem for a wavy wall. *Journal of Fluid Mechanics* 77, 243–256.

Sheffield, R.E., Metzner, A.B., 1976. Flow of non linear fluids through porous media. *AIChE Journal* 22, 738–744.

Sheik, I.L., Ranganayakulu, C., Shah, R.K., 2009. Numerical study of flow patterns of compact plate-fin heat exchangers and generation of design data for offset and wavy fins. *International Journal of*

Heat and Mass Transfer 52, 3972–3983.

Sheikholeslami, M., Bandpy, M.G., Ellahi, R., Zeeshan, A., 2014. Simulation of MHD CuO–water nanofluid flow and convective heat transfer considering Lorentz forces. *Journal of Management and Magnetic Materials* 369, 69–80.

Shenoy, A.V., 1986. Natural convection heat transfer to power-law fluids. *Handbook of Heat and Mass Transfer*, Vol. 1. Gulf Publishing Company, Houston.

Shenoy, A.V., 1992. Darcy-Forchheimer natural, forced and mixed convection heat transfer from an isothermal vertical flat plate embedded in a porous medium saturated with an elastic fluid of constant viscosity. *International Journal of Engineering Science* 30, 455–467.

Shenoy, A.V., 1993a. Forced convection heat transfer to an elastic fluid of constant viscosity flowing through a channel filled with a Brinkman-Darcy porous medium. *Warme-und Stoffubertragung* 28, 295–297.

Shenoy, A.V., 1993b. Darcy-Forchheimer natural, forced and mixed convection heat transfer in non-Newtonian power-law fluid-saturated porous media. *Transport in Porous Media* 11, 219–241.

Shenoy, A.V., 1994. Non-Newtonian fluid heat transfer in porous media. *Advances in Heat Transfer* 24, 101–190.

Shenoy, A.V., 1999. *Rheology of Filled Polymer Systems*. Springer, New York.

Shercliff, J.A., 1965. *A Textbook of Magnetohydrodynamics*. Pergamon Press, Oxford, UK.

Sheremet, M.A., Pop, I., 2014a. Thermo-bioconvection in a square porous cavity filled by oxytactic microorganisms. *Transport in Porous Media* 103, 191–205.

Sheremet, M.A., Pop, I., 2014b. Natural convection in a square porous cavity with sinusoidal temperature distributions on both side walls filled with a nanofluid: Buongiorno's mathematical model. *Transport in Porous Media* 105, 411–429.

Sheremet, M.A., Pop, I., 2014c. Conjugate natural convection in a square porous cavity filled by a nanofluid using Buongiorno's mathematical model. *International Journal of Heat and Mass Transfer* 79, 137–145.

Sheremet, M.A., Pop, I., 2015a. Natural convection in a wavy porous cavity with sinusoidal temperature distributions on both side walls filled with a nanofluid: Buongiorno's mathematical model. *ASME*

Journal of Heat Transfer 137, 072601.

Sheremet, M.A., Pop, I., 2015b. Free convection in a triangular cavity filled with a porous medium saturated by a nanofluid: Buongiorno's mathematical model. *International Journal of Numerical Methods for Heat and Fluid Flow* 25, 1138–1161.

Sheremet, M.A., Pop, I., 2015c. Natural convection in a horizontal cylindrical annulus filled with a porous medium saturated by a nanofluid using Tiwari and Das' nanofluid model. *The European Physical Journal Plus* 130, 107 (12 pages).

Sheremet, M.A., Pop, I., 2015d. Mixed convection in a lid-driven square cavity filled by a nanofluid: Buongiorno's mathematical model. *Applied Mathematics and Computation* 266, 792–808.

Sheremet, M.A., Pop, I., 2015e. Free convection in a porous horizontal cylindrical annulus with a nanofluid using Buongiorno's model. *Computers & Fluids* 118, 182–190.

Sheremet, M.A., Groşan, T., Pop, I., 2014. Free convection in shallow and slender porous cavities filled by a nanofluid using Buongiorno's model. *ASME Journal of Heat Transfer* 136, 082501-1-082501-5.

Sheremet, M.A., Groşan, T., Pop, I., 2015a. Free convection in a square cavity filled with a porous medium saturated by nanofluid using Tiwari and Das' nanofluid model. *Transport in Porous Media* 106, 595–610.

Sheremet, M.A., Groşan, T., Pop, I., 2015b. Natural convection in a cubical porous cavity saturated with nanofluid using Tiwari and Das' nanofluid model. *Journal of Porous Media* 18, 585–596.

Sheremet, M.A., Groşan, T., Pop, I., 2015c. Steady-state free convection in right-angle porous trapezoidal cavity filled by a nanofluid: Buongiorno's mathematical model. *European Journal of Mechanics - B/Fluids* 53, 241–250.

Sheremet, M.A., Pop, I., Ishak, A., 2015d. Double-diffusive mixed convection in a porous open cavity filled with a nanofluid using Buongiorno's model. *Transport in Porous Media* 109, 131–145.

Sheremet, M.A., Pop, I., Rahman, M.M., 2015e. Three-dimensional natural convection in a porous enclosure filled with a nanofluid using Buongiorno's mathematical model. *International Journal of Heat and Mass Transfer* 82, 396–405.

Sheremet, M.A., Dinarvand, S., Pop, I., 2015f. Effect of thermal stratification on free convection in a square porous cavity filled with a nanofluid using Tiwari and Das' nanofluid model. *Physica E* 69,

332–341.

Sheremet, M.A., Pop, I., Shenoy, A., 2015g. Unsteady free convection in a porous open wavy cavity filled with a nanofluid using Buongiorno's mathematical model. *International Communications in Heat and Mass Transfer* 67, 66–72.

Sheremet, M.A., Pop, I., Bachok, N., 2016a. Effect of thermal dispersion on transient natural convection in a wavy-walled porous cavity filled with a nanofluid: Tiwari and Das' nanofluid model. *International Journal of Heat and Mass Transfer* 92, 1053–1060.

Sheremet, M.A., Pop, I., Shenoy, A., 2016b. Natural convection in a wavy open porous cavity filled with a nanofluid: Tiwari and Das' nanofluid model. *The European Physical Journal Plus* 131, 62 (12 pages).

Sheremet, M.A., Pop, I., Roşca, N.C., 2016c. Magnetic field effect on the unsteady natural convection in a wavy-walled cavity filled with a nanofluid: Buongiorno's mathematical model. *Journal of the Taiwan Institute of Chemical Engineers* 61, 211–222.

Sheremet, M.A., Trifonova, T.A., 2013. Unsteady conjugate natural convection in a vertical cylinder partially filled with a porous medium. *Numerical Heat Transfer, Part A* 64, 994–1015.

Shih, T.M., 1984. *Numerical Heat Transfer.* Hemisphere, Washington DC.

Siddiqa, S., Hossain, M.A., 2013. Natural convection flow over wavy horizontal surface. *Advances in Mechanical Engineering* 7, 2013. Article ID 743034.

Siddiqa, S., Hossain, M.A., Gorla, R.S.R., 2015. Natural convection flow of viscous fluid over triangular wavy horizontal surface. *Computers & Fluids* 106, 130–134.

Siddiqa, S., Hossain, M.A., Saha, S.C., 2013. Natural convection flow with surface radiation along a vertical wavy surface. *Numerical Heat Transfer, Part A* 64, 1–16.

Siddiqa, S., Hossain, M.A., Saha, Suvash. C., 2014. The effect of thermal radiation on the natural convection boundary layer flow over a wavy horizontal surface. *International Journal of Thermal Sciences* 84, 143–150.

Singh, R.R., Singh, A.K., Singh, U., Singh, A.K., Singh, N.P., 2014. Hydromagnetic convection flow in a porous medium bounded between vertical wavy wall and parallel flat wall: analysis using Darcy-Brinkman-Forchheimer model. *Proceedings of the National*

Academy of Sciences, India Section A: Physical Sciences 84, 409–431.

Sobey, I.J., 1980. On flow through furrowed channels. Part I: Calculated flow patterns. *Journal of Fluid Mechanics* 96, 1–26.

Sorenson, J.P., Stewart, W.E., 1974. Computation of forced convection in slow flow through ducts and packed beds-II. Velocity profile in a simple cubic array of spheres. *Chemical Engineering Science* 29, 819–825.

Sparrow, E.M., Comb, J.W., 1983. Effect of interwall spacing and fluid flow inlet conditions on a corrugated-wall heat exchanger. *International Journal of Heat and Mass Transfer* 26, 993–1005.

Sparrow, E.M., Prata, A.T., 1983. Numerical solutions for laminar flow and heat transfer in a periodically converging diverging tube, with experimental confirmation. *Numerical Heat Transfer* 6, 441–461.

Sreeremya, T.S., Krishnan, A., Mohamed, A.P., Hareesh, U.S., Ghosh, S., 2014. Synthesis and characterization of cerium oxide based nanofluids: An efficient coolant in heat transport applications. *Chemical Engineering Journal* 255, 282–289.

Stephanoff, K.D., Sobey, I.J., Bellhouse, B.J., 1980. On flow through furrowed channels. Part 2. Observed flow patterns. *Journal of Fluid Mechanics* 96, 27–32.

Straus, J.M., 1974. Large amplitude convection in porous media. *Journal of Fluid Mechanics* 64, 51–83.

Sun, Q., Pop, I., 2011. Free convection in a triangle cavity filled with a porous medium saturated with nanofluids with flush mounted heater on the wall. *International Journal of Thermal Sciences* 50, 2141–2153.

Sun, Q., Pop, I., 2014. Free convection in a tilted triangle porous cavity filled with Cu–water nanofluid with flush mounted heater on the wall. *International Journal of Numerical Methods for Heat and Fluid Flow* 24, 2–20.

Taneda, G., Vittori, G., 1996. Fluid flow and heat transfer in a two-dimensional wavy channel. *Heat and Mass Transfer* 31, 411–418.

Tao, Y.B., He, Y.L., Wu, Z.G., Tao, W.Q., 2007. Numerical design of an efficient wavy fin surface based on the local heat transfer coefficient study. *Journal of Enhanced Heat Transfer* 14, 315–322.

Tashtoush, B., Al-Odat, M., 2004. Magnetic field effect on heat and fluid flow over a wavy surface with variable heat flux. *Journal of Management and Magnetic Materials* 268, 357–363.

Taylor, G.I., 1953. Dispersion of soluble matter in solvent flowing slowly through a table. *Proceedings of the Royal Society of London* A219, 186–203.

Telionis, D.P., 1981. *Unsteady Viscous Flows*. Springer, New York.

Telles, R.S., Trevisan, O.V., 1993. Dispersion in heat and mass transfer natural convection along vertical boundaries in porous media. *International Journal of Heat and Mass Transfer* 36, 1357–1365.

Temam, R., 1977. *Navier–Stokes Equations. Theory and Numerical Analysis*. North-Holland, Amsterdam.

Thien-Phan, N., 1980. On the Stokes flow of a viscous fluid through corrugated pipes. *Journal of Applied Mechanics* 47, 961–963.

Thien-Phan, N., 1981. On Stokes flows in channels and pipes with parallel stationary random surface roughness. *Journal of Applied Mathematics and Mechanics (ZAMM)* 61, 193–199.

Tilton, J.N., Payatakes, A.C., 1984. Collocation solution of creeping Newtonian flow through sinusoidal tubes: a correction. *AIChE Journal* 30, 1016–1021.

Tiwari, R.K., Das, M.K., 2007. Heat transfer augmentation in a two-sided lid-driven differentially heated square cavity utilizing nanofluids. *International Journal of Heat and Mass Transfer* 50, 2002–2018.

Tsangaris, S., Leiter, E., 1984. On laminar steady flow in sinusoidal channels. *Journal of Engineering Mathematics* 18, 89–103.

Tsangaris, S., Potamitis, D., 1986. On laminar small Reynolds number flow over wavy walls. *Acta Mechanica* 61, 109–115.

Umavathi, J.C., Shekar, M., 2015. Unsteady mixed convective flow confined between vertical wavy wall and parallel flat wall filled with porous and fluid layer. *Heat Transfer Engineering* 36, 1–20.

Vadasz, P. (ed.), 2008. *Emerging Topics in Heat and Mass Transfer in Porous Media*. Springer, New York.

Vafai, K. (ed.), 2005. *Handbook of Porous Media*. Taylor and Francis, New York.

Vafai, K., Hadim, H.A., 2000. *Handbook of Porous Media*. Marcel Dekker, New York.

Vajravelu, K., 1980. Fluid flow and heat transfer in horizontal wavy channels. *Acta Mechanica* 35, 245–258.

Vajravelu, K., 1989. Combined free and forced convection in hydromagnetic flow in vertical wavy channels with traveling thermal waves. *International Journal of Engineering Science* 27, 289–300.

Vajravelu, K., Sastri, K.S., 1978. Free convective heat transfer in a viscous incompressible fluid confined between a long vertical wavy wall and a parallel flat wall. *Journal of Fluid Mechanics* 86, 365–383.

Vajravelu, K., Sastri, K.S., 1980. Natural convective heat transfer in vertical wavy channels. *International Journal of Heat and Mass Transfer* 23, 408–411.

Vyas, S., Zhang, J., Manglik, R.M., 2004. Steady recirculation and laminar forced convection in a sinusoidal wavy channel. *ASME Journal of Heat Transfer* 126, 500.

Wakao, N., Kaguei, S., 1982. *Heat and Mass Transfer in Packed Beds.* Gordon & Breach Science Publishers, New York.

Walker, K.L., Homsy, G.M., 1978. Convection in a porous cavity. *Journal of Fluid Mechanics* 87, 338–363.

Wang, C.C., Chen, C.K., 2002. Forced convection in a wavy-wall channel. *International Journal of Heat and Mass Transfer* 45, 2587–2595.

Wang, C.C., Chen, C.K., 2005a. Forced convection in micropolar fluid flow through a wavy-wall channel. *Numerical Heat Transfer, Part A* 48, 879–900.

Wang, C.C., Chen, C.K., 2005b. Mixed convection boundary layer flow on inclined wavy plates including the magnetic field effect. *International Journal of Thermal Sciences* 44, 577–586.

Wang, C.Y., 1987. Flow between longitudinally corrugated cylinders. *Applied Scientific Research* 44, 277–286.

Wang, C.Y., 2006a. Effect of helical corrugations on the low Reynolds number flow in a tube. *AIChE Journal* 52, 2008–2012.

Wang, C.Y., 2006b. Stokes flow through a tube with bumpy wall. *Physics of Fluids* 18, 078101.

Wang, G., Vanka, S.P., 1995. Convective heat transfer in periodic wavy passage. *International Journal of Heat and Mass Transfer* 38, 3219–3230.

Wang, X.Q., Mujumdar, A.S., 2007. Heat transfer characteristics of nanofluids: a review. *International Journal of Thermal Sciences* 46, 1–19.

Watanabe, T., Pop, I., 1993. Magnetohydrodynamic free convection flow over a wedge in the presence of a transverse magnetic field. *International Communications in Heat and Mass Transfer* 20, 871–881.

Watson, A., Poots, G., 1971. The effect of sinusoidal protrusions on laminar free convection between vertical walls. *Journal of Fluid Mechanics* 49, 33–48.

Wen, D., Lin, G., Vafaei, S., Zhang, K., 2011. Review of nanofluids for heat transfer applications. *Particuology* 7, 141–150.

Wenming, Y., Dongyue, J., Kenny, C.K.Y., Dan, Z., Jianfeng, P., 2015. Combustion process and entropy generation in a novel microcombustor with a block insert. *Chemical Engineering Journal* 274, 231–237.

Whitaker, S., 1989. Heat transfer in catalytic packed bed reactors. *Handbook of Heat Mass Transfer*. Gulf Publishing Co., Houston, Texas 3, 361–417.

White, F., 2006. *Viscous Fluid Flow* (3rd *edition*). McGraw-Hill, New York.

Williams, G.P., 1969. Numerical integration of the three-dimensional Navier–Stokes equations for incompressible flow. *Journal of Fluid Mechanics* 37, 127–750.

Wong, K.V., Leon, O.D., 2010. Applications of nanofluids: current and future. *Advances in Mechanical Engineering*. Article ID 519659, 11 pages.

Woods, L.C., 1975. *Thermodynamics of Fluid Systems*. Oxford University Press, Oxford, UK.

Wright, S., Rawson, H., 1973. Calculation of natural convection in a rectangular cell containing glass with specified temperatures on the boundaries. *Glass Technology* 14, 42–49.

Yang, L.C., Asako, Y., Yamaguchi, Y., Faghri, M., 1997. Numerical prediction of transitional characteristics of flow and heat transfer in a corrugated duct. *ASME Journal of Heat Transfer* 119, 62–69.

Yang, Y.-Z., Chen, C.-K., Lin, M.-T., 1996. Natural convection of non-Newtonian fluids along a wavy vertical plate including the magnetic field effect. *International Journal of Heat and Mass Transfer* 39, 2831–2842.

Yao, L.S., 1983. Natural convection along a vertical wavy surface. *ASME Journal of Heat Transfer* 105, 465–468.

Yih, K.A., 1999. MHD forced convection flow adjacent to a non-isothermal wedge. *International Communications in Heat and Mass Transfer* 26, 819–827.

Young, D.F., 1968. Effect of a time-dependent stenosis on flow through a tube. *Journal of Engineering for Industry* 90, 248–254.

Yu, W.H., France, D.M., Routbort, J.L., Choi, S.U.S., 2008. Review and comparison of nanofluid thermal conductivity and heat transfer enhancement. *Heat Transfer Engineering* 29, 432–460.

Zargartalebi, H., Noghrehabadi, A., Ghalambaz, M., Pop, I., 2015. Natural convection boundary layer flow over a horizontal plate embedded in porous medium saturated with a nanofluid: case of variable thermophysical properties. *Transport in Porous Media* 107, 153–170.

Zhang, J., 2005. Numerical simulations of steady low-Reynolds-number flows and enhanced heat transfer in wavy plate-fin passages. Ph.D. Thesis. University of Cincinnati.

Zhang, J., Kundu, J., Manglik, R.M., 2004. Effect of fin waviness and spacing on the lateral vortex structure and laminar heat transfer in wavy-plate-fin cores. *International Journal of Heat and Mass Transfer* 47, 1719–1730.

Zhang, J., Muley, A., Borghese, J.B., Manglik, R.M., 2003. Computational and experimental study of enhanced laminar flow heat transfer in three-dimensional sinusoidal wavy-plate-fin channels. In: *Proceedings of ASME Summer Heat Transfer Conference*. Vol. 1, 665–672.

Index